French-Swedish Winterschool on Neutron Scattering 2016 (JDN 23)

Applications to Soft Matter

EPJ Web of Conferences Volume 188 (2018)

Uppsala, Sweden
6-9 December 2016

Editors:

M. Wolff
F. Cousin

ISBN: 978-1-5108-7120-5

Printed from e-media with permission by:

Curran Associates, Inc.
57 Morehouse Lane
Red Hook, NY 12571

Some format issues inherent in the e-media version may also appear in this print version.

Printed by Curran Associates, Inc. (2018)

For additional information, please contact EDP Sciences – Web of Conferences at the address below.

EDP Sciences – Web of Conferences
17, Avenue du Hoggar
Parc d'Activité de Courtabœuf
BP 112
F-91944 Les Ulis Cedex A
France

Phone: +33 (0) 1 69 18 75 75
Fax: +33 (0) 1 69 28 84 91

contact-edps@webofconferences.org

Additional copies of this publication are available from:

Curran Associates, Inc.
57 Morehouse Lane
Red Hook, NY 12571 USA
Phone: 845-758-0400
Fax: 845-758-2633
Email: curran@proceedings.com
Web: www.proceedings.com

TABLE OF CONTENTS

Foreword

The 2016 edition of the école thematique of the Société Française de la Neutronique (SFN) was organized together with the Swedish neutron scattering society (SNSS) as a four days workshop at Uppsala University from Dec. 6 to 9 and focused on Neutron Scattering: Applications to soft matter. Organizing the school in Sweden emerged from a series of joint teaching events during the last years and the fact that Sweden will host the forthcoming European Spallation Source (ESS). 23 students and twelve teachers gathered at temperatures well below zero degrees at Ångstöm laboratory in Uppsala, about 80 km north of Stockholm. The first day of the winterschool was dedicated to the general introduction of soft matter and neutron scattering presented by Julian Oberdisse, Montpellier, and Thomas Ederth, Linköping. The rest of the school introduced the different experimental methods which are themost relevant to soft matter research as well as recent scientific challenges in the field. Thursday, Dec. 8, was almost exclusively dedicated to a journey at the Nobel lectures in Stockholm. The participants left Uppsala early in the morning by bus to arrive in time for the best seats in Magna Aula to listen to the Physics, Chemistry and Economy 2016 laureates. Inspired by the presentations on topological effects, molecular machines and limitations of contracts, the students returned to Uppsala in the late afternoon to listen to lectures on neutron scattering applied to biological systems. This issue is dedicated to the lecture notes of the school.

edp sciences

Web of Conferences

Statement of Peer review

In submitting conference proceedings to *Web of Conferences*, the editors of the proceedings certify to the Publisher that

1. They adhere to its **Policy on Publishing Integrity** in order to safeguard good scientific practice in publishing.
2. All articles have been subjected to peer review administered by the proceedings editors.
3. Reviews have been conducted by expert referees, who have been requested to provide unbiased and constructive comments aimed, whenever possible, at improving the work.
4. Proceedings editors have taken all reasonable steps to ensure the quality of the materials they publish and their decision to accept or reject a paper for publication has been based only on the merits of the work and the relevance to the journal.

Title, date and place of the conference

2016 French - Swedish Winter School in Neutron Scattering: Applications for Soft Matter
2016 - Dec - 6 to 2016 - Dec - 9, Uppsala

Proceedings editor(s): Max Wolff, Fabrice Cousin

Date and editor's signature 2018-06-26

17, avenue du Hoggar - PA de Courtabœuf – BP 112 - 91944 Les Ulis Cedex A (France)
Tél. : 33 (0)1 69 18 75 75 - Fax : 33(0)1 69 07 45 17 – www.edpsciences.org

Introduction to soft matter and neutron scattering

Julian Oberdisse*

Laboratoire Charles Coulomb (L2C), University of Montpellier, CNRS, Montpellier, France

Abstract. As an opening lecture to the French-Swedish neutron scattering school held in Uppsala (6th to 9th of December 2016), the basic concepts of both soft matter science and neutron scattering are introduced. Typical soft matter systems like self-assembled surfactants in water, microemulsions, (co-)polymers, and colloids are presented. It will be shown that widely different systems have a common underlying physics dominated by the thermal energy, with astonishing consequences on their statistical thermodynamics, and ultimately rheological properties – namely softness. In the second part, the fundamentals of neutron scattering techniques and in particular small-angle neutron scattering as a powerful method to characterize soft matter systems will be outlined.

1 Introduction

Soft condensed matter has grown into a rather wide and mature scientific discipline over the past 50 years, unifying different branches of physics, chemistry, and biology – with their interfacial disciplines physical chemistry, biophysics, etc. – into a common approach mainly characterized by the importance of entropy. There are many examples of initially independent fields, like surfactant science or liquid crystals, which have existed for more than one hundred years as a true 'nano'-science, or polymer physics, which really started with Staudinger and his macromolecular hypothesis between the world wars of the 20th century. Some major progress was made by giants of science, sometimes rewarded – but not always – by a Nobel prize, like the ones of Staudinger†, Ziegler and Natta in polymer chemistry‡, as well as Flory in polymer physics and chemistry§, or De Gennes for several contributions including polymer

* Corresponding author: julian.oberdisse@umontpellier.fr

† The Nobel Prize in Chemistry 1953 was awarded to Hermann Staudinger *"for his discoveries in the field of macromolecular chemistry"*.
‡ The Nobel Prize in Chemistry 1963 was awarded jointly to Karl Ziegler and Giulio Natta *"for their discoveries in the field of the chemistry and technology of high polymers"*.
§ The Nobel Prize in Chemistry 1974 was awarded to Paul J. Flory *"for his fundamental achievements, both theoretical and experimental, in the physical chemistry of the macromolecules"*.

physics[**], a prize the late Sir Sam Edwards might have merited as well. The reason I mention these Nobel prize winners explicitly is that they enter in a particular resonance with the French-Swedish winter school held in Uppsala in December 2016, during the week of Nobel celebrations. The organizers had the brilliant idea to take us to the Nobel lectures in physics[††] (D.J. Thouless, M. Kosterlitz, D. Haldane) and chemistry[‡‡] (J.P. Sauvage, F. Stoddart, B. Feringa) in Stockholm, which was truly an extraordinary event.

In this short introduction – short with respect to entire textbooks available on the field of soft matter and neutron scattering [1, 2], see also ref. [3, 4] from previous neutron schools – , I will outline the fundamental concepts and properties of soft condensed matter, namely viscoelasticity, and its link to microstructure and thermal motion. These concepts are then illustrated with the main sub-disciplines of soft matter: from surfactants and microemulsions, to polymers and copolymers, colloids, and liquid crystals. I will then show how neutron scattering can be particularly appropriate to study such systems. Indeed, soft matter displays a wide range of different structures, as well as dynamical properties, and neutron scattering is a technique which is well-suited to unravel both their structure and dynamics, due to the unique properties of the neutron in terms of penetration power, wavelength, and energy. While different aspects of neutron scattering will be presented in detail in the following lectures, we restrict ourselves here to a discussion of small-angle scattering as a tool to characterize the distribution of both volumes and surfaces in space, with some hints to application of other scattering techniques to soft matter, like reflectometry, GISANS (grazing incidence small-angle neutron scattering), and QENS (quasi-elastic neutron scattering), all three techniques are also discussed in dedicated chapters in this volume.

2 Fundamental concepts of soft condensed matter

It will not surprise the reader to learn that the first and main property of soft condensed matter is its … softness (i). But why is soft matter soft? Before going into microscopic detail, we need to turn to the branch of science which provides tools to characterize softness, called rheology, the science of flow. The basic rheological experiment – many different versions, time dependencies, and geometries exist, which we ignore here – is to exert a stress σ (a force per unit surface) on the surface of a sample, and observe its deformation ε, or strain, given by the relative displacement of the surface, relative with respect to sample size. A weak stress inducing a large strain is characteristic of a soft material, and this is commonly described by a low elastic modulus $G = \sigma/\varepsilon$. A second property of many soft matter systems is that they can relax a stress they have undergone (ii), via more or less slow internal motions, introducing a characteristic relaxation time τ. The latter is related to flow of the samples described by viscosity η, and η equals $G\tau$ in the simplest case of a Maxwellian fluid [5]. Of course many complications exist, and these material 'constants' may be very complicated functions of time, or equivalently frequency, or history of deformation. But we may retain that soft samples are described by a modulus indicating elasticity, and by viscosity representative of flow: soft matter systems are thus usually viscoelastic. In practice, their linear rheological response is given by G' and G"(ω), where ω is the frequency $2\pi f$ of a low-amplitude

** The Nobel Prize in Physics 1991 was awarded to Pierre-Gilles de Gennes *"for discovering that methods developed for studying order phenomena in simple systems can be generalized to more complex forms of matter, in particular to liquid crystals and polymers"*.
†† The Nobel Prize in Physics 2016 was divided, one half awarded to David J. Thouless, the other half jointly to F. Duncan M. Haldane and J. Michael Kosterlitz *"for theoretical discoveries of topological phase transitions and topological phases of matter"*.
‡‡ The Nobel Prize in Chemistry 2016 was awarded jointly to Jean-Pierre Sauvage, Sir J. Fraser Stoddart and Bernard L. Feringa *"for the design and synthesis of molecular machines"*.

oscillatory shear experiment, or equivalently their Fourier transform G(t). The latter is very useful, as it allows to discriminate solid from liquid behavior, G(t) tending towards zero for liquids – if one waits long enough, illustrating that viscoelasticity is also a question of characteristic relaxation time. The poor person studying the bitumen drop experiment might very well be convinced to look at a solid over the time of a PhD – there is one drop about every ten years[§§].

The origin of the viscoelastic properties of soft matter systems reside in their microscopic structure. Indeed, stress is transmitted across the sample by the building blocks of the specific soft matter system under study. These may be interacting neighboring colloidal particles, i.e. nanoparticles, or polymer molecules, etc. The more of them there are, the more stress is transmitted, i.e. the modulus G is directly proportional to the density of microscopic objects n, which is related to the characteristic size of objects ξ by $n = 1/\xi^3$. The third fundamental property (iii) of soft matter is that these building blocks are very big with respect to atoms making up atomic systems, like e.g. metals. Of course the nature of the cohesive forces is different there, we just wish to underline the enormous difference in strength related to the different sizes of the building block of the material. A polymer molecule, for instance, can be made of tens or even hundreds of thousands of atoms. It may be concluded immediately that polymeric systems are thousands of times weaker than, say, metals, and softness is thus related to microstructure (as well as to details of interactions). Similar arguments hold for any soft matter systems discussed in this article. Colloids, e.g., cover a typical size range from about a nanometer (below one would talk of an atomic cluster), to a micron. This upper limit is somewhat magic, as it represents an upper bound for several phenomena typical for soft matter, and deserves discussion. First, too big particles with respect to the wavelength of visible light scatter light strongly: suspensions become opaque, because of multiple light scattering, unless special index-matching conditions are met. Moreover, sedimentation sets in for big particles, because they become too heavy. This brings us to the fourth key property, (iv) Brownian or thermal motion. Its microscopic origin is due to the many random collisions with the much smaller solvent (or any matrix) molecules, and it results in a thermal energy per object of approximately kT – the exact value being not important for comparison, and depending on the number of degrees of freedom, each contributing on average ½kT. If particles are too big, gravitational energy may overcome thermal energy, and particles sediment (or cream, like oil droplets in milk, depending on relative density). To find the critical size of particles below which Brownian motion dominates, one may either argue in terms of gravitational energy for a height of R equaling the thermal energy, or by comparing the particle speed due to thermal diffusion over a length of R to the speed of sedimentation. The critical size is somewhat magically fixed, as comparing displacement of particles by shear to Brownian motion – the ratio is called the Péclet number – gives the same result. The outcome is one micron, which delimits the colloidal domain. The phenomenon of thermal motion which is fundamental to colloid science and soft matter in general is thus linked to microscopic properties, namely collisions with much smaller particles, molecules and atoms, and therefore microstructure and size. Let's add that sedimentation measurements and their analysis, as well as confirmation of Einstein's theory of Brownian motion in 1905 [6], led to the precise determination of the Avogadro number, and to the Nobel prize of Jean Perrin in 1926[***]. It is a noteworthy coincidence in time and place that Theodor Svedberg received the

---

[§§] https://en.wikipedia.org/wiki/Pitch_drop_experiment
[***] The Nobel Prize in Physics 1926 was awarded to Jean Baptiste Perrin *"for his work on the discontinuous structure of matter, and especially for his discovery of sedimentation equilibrium"*.

Noble Price in chemistry the same year for his work on dispersed systems at University of Uppsala[†††].

Coming back to softness, the alert reader might argue that it is not only the number density, but also the strength of the interaction between the building blocks which sets the modulus. Soft matter interaction potentials may be of very different nature: colloids, e.g., may interact through a DLVO potential [7] summing up electrostatic and van der Waals terms, whereas polymer chains may pull on anchor points or induce depletion interaction for entropic reasons, or surfactant molecules might stick together following the hydrophobic effect, or form layers which can possess bending rigidity. The wealth of soft matter science actually comes from the endless possibilities of combinations of these interactions, like polymer molecules with hydrophobic stickers linking micelles, which mutually interact through electrostatic interactions, and so on. However, they all have one thing in common: the typical energy scale of interactions is close to the thermal energy, kT. Coming back to rheology, this property provides us with an estimate of the modulus of soft matter systems: $G = kT/\xi^3$.

As we will not speak much about colloidal interactions in this introductory chapter, we would like to mention that nanoparticles are often hard matter (inorganic) objects, or possibly polymeric latex beads, which usually cannot be compressed. Then the excluded volume interaction needs to be taken into account, opening the road to the transdisciplinary chapter of the (colloidal) glass transition. As often with soft matter, intermediate states exist, like star polymers which may eventually interpenetrate, depending on their architecture, and density.

On the other hand, systems which will be discussed in detail are polymer chains, as they illustrate perfectly the importance of entropy. Due to their many internal degrees of freedom, individual polymer molecules are statistical systems by themselves. As there is no energetical cost involved with going from one conformation to another, they are virtually all energetically equivalent. Thermal energy then generates Brownian motion between the different states. This leads to structure being averaged over large configurational spaces to be explored by the systems, and to dynamical properties as transitions between different conformations. Such dynamics are in the focus of neutron studies. From a thermodynamic point of view, the free energy $F = U - TS$ is essentially given by the entropy term, as U is mostly the same for the different conformations. Ludwig Boltzmann taught us towards the end of the 19th century how to link entropy S to configurations (here polymer conformations, denoted W, but this holds for any system): $S = k \ln W$, k being the famous Boltzmann constant. This is why soft matter systems are entropy-driven, making the exploration of configurational spaces rather than enthalpic ones a key endeavour of soft matter science.

It is not easy to characterize what soft matter really is in one sentence. We have tried here to summarize its main properties (i-iv), which may seem rather disconnected. However, the particular rheological properties, softness and viscoelasticity, are linked to microscopic features common to all systems, which are the importance of entropy, object size and type of interactions – all of them setting the scale of transmitted constraints, as well as the reversibility of links between objects.

The outline of this chapter is the following: various soft matter systems will be introduced, starting with surfactant self-assembly and microemulsions. We then go to polymers, colloids and aggregates, and finally nanocomposite systems. In the second part, the fundamental

[†††] The Nobel Prize in Chemistry 1926 was awarded to The Svedberg *"for his work on dispersed systems"*, namely in physical and colloid chemistry, and the development of the ultracentrifuge.

scattering process of neutrons is introduced, and then discussed for the technique of small angle scattering, with special attention to applications for soft matter science.

3 Physical properties of soft matter systems

3.1 Self-assembled surfactant systems and microemulsions

Self-assembly. Surfactant molecules are amphiphilic molecules, i.e. they possess a hydrophobic and a hydrophilic moiety, which are usually an aliphatic chain (or several chains), and a hydrophilic head group, respectively. According to the nature of the head group, different classes exist, such as ionic and non-ionic surfactant molecules. The imagination and capacities of organic chemists being virtually without limit, many different types of molecules have been synthesized, including zwitterionic surfactants, double chain, or double (called gemini) or multiple head groups. There are two interrelated key properties of these molecules. On the microscopic side, they tend to self-assemble in a selective solvent, here usually water. The hydrophobic parts cluster together to avoid contact with water, and the hydrophilic parts decorate the interface towards water. The simplest aggregate is the micelle, which usually contains some tens, up to about a hundred molecules. This aggregation is the result of a thermodynamic balance between the translational entropy of the molecules, which tends to have them dispersed, and the hydrophobic effect, and it takes place at a well-defined (temperature-dependent) concentration, the critical micellar concentration, or cmc. The second, macroscopic, property which gave its name to the molecule, is that they are surface-active: as one adds them to water, the air-water surface tension is reduced considerably. The surface tension curve plotted against concentration presents a kink, located at the cmc, demonstrating that microstructure and macroscopic properties are intimately related.

Micelles are globular objects of radius approximately given by the aliphatic chain length. It is important to recognize that this radius is actually the radius of curvature of the interface. The exact geometry depends on a second parameter, the head group interactions, which might favor more or less curved interfaces, as described by the packing parameter p defined by $p = V/l\sigma$ [8, 9]. Here V is the molecular volume, σ the head group area resulting from their interactions, and l the extended chain length. The compromise between the two parameters might lead to more spherical or more ellipsoidal surfaces, i.e. the higher the packing parameter, the lower the spontaneous interfacial curvature. This concept can be used to understand the sequence of morphology of micellar phases (called L_1) shown at the bottom of Figure 1: the most bulky head groups give the lowest packing parameter, thus the highest curvature, and globular micelles are obtained. By either adding cosurfactant molecules having a smaller head group (for non-ionics), or screening electrostatic head group repulsion by adding salt (for ionics), the head group area is decreased, the packing parameter is increased, the curvature thus decreased, and micelles (highly curved monolayers in the two principal directions) deform into ellipsoidal structures, and finally cylindrical or wormlike micelles, with the monolayers curved on average only in one direction.

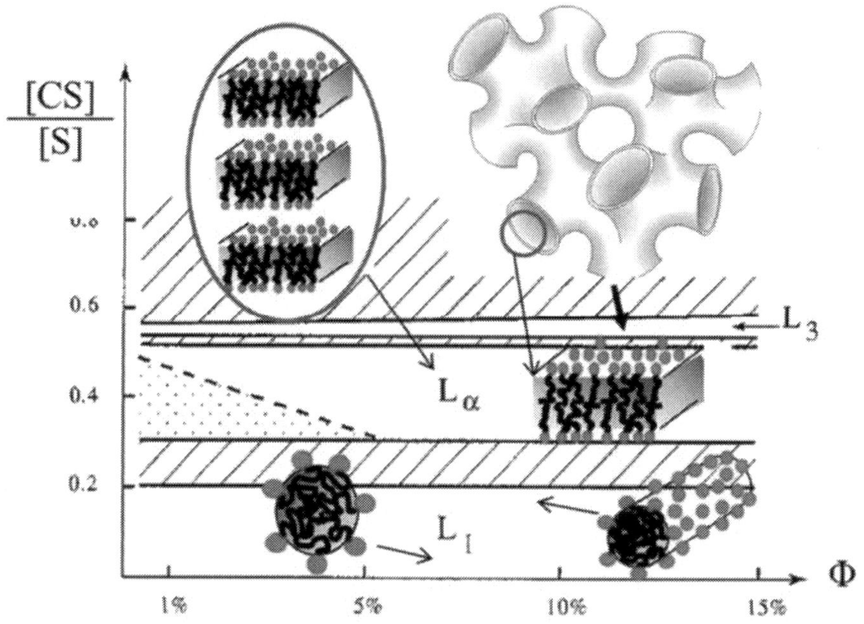

Fig. 1. Phase diagram of Triton X-100/octanol/water system. The x-axis is the total amphiphile weight fraction, the y-axis represents the cosurfactant to surfactant ratio, [Oct]/[TX100]. The different surfactant aggregate morphologies belonging to the different phases are also indicated. Both L_α and L_3 are bilayer phases. Phase diagram first published in [10].

Decreasing the curvature even more leads to bilayer phases made of two on average flat monolayers sticking on each other, as illustrated in the middle part of Figure 1, for a Triton X-100/octanol system. Micelles can thus be seen as a surfactant monolayer which is wrapped up on itself. The physics of surfactant systems is therefore given to a great extent by the bending properties of the monolayer, and of the bilayer for bilayer phases.

The formation of liquid crystalline phases is often encountered in soft condensed matter at high concentration, in particular with anisotropic molecules, or rods. As Lars Onsager – yet another Nobel laureate, although not explicitly for phase transitions[‡‡‡] – understood, this is due to the entropic optimization between individual rotation and collective translational motion. Above a critical concentration, the system prefers arranging into ordered materials, with positional or orientational order (described by the director) typical for solid crystals, while still maintaining its liquid aspects, thus forming liquid crystals. In surfactant or copolymer phases, the formation of liquid crystalline structures can be understood with the physics of monolayers outlined above [11-13]. In particular, the lamellar phase depicted in Figure 1 is already a (smectic A) liquid crystal with positional order in the direction perpendicular to the on average parallel and thus oriented bilayer planes. The physical mechanism underlying the sequence of liquid crystalline phases encountered at high concentrations is that surfactant molecules compete for hydration water of the head groups. As the concentration increases, there is less and less water, and therefore the packing parameter increases, triggering an evolution of aggregate morphology, from globular objects, to cylinders, to bilayers. This results in the formation of cubic micellar phases, followed by columnar liquid crystals. Then, as the monolayer curvature decreases further, branched cylinders are formed, which can arrange in bicontinuous cubic structures also referred to as

[‡‡‡] The Nobel Prize in Chemistry 1968 was awarded to Lars Onsager *"for the discovery of the reciprocal relations bearing his name, which are fundamental for the thermodynamics of irreversible processes"*.

 https://doi.org/10.1051/epjconf/201818801001

"plumber's nightmare". The full zoology of liquid crystalline phases cannot be discussed here in detail, but for the sake of scattering analysis, we note that such ordered structures induce scattering peaks, which can be indexed and used to analyze the structure, as done for example with the famous isasomes [14-18].

In the following section, the most successful theoretical approach – the Helfrich Hamiltonian – to surfactant phases is discussed. It is also directly relevant to microemulsions, which are also presented. For didactical reasons, the introductory parts of a recent review on microemulsions [19] have been included in this article.

Bending properties and morphology of surfactant phases. The bending properties of surfactant monolayers were described in the 1970s by Helfrich et al [20], originally for lipid bilayers. These properties are commonly described by a series expansion of the bending energy for the lowest curvatures. The free energy dF needed to bend a piece of monolayer of surface dA can then be expressed in terms of the two principal curvatures c_1 and c_2, which are related to the principal radii of curvature by $c_i = 1/R_i$. The so-called Helfrich Hamiltonian of the bending free energy reads:

$$dF = \left[\frac{\kappa}{2}(c_1 + c_1 - 2c_0)^2 + \bar{\kappa} c_1 c_2\right] dA \qquad (1)$$

The spontaneous curvature, c_0, describes the state of lowest bending energy of the monolayer - note that notations with and without the factor 2 exist in the literature; here we chose to write the total spontaneous curvature as $2c_0$. Due to symmetry reasons, this term is zero for bilayers or balanced monolayers. Alternatively, the free energy can be formulated based on the total curvature $H = c_1 + c_2$, and the Gaussian curvature given by the product $K = c_1c_2$. The above expression for the bending energy serves as definition of two material constants of the monolayer, the bending moduli, κ and $\bar{\kappa}$, which are sometimes called bending rigidity and saddle-splay modulus, respectively. Both moduli are in units of energy, i.e. do not refer to any length scale, which has the surprising consequence that the bending energy of a spherical shell as depicted in Figure 2 (with $c_0 = 0$) is independent of its radius: $8\pi (\kappa + \bar{\kappa}/2)$.

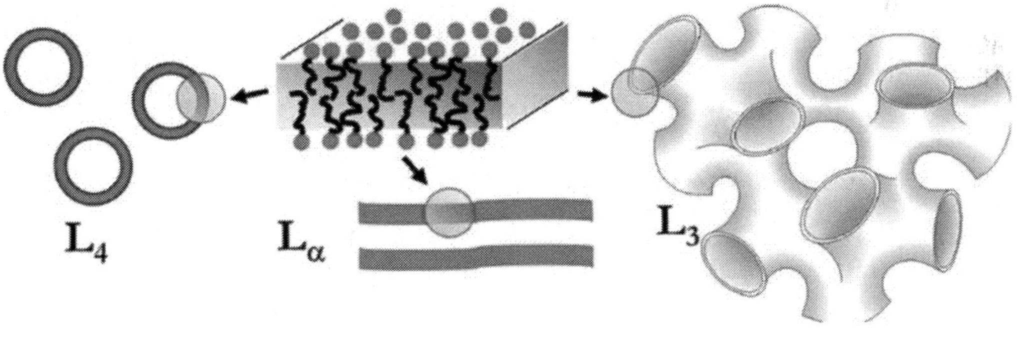

Fig. 2. Surfactant bilayer phases – vesicle, lamellar, and sponge phase.

The roles of κ and $\bar{\kappa}$ are different. It can be shown with the Gauss-Bonnet theorem, that the second term in eq.(1), which is related to the Gaussian curvature, depends only on the topology of the surface, and in particular on the number of handles and independent pieces of membrane or monolayer [21]. As an example for these terms describing the topology of a membrane, one can think of a cup, which has one handle, and is topologically equivalent to a torus (e.g., like a doughnut). In order to have a handle (resp., many), a piece of matter needs to be curved in a particular way, and this is picked up by the Gauss-Bonnet theorem. The dependence on the Gaussian curvature is particularly obvious in the phase sequence of non ionic bilayer phases governed by the surfactant-cosurfactant ratio, where $\bar{\kappa}$ can be shown to trigger morphology changes from vesicular (called L_4) [22], to lamellar (L_α), and finally to sponge (L_3) phases [23, 24], a sequence partially present in the phase diagram in Figure 1, and shown in Figure 2. Concerning this last and most intriguing phase, it should be noted that there is a close relationship between the sponge phase and bicontinuous microemulsions to be discussed below [25]. Having understood the influence of $\bar{\kappa}$, it follows that for a fixed topology of any given structure, local topology-preserving bending fluctuations are entirely determined by κ, and can in principle be determined by dynamical studies [26, 27] .

Alternatively to eq.(1), the bending energy can also be expressed in terms of the deviation of the packing parameter (defined above) of the molecules from their spontaneous value p_0. Indeed, if one integrates over the monolayer thickness λ, the packing parameter is found to be linked to the curvature in order to comply to complete surface coverage: $p = 1 + H\lambda + \frac{1}{3} K\lambda^2$ [28], as long as the topology remains unchanged. This covering relationship allows rewriting the Helfrich Hamiltonian as $\frac{1}{2}\kappa^* (p-p_0)^2$, with $\kappa^* = 2\kappa + \bar{\kappa}$ [29]. One sees that monolayers adopting configurations close to spontaneous packing can only be stable with respect to bending if κ^* is positive, i.e. if there is an energy penalty for spherical geometries.

3.2 Microemulsions

Microemulsions are surfactant systems which contain not only water, but also an immiscible oil. The role of the surfactant molecules is to allow a thermodynamically stable mixture of the two solvents. The canonical definition of microemulsions states that they are thermodynamically stable, isotropic, and liquid solutions of two immiscible fluids, typically oil and water, and surfactant [30]. Such systems present a low interfacial tension, combined with a high interfacial area between the immiscible fluids. They have many applications, basically everything which concerns mixing oily and aqueous matter, from food science to enhanced oil recovery. The properties of microemulsions are intimately related to their microstructure, and as this is a school on neutron scattering, you will be delighted to learn that neutron scattering is perfectly suited for the structural analysis of microemulsions. Besides strong and easily varied contrast, this is due to the essence of microemulsions: their structures are spatial distributions of domains of oil and water, respectively, delimited by a surfactant monolayer. Their geometry might be very complicated, but the measurement of the statistical properties of volumes and surfaces is exactly what neutrons, and in particular small-angle neutron scattering, can do for you. Moreover, by varying the contrast by deuteration, one can highlight either the bulk, or the interfacial layer, the latter being called film contrast. And last but not least, the Porod law gives a direct access to the specific (surfactant) surface, and thus the head group area σ. As already mentioned, the following paragraphs may also be found in a recent review [19].

The size and shape of the oil and water domains depend on the relative amount of oil and water, and surfactant. The usually encountered morphology for minority compounds, like a little oil with much water, are (discontinuous) oil droplets suspended in a continuous water phase, or vice versa. In approximately equal mixtures, both phases may become continuous, and bicontinuity will be discussed below. For more detailed information, we refer the reader to a general introduction to both physics and characterization of microemulsions proposed by Chevalier and Zemb [31], as well as to the bookchapter by Zemb [2].

Besides the volumes of the phases, microemulsion structure depends crucially on the quantity and properties of the surfactant. Indeed, the amount of the latter defines the available specific surface S/V, as all oil-water interfaces need to be covered for energetical reasons: S/V is given by the product of surfactant concentration (number density) and head group area σ. Moreover, if we ignore solubility issues, the bending properties of the surfactant layer determine which type of morphology is accessible, and the strength of thermal fluctuation around these morphologies.

The existence of different phases in the phase diagram of microemulsions can thus be understood from the monolayer bending properties. A high curvature towards oil may set a typical droplet size, and the surfactant concentration the number of droplets. This defines how much oil can be solubilized, and emulsification failure with supernatant oil will in general be found for higher oil concentrations – forming what is called a Winsor I domain in the phase diagram, see Figure 3. There the number of phases of this domain are reported as $\underline{2}$, with the lower bar indicating the droplet phase below in the test tube. If the spontaneous curvature is of opposite sign, inverse systems with solubilized water in a continuous oil phase will be found – in case of emulsification failure as supernatant to an excess water phase, forming a Winsor II domain, noted here $\overline{2}$ in a similar way. Close to zero curvature, bicontinuous structures have been conjectured [32] and been proven to exist by NMR self-diffusion and conductivity experiments [33-36]. In these Winsor III domains, both excess oil and water exist, denoted by the three phases "3" in Figure 3. If one now increases the amount of surfactant, more and more oil and water can be solubilized, up to a point where the excess phases disappear – forming a single bicontinuous microemulsion phase, denoted "1". The phase behavior described here in the curvature-surfactant phase plane for equal quantities of oil and water has the shape of a fish and is called the Kahlweit-Strey fish diagram [37, 38]. It is shown in Figure 3. It gives a direct illustration of the fact that the topology of the phases can be tuned and even inverted (the famous Shinoda cut) [39, 40], opening the road to chemical reactions either in droplets of either type used as microreactors, or at the (giant) interface of bicontinous microemulsions, benefitting of both available solubility channels in the samples[41, 42], also for decontamination[43, 44].

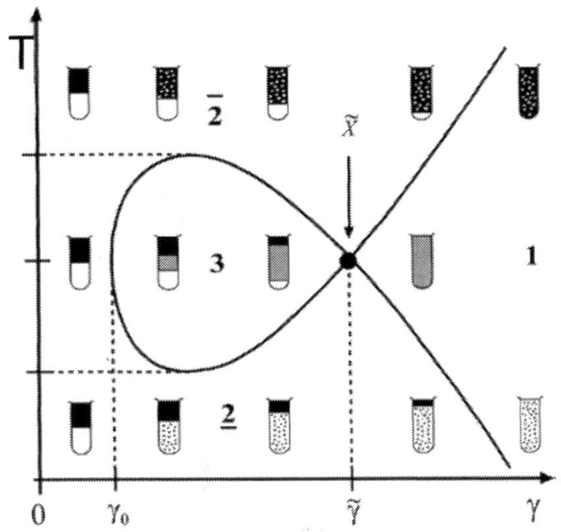

Fig. 3. Typical microemulsion phase diagram as a function of surfactant concentration and temperature. The Winsor phases are indicated, the numbers indicating the number of phases, forming the Kahlweit-Strey fish diagram, see text for details. Picture taken with permission from "Phase behaviour, Interfacial Tension and Microstructure of Microemulsions", T. Sottmann and C. Stubenrauch; pp. 1-47. Blackwell Publishing (Wiley-Blackwell) 2009.

3.3 Polymers

As already mentioned, it was recognized by Staudinger about one hundred years ago that polymers are macromolecules made of many small subunits called monomers. We start the discussion with a polymer molecule suspended in a solvent. Due to the many degrees of freedom, without any enthalpic preference, a polymer molecule is a statistical object in itself, possessing entropy, S = k ln W. The number of configurations depends on external constraints, like confinement, or stress on the molecule. If for instance we pull the chain ends apart, the number of configurations decreases, and so does S, as the logarithm is a monotonic function. In absence of enthalpic contributions, the free energy F = - TS thus increases, meaning that we have to provide work to pull the chain ends apart. This is illustrated in Figure 4, where two out of many possible chain conformations are drawn for an identical end-to-end distance on the left and in the middle, whereas the stretched chain on the right has almost no degrees of freedom left. The corresponding force on the chain ends is thus of entropic origin, and there is an entropic spring constant describing this process. Not surprisingly, it is found to be proportional to kT, as the thermal motion is at the origin of this elasticity. Together with the density argument outlined in section 2, the modulus of a network of chains, called elastomer or rubber, is thus also given by $G = kT/\xi^3$, where the characteristic size ξ now describes the distance between chain connections (called crosslinks, or possibly entanglements).

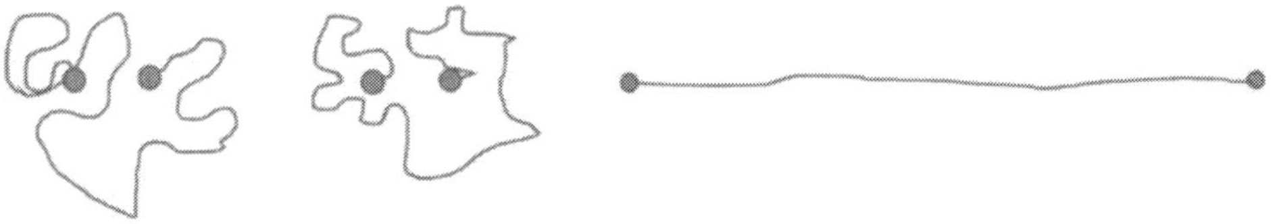

Fig. 4. Illustration of two polymer chains having the same end-to-end vector represented by the blue dots (left and middle). On the right, a stretched chain.

Coming back to individual chains, the distribution of chain ends can be described for an ideal chain, i.e. ignoring monomer interactions and excluded volume, by a Gaussian [45, 46], which is why these chains are also called Gaussian chains. The distribution function for the end-to-end vector r of such a chain made of N segments of length b reads:

$$P(r) = \left(\frac{3}{2\pi N b^2}\right)^{3/2} exp\left(-\frac{3r^2}{2Nb^2}\right) \tag{2}$$

The Gaussian distribution function can be derived from a random walk describing the polymer conformation, and it thus also describes unperturbed Brownian motion of particles, by treating the number of monomers N and the time as analogous quantities. If, however, the excluded volume of the monomers needs to be taken into account, then a self-avoiding random walk is appropriate. The difference between the two is that the self-avoiding chain is bigger due to the monomer repulsion. This is the case for particular solvents, called 'good solvents', whereas the ideal case is only found in so called Θ–solvents, with reference to a specific temperature Θ where monomeric interactions are screened.

The typical size of a macromolecule is described by the average radius of gyration R_g, which is also the quantity directly measured by SANS, see the next section. The scaling laws of this radius with the number of monomers are characteristic: R_g is proportional to $N^{1/2}$, and $N^{3/5}$, for Gaussian and self-avoiding chains, respectively. The exponent is also measurable by SANS, it appears as the inverse of the average fractal dimension of these objects, cf. below. One might add that a single macromolecule will of course never possess exactly the measured radius of gyration: it is the strength of scattering techniques to average over the many molecules evolving in the beam. The type of average, however, depends on polydispersity, which is why some care is necessary for data analysis. On the other hand, the situation is analogous to measuring the temperature in a room: usually no one is interested in the speed of all the air atoms, we are all happy with a single number, T, describing the average energy of all of them.

In order to get an idea of the virtually infinite number of possible conformations of large molecules, we take the example of a long molecule, say one carrying 10000 monomers. If there are three possible bond angles between two of them, and if we ignore excluded volume, then there are 3^{10000} possibilities – an enormous number, approximately 10^{5000}. If it takes one picosecond to explore one configuration, then it takes much much much longer than the age of the universe (!) to explore all individual conformations of a molecule, between completely stretched, and a small coil. Also, the ratio in size between the two extremes is roughly a factor of one hundred, and one can try to imagine all the intermediate possibilities. Luckily, the Gaussian distribution in eq.(2) also tell us that most conformations correspond to approximately the same typical size of the random coil given by Nb^2, leading to a rather well defined average value.

The exploration of the many possible configurations of a large molecule immediately introduces dynamics occurring as the transitions of one microstate to another. Measurement of dynamics with neutrons which are an ideally suited probe for soft matter due to the available q- and energy-range, will be described in the lecture by Jean-Marc Zanotti [26]. We will thus not go into detail here, but would like to mention at least the fundamental concepts of the broad field of polymer dynamics. First, polymers as individual molecules undergo thermal motion, typically described by Rouse and Zimm modes [45]. Experimentally, these motions lead to a loss of correlation with respect to some initial state with time, and correlation functions are exactly what is measured by neutrons with quasi-elastic methods, called QENS [47-50]. With these methods, it is possible to either follow individual protons using incoherent scattering, or entire chains if appropriately labelled, using coherent measurements. In dense polymer systems, like melts, i.e. in absence of solvent, polymer molecules are constrained by their neighboring molecules, as illustrated in Figure 5. It was recognized by De Gennes in the early 1970s that they could only move away from their current conformation by moving along the chain direction. This peculiar motion is called reptation, in analogy with the motion of reptiles, like snakes. Reptation can be imagined as the back- and-forth motion of a chain in a tube formed by the neighboring chains. This generates a very particular 'tube-escape' dynamics, and the tube model has been highly successful in describing microscopic polymer dynamics and rheology. Interestingly, the link between the last two a priori independent measurements is given by the underlying relaxation mechanism of the polymer chains, which are also stress carrying entities. The 'softness' and slow relaxation of soft condensed matter is thus once again based on thermal motion governing interactions and reorganization of large objects, over large length scales.

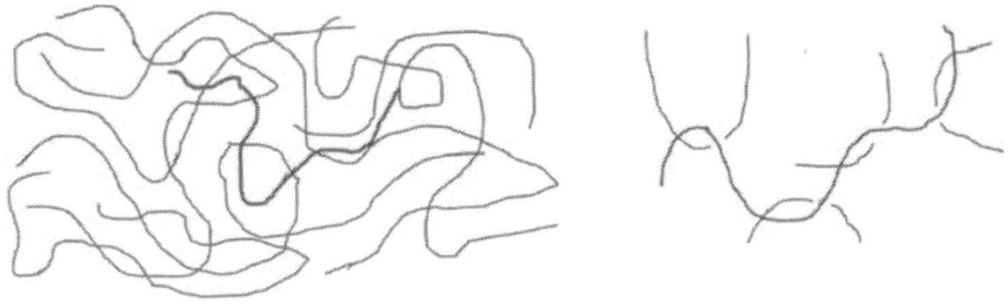

Fig. 5. On the left, illustration of a chain in a melt of chains. On the right, illustration of constrained topology of the red chain by the neighboring chains forming the tube.

As a last comment in this section, we may note that polymer physics did not stop with linear chains, which is mainly the chemists' "fault". Other architectures, and other interactions, exist, of arbitrary complexity, like the already mentioned copolymers, but this is outside the scope of this chapter. The lectures by Combet [51] and Rubatat [52] provide an idea of the infinite possibilities.

3.4 Colloids and aggregates

Colloids are another major building block of soft condensed matter. The word is of Greek origin, related to stickiness. Today one speaks mostly of nanoparticles, but they have been used spontaneously for a long time by mankind. Gold particles studied by M. Faraday in the 19th century are generally cited as the first scientific approach to colloids.

In spite of the many funny shapes nanoparticles may have, we will restrain ourselves here to the discussion of spherical and mostly monodisperse nanoparticles. The dispersion state of nanoparticles suspended in a solvent depends on the interaction between the particles. The translational (or mixing) entropy of the particles tends to a random dispersion, whereas attractive interactions may induce aggregation and phase separation, and repulsive interactions may structure the suspension, and stabilize it. The most important interparticle interactions are excluded volume interactions of the particles themselves, followed by long range electrostatic interactions for charged particles [7], or steric repulsion due protective layers formed by molecules adsorbed or attached (grafted) to the particle surface. On the other hand, attractive van der Waals forces are always present but short-ranged, and depletion forces due to exclusion of smaller particles – they create osmotic pressure – may have intermediate ranges given by the size of the excluded object. Depending on particle concentration, very different phenomena might occur, like the colloidal glass transition at high concentration [53, 54], the discussion of which is however outside the scope of this chapter.

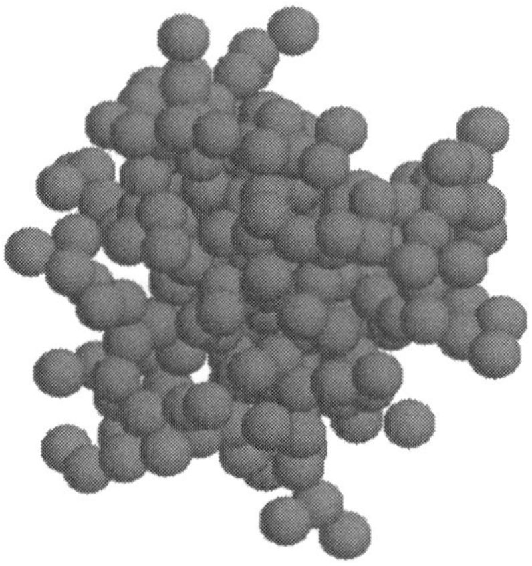

Fig. 6. Illustration of a simulated aggregate of colloidal spheres.

Colloidal aggregates as illustrated in Figure 6 may form due to any of the above-mentioned interactions. They are encountered as soon as colloidal stability is weakened, and their shapes and in particular fractal dimensions have been extensively discussed in the literature. Some thirty years ago, there was a major hype on relating fractal dimensions to the type of diffusion process leading to collisions and 'reaction' in the sense of stickiness, like reaction or diffusion-limited aggregation (RLA or DLA, resp.), or similarly cluster-aggregation on a larger scale. For the purpose of the present chapter, we dare the comment that the hope of universality was somewhat disappointed. It is true, however, that fractal dimensions are the appropriate property for certain classes of aggregates, and that the latter can be easily accessed by scattering methods as explained below. Finally, aggregates are of major importance for some complex applications of soft matter, like polymer nanocomposites, as outlined in the following section.

3.5 Introducing complexity by combining soft matter systems

In the above paragraphs, we have outlined the most fundamental properties of soft matter, with special emphasis on surfactant systems and microemulsions, polymers, colloids, and colloidal aggregates. As already indicated in some cases, these properties can be coupled, and mixed systems are constantly invented. Needless to say, they all cannot be presented here, but some may be mentioned as examples: copolymers and their mixtures with micellar systems [55-57], forming sometimes complex colloids [58-66], or polyelectrolyte complexes [67-69], copolymers added to microemulsions [70, 71], and of course many systems of biophysical [72] or agronomical interest [73, 74]. Moreover, polymer molecules may have interesting thermodynamic properties, like microgel particles displaying a volume phase transition [75-77]. Colloids may form structures, but they may also be studied for their individual motion, like self-propelled particles [78]. Other sub-disciplines of soft condensed matter have also been excluded from this review focused on microscopic aspects, like wetting, or textures of thermotropic liquid crystals.

Polymer nanocomposites are a famous application of both colloids and polymers. These are materials made of hard and usually inorganic nanoparticles embedded in a soft polymer matrix. Car tires are famous examples, with carbon black or silica playing the role of the filler nanoparticles. Their role is to reinforce the polymer, in order to decrease abrasion and increase stiffness and strength, while keeping wet skid properties – the tire sticking to the road – as good as possible, without compromising the lowest possible rolling resistance. The latter is responsible for the about 20% of the fuel consumption, and thus CO_2 production. From a fundamental point of view, many scientific questions arise, related to the rheology of the material, and its non-linear mechanical properties. It is thus important to study both the microscopic structure and the dynamics of such systems, in order to understand 'more useful' properties. The microstructure includes chain conformation and filler dispersion, and the dynamics can be studied on any scale from atoms, to molecules, to nanoparticles, and finally to NP aggregates and networks. As you may doubt, neutron (and also x-ray) scattering is well suitable for such studies. For further reading, the reader is referred to our recent work on model [79-83] and industrial [84-88] systems, including dynamics [89, 90], as well as reviews [91, 92].

After having reviewed the main disciplines and systems of soft condensed matter, we now turn to neutron scattering, and in particular structural characterization by SANS. It is recalled that more details can be found in the following chapters, which corresponds to the lectures given in this school: after an introduction to the properties of neutrons by T. Ederth [93], a variety of applications of neutrons to soft matter systems is presented: Jérôme Combet focuses on polyelectrolytes and small-angle scattering [51], Camille Loupiac et al discuss neutron imaging and tomography applied to food science [94], and Laurent Rubatat introduces block copolymer electrolytes for fuel cells [52]. Then three chapters are devoted to dynamics and two surface techniques, respectively. Jean-Marc Zanotti et al present inelastic and quasi-elastic scattering applied to soft matter [26]. Finally, Fabrice Cousin and Alexis Chennevière discuss neutron reflectivity aiming at density profiles perpendicular to surfaces [95], and Max Wolff, the main organizer of this winter school with Fabrice Cousin, introduces grazing incidence scattering capable of revealing in-surface structures[96].

4 Introduction to neutron scattering and SANS

Neutrons as a probe to explore the structure and dynamics will be presented in detail in the chapter of Thomas Ederth [93]. They are highly suitable for soft matter studies, primarily for four reasons. First, they have a high penetration power of matter, which enables one to study thick samples – typically in the millimeter range, which is a lot more convenient than, say, one hundred nanometer slices necessary for TEM for instance. This does not only limit the introduction of artefacts with sample preparation, but also favors the measurement of true averages over large and thick samples. Secondly, the wavelength of cold neutrons is in the 5 to 20 Å range, and the scattering wave vector q is thus appropriate for the study of nano-structures. Here one also understands immediately that the wavelength of typically 10 Å is still a bit too small as compared to large soft matter structures, which may reach up to thousands of Ångström, with the consequence that one needs to measure at small angles (typically well below a few degrees) in order to characterize these structures [96]. Thirdly, the energy of the neutron is comparable to the thermal energy kT, which we have seen to be characteristic for soft matter systems. It can thus efficiently exchange energy with the sample, and the measurement of these energy changes in inelastic or quasi-elastic scattering experiments give information on the system's dynamics [26]. Finally, the nuclear interaction between the neutrons and the nuclei may be very different for isotopes, and in particular for protons and deuterium. This leads to the possibility of isotopic substitution, i.e. the introduction or modification of scattering contrast via the chemistry. As water is ubiquitous in soft matter studies, the (possibly partial) replacement of H_2O by D_2O can be used to highlight the scattering contributions of parts of a sample, giving access to otherwise hidden information. Another famous example to be discussed below is the structure of polymer chains in a melt.

4.1 Basic scattering process of neutrons

When cold or thermal neutrons interact with matter, their typical wavelength is in the Å to nm range, whereas nuclear interactions take place at a femtometer scale. As a result of this point-like interaction, a spherical secondary neutron wave is created, centered at the position of the scattering nucleus. In case of elastic interactions, this spherical wave has the same wavelength and energy as the incoming wave, but spreads in all directions of space, contrarily to the incoming wave, which is directed towards the beamstop, see Figure 7.

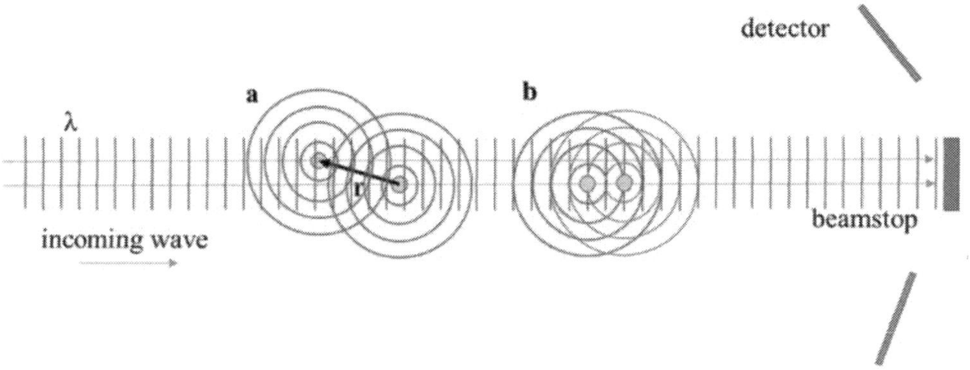

Fig. 7. **(a)** Illustration of spherical secondary waves produced by two scattering centers located at a distance r. Their interference may be measured on a detector. The incoming wave of wavelength λ comes from the left, and the forward direction is to the right. Note that the secondary waves are in phase in the forward direction ($\Delta\Phi$=0) independently of r – a particular configuration r has been chosen to illustrate this in **(b)**, but it is true for all r.

By measuring a non-vanishing neutron flux outside the beamstop, we can thus conclude that there must be at least one atom in the beam! Moreover, the amplitude of the scattered wave is proportional to the scattering power of the nucleus. The latter has the unit of length, and is called the (coherent) scattering length of this nucleus i, usually denoted b_i. The situation becomes more interesting for two atoms, located at a distance r, also shown in Figure 7. Now the two secondary waves can interfere, i.e. their amplitudes add up, and the result of this addition depends on their phase shift. The latter depends on the distance between the atoms r, and on the direction in which the interference is measured. This direction is described by the scattering angle Θ, which is usually defined as the angle between the primary beam and the scattering direction. A straightforward calculation shows that the norm of the scattering vector q is given by:

$$q = 4\pi/\lambda \sin(\theta/2) \tag{3}$$

And the phase shift is given by:

$$\Delta\Phi = q\ r \tag{4}$$

Note that q r is a scalar product of two vectors, but that for isotropic samples the final result is usually expressed in terms of the norm of q given by eq.(3). The phase shift is thus zero for scattering detected in the forward direction (Θ = 0), which will be of importance below. This is also represented graphically in Figure 7. For the moment, it is noted that a measurable impact of a phase shift is possible if the phase shift is of the order of one (or π to be exact). If atoms are close (r small), i.e. if one looks at local structures, then q has to be large in order for the product to be non-zero. On the contrary, if one wishes to study large structures (r big), then, one has to go to small angles. As said already, typical soft matter systems with sizes greater than the neutron wavelength thus imply measurements at small angles, which gives the name to the technique, SANS.

Finally, the alert reader might have noticed that samples usually contain more than only two atoms. This has two consequences. First, the sum over all the amplitudes of pairs of atoms has to be extended to all pairs in order to get the complete interference. Secondly, individual atoms may be regrouped and described by a density, in a continuum approach. This is justified as the size of the atoms are not probed anyway in small-angle experiments. Then the material is not described any more as a sum of many atoms, but as a continuum of a given density. The scattering length introduced above then becomes a scattering length density:

$$\rho = \Sigma\ b_i/V \tag{5}$$

This density is usually calculated over a 'small' volume V, where V contains the atoms taken in the sum, each contributing b_i. In soft matter systems, a typical 'small' volume may be the one of a monomer, or of a homogeneous particle. Heterogeneous systems are then described as a spatial modulation of ρ(r), and in the presence of a matrix of scattering length density ρ_m (which may be a solvent), the contrast Δρ is the origin of small-angle scattering:

$$\Delta\rho(r) = \rho(r) - \rho_m \tag{6}$$

Microstructure is thus described by Δρ(r), and in a small angle scattering experiment, the Fourier transform of its auto-correlation function is measured. An intuitive approach to the meaning of this mathematical transformation is presented below. If objects (atoms, molecules, particles…) move, then a time-dependence is introduced, to be picked up by quasi- or inelastic neutron scattering experiments, see [26].

4.2 Small angle (neutron) scattering

Note again that textbooks and chapters of previous neutron schools exist [2, 4]. Here we develop an argument allowing an intuitive understanding of form factors. It was explained to me some twenty years ago by Loic Auvray, during a long night of SANS measurements, and has helped me ever since. In a small-angle scattering experiment, the intensities coming from the superposition of secondary waves interfering on the detector position are counted. If we are interested in the (only theoretically observable) intensity at zero scattering angle, then the amplitudes of the secondary waves add up, because there is no phase shift as said before. Therefore, the total amplitude is proportional to the number of scattering atoms, possibly weighted by their scattering length density. As the intensity is given by the square of the amplitude, a square of this number arises. One of the two factors is usually converted into the volume fraction, via the product with the number density. The remaining factor counts the number of scattering atoms, and it allows us to recognize that very small angle measurements – like with light scattering –, or better, very low-q measurements, give access to the mass of (independent, uncorrelated) objects, like molecules or particles.

Real measurements are done at finite q-values, and interference must lead to a decrease of the intensity with increasing q, from its maximal value at $q \to 0$. For individual, finite-sized objects, this is called the Guinier domain, and corresponds to a Gaussian in q:

$$I(q) = I_0 \exp(-q^2 R_g^2/3) \tag{7}$$

Here R_g denotes the radius of gyration of the object. It is the most important quantity describing the dimension of any object of finite size. It is related (and can be calculated accordingly) to the distribution of mass around the center of mass, giving a higher weight to masses which are located far from the center, just as in the mechanics of rotation (gyration). For known geometries, like surfactant micelles, directly usable forms of this equation will be given below.

The parameter of the low-q decrease with q is thus given by the total size characterized by R_g. This can be generalized to arbitrary q: if we draw a sphere of diameter q^{-1} around the particle, then all atoms distant by r within that sphere are closer than q^{-1}, leading to q*r being a small number. From eq.(4) we know that the phase shift is thus small, and interference will be constructive, leading to the same counting of atoms (or nuclei) as in the above $q \to 0$ argument. For particles at larger distances, thus located outside the sphere of diameter q^{-1}, there is no constructive interference any more. For form factors at finite q, the mass (or volume) inside a sphere of diameter q^{-1} gives thus the measured intensity at that q-vector:

$$I(q) \approx \text{mass inside sphere of radius } q^{-1} \tag{8}$$

Of course this “equation” can only account for the statistical properties of the mass distribution of an object, which is usually something we are interested in. Once one understands the type of object one observes, one can use more specific geometric form factors, like the ones listed by J.S. Pedersen in ref. [2].

An example of an application of eq.(8) is shown in Figure 8. A cylinder is observed at low and high q, leading to a large and a small sphere of diameter q^{-1}. The volume inside the sphere is given by the length times the cross section S of the cylinder. The intensity is thus large at small q, and decreases with q. As the length is given by q^{-1}, the final intensity reads in the intermediate scattering regime:

$$I(q) = \pi\, S\, \Delta\rho^2\, \Phi\, q^{-1} \tag{9}$$

where $\Delta\rho$ is the contrast of the homogeneous cylinder of volume fraction Φ. The prefactor π comes from the exact Fourier transform. If the cylinder is of finite size, then there exists a low q-value corresponding to a very large sphere containing the entire cylinder: the intensity reaches a finite plateau, described by eq.(7), with I_0 containing the information of the total cylinder mass, or volume. At high q, one reaches the scattering characteristic of the cross section (see below), and finally the Porod regime of the cylinder surface. Here one may note that the q^{-4} Porod law and its generalization to surface fractals, $q^{-(6-Df)}$ is easily obtained with the same argument, which is left as an exercise to the reader.

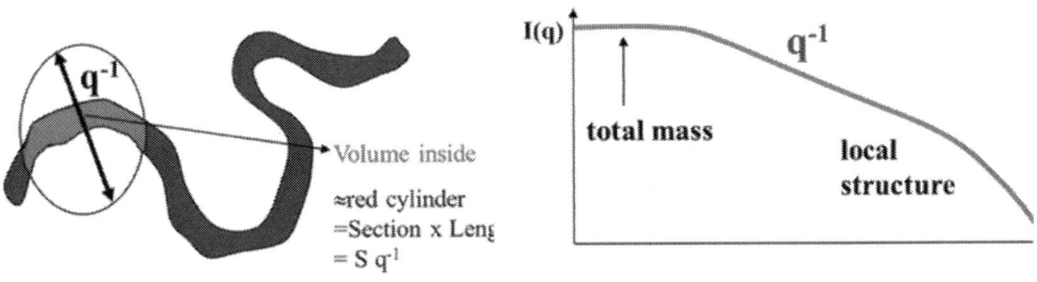

Fig. 8. Example of form factor of a smooth and mostly straight cylinder. The low-q regime gives the total mass and R_g, the intermediate one is calculated using eq.(8) following the sketch on the left, and the high-q domain is given by the Porod law. Undulations of the cylinder as shown on the left would increase the fractal dimension, and thus make the power law steeper.

To summarize, the intermediate scattering regime thus contains information about the spatial distribution of matter. It is closely related to the fractal dimension d – defined by $N = (R/a)^d$ where a is a microscopic scale. Indeed, using this definition, one recognizes that the exponent of minus one in eq.(9) refers to the dimension of one of the cylinder: it is a linear object. If the cylinder were randomly bent, then there would be more mass, the fractal dimension would be higher, and so would the exponent. The latter argument would bring us to the definition of persistence length of wormlike chains from a crossover between fractal regimes, it is also left to the reader. Similarly, the fractal dimension of flat bilayers will lead to a q^{-2} dependence, which is also the case for Gaussian chains. Of course, bilayer scattering (or of any large disk) can be calculated using eq.(8), giving a result similar to eq.(9), but with a prefactor 2π and a $1/q^2$ dependence.

For lack of space, a discussion of structure factors and their peaks cannot be done here. Its thorough treatment necessitates the introduction of concepts of statistical thermomechanics, cf. the chapter by R. Klein in ref. [2]. A simple approach may nonetheless be mentioned, namely dilution lines. It is always beneficial to observe structures as a function of concentration, as scattering peaks move in a particular way depending on the type of order (like liquid crystals, section 3.1), and in particular on its dimensionality. Lamellar phases, e.g., present a one-dimensional order, and the peak position q_0 is proportional to the concentration. Deviations appear when the one-dimensional order breaks-down, in particular at low concentrations. Then lamellae are far apart, and can undulate. The measurement of the peak position thus provides a measurement of the bending modulus of the bilayer allowing these undulations, which brings us to the first specific application of scattering, surfactant phases.

4.3 Small-angle scattering of surfactant phases

We have seen in section 3 that surfactant molecules self-assemble in geometrically well-defined objects, namely spheres, cylinders, and flat bilayers. The form factor scattering of these objects can be calculated analytically. Note that for anisotropic objects, the orientation needs to be specified, or on the contrary an isotropic average over randomly oriented objects may be appropriate. The Guinier expression for the radius of gyration, eq. (7), can be reformulated for these three geometries, i.e. spheres, cylinders, and flat bilayers, assuming random orientations:

$$I(q) = V\ \Phi\ \Delta\rho^2 \exp(-q^2 R_s^2/5) \tag{10a}$$

$$I(q) = \pi\ \Phi\ \Delta\rho^2\ S\ q^{-1} \exp(-q^2 R_c^2/4) \tag{10b}$$

$$I(q) = 2\ \pi\ \Phi\ \Delta\rho^2\ \delta\ q^{-2} \exp(-q^2 \delta^2/12) \tag{10c}$$

Note that the q-dependent prefactors take the fractal dimension of the objects into account. R_s, R_c, and δ refer to the geometric radius of the spheres, the cylindrical cross section, and the bilayer thickness, respectively. Fitting the q-dependence of these expressions thus gives both the fractal dimension and the radius or thickness, while the prefactor gives a measure of the amount of scattering material. These two pieces of information should be treated separately: for a homogenous monodisperse spherical mineral particle, e.g., the radius R_s in eq.(10a) should also give the volume in the prefactor. On the other hand, a polymeric particle might take up solvent, and then extend further into the solution, giving a high "wet" R_s, whereas the scattering volume is still given by the "dry" particle radius. Comparing the two gives thus an estimation of the solvation. A word of caution about fits might also be necessary: as the intensity is the square of the amplitude, the phase information is lost, and a good fit in itself does not give any guarantee of the geometry. As an example, if the fractal dimensions are unavailable, any high-q fit of the exponentials in eq. (10) might fit the data, or if one does, all do, not allowing one to discriminate between the geometries.

Other geometries exist. Bilayers may bend into spherical objects, called vesicles, see Figure 2. Due to the symmetry, the form factor calculation is straightforward, and has been used to determine electrostatic contributions to bending rigidities, for neat and fuzzy interfaces [97]. As additional information, one might add that some structure factors are particularly suited for surfactant systems. Non-ionic micelles or vesicles are usually well described using the Percus Yevick structure factor [98, 99], and charged ones by RMSA (rescaled mean spherical approximation) [100, 101]. Without going into detail, one should at least know that repulsive interactions lead to a scattering peak, of approximate position $2\pi/D$, where D is the typical concentration-dependend center-to-center distance, whereas hard sphere interactions show up only at close contact, i.e. $D = 2R$.

4.4 Small-angle scattering of microemulsions

Neutrons are particularly suited for the structural characterization of microemulsions as isotopic substitution can be used to highlight either the domains, or the surfactant layer. Again for didactical reasons, we reproduce here the scattering part of a recent review on microemulsions [19]. For simple geometries of domains, like droplets, including non spherical or polydisperse droplets, analytical or quasi-analytical descriptions of the scattered intensity as a function of scattering vector q exist, just like the low-q expressions given in eq.(10). One may also note that a straightforward description of the high-q Porod scattering – if interfaces are well-defined – allows the determination of the specific surface S/V, which as already mentioned gives access to the head group area σ of the surfactant molecules. Finally, as with any surfactant system, we would like to add that it is possible to describe fuzzy interfaces [25, 102].

Bicontinuous structures have been proven to be much harder to describe theoretically, although freeze-fracture electron microscopy gave a clear hint at the structure [103]. Loic Auvray et al have shown early that balanced bicontinuous phases may possess a surfactant layer of zero mean curvature using contrast variation [104]. In this context, one may note the discussion of the Kirste-Porod theorem [105] by Auvray and Auroy [106], a theorem which relates the shape of the high-q intensity in Porod presentation $q^4I(q)$ to the interfacial curvature. The small-angle scattering of interconnected domains has been approached in different ways. First, it was recognized that scattering often displays a peak characterizing a nanometric domain size. It is therefore tempting to divide space in cubes (in 3D), and evaluate the amount of surface between (not-identically filled) neighboring cubes, as a function of the volume fraction Φ of for instance the water phase. This approach leads to the determination of a dimensionless number A, given by the product of the specific surface S/V and the characteristic length D determined from the peak position:

$$D^*S/V = A\,\Phi\,(1-\Phi) \qquad (11)$$

In such models, the essential difference between different topologies amounts to different values of the dimensionless number A. For cubes, A = 6. The statistical mechanics of more disordered systems are described by the Talmon-Prager model, which predicts A = 5.82 using a Voronoi tessellation of space [107-109]. Voronoi-type models are based on a geometrical and random subdivision of space, which is also underlying the family of DOC-models proposed by Zemb et al [28, 31, 110, 111]. Obvious difficulties with all of these models are the non physical description of highly bent states at the edges of the cubes or volumes, or at least the non homogeneous distribution of curvature across the samples. Needless to say, fluctuations are entirely absent from such models.

Secondly, analytical models of random mixtures with nanometric domain formation have been proposed, in particular by Debye for inhomogeneous solids [112, 113]. The Debye model for the scattered intensity is simply the Fourier transform of a correlation function $\rho(r)$ with only one characteristic length scale, the domain size D: $\rho(r) \approx \exp(-r/D)$. There is thus no correlation between neighboring domains. In reciprocal space, the intensity reads:

$$I(q) = I_0/(1+q^2D^2)^2 \qquad (12)$$

If one multiplies out the square in eq.(4), one immediately sees that the dominant high-q term obeys a Porod behavior. The prefactor I_0 is related to the product $8\pi D^3\,\Delta\rho^2\,\Phi_1\Phi_2$, where $\Delta\rho$ is the scattering contrast, and Φ_i are the volume fractions. The surface-to-volume ratio is given by $4\,\Phi_1\Phi_2/D$. Due to the absence of inter-domain correlations, however, there is no peak predicted by eq.(12), which is linked to the fact that in the expansion of the square, the quadratic and the biquadratic terms have prefactors based on one length only, D. One may note that a similar equation, called the Ornstein-Zernike equation, exists for the description of rapidly decaying thermal fluctuations. Then the correlation function has the Yukawa form $\rho(r) \approx 1/r \exp(-r/D)$, and the intensity is proportional to $1/(1+q^2D^2)$. In this case, there is no Porod scattering at high q, and no scattering peak indicative of domain size.

The extension of the Debye scattering law to correlated domains leads to the famous Teubner-Strey formula [114]. The correlation function is now proportional to $1/r \sin(2\pi r/D) \exp(-r/\xi)$, and the corresponding intensity may be written as:

$$I(q) = I_0/\,(1 + B\,q^2 + C\,q^4) \qquad (13)$$

Note that fitting eq.(5) to the scattering peaks thus provides the constants B and C, which can be directly transformed into the correlation length ξ and domain size D [115]. Here the prefactors of the q^2 and q^4 term are now different, which allows the description of the scattering peak. The position of the peak is related to domain size D, whereas its width is related to the correlation length. The more ordered the microemulsion, the higher the correlation length, and thus the thinner the peak. The Teubner-Strey function is extremely successful in describing bicontinous microemulsions. A word of caution should however be added: the presence of a peak is no guarantee of bicontinuity, and indeed it is impossible to prove bicontinuity by scattering only, as particulate models might very well work for bicontinuous systems [116]. A peak arising from, say, repelling droplets might be well described by the TS-formula over a limited q-range. It is recommended to always check Porod scattering and dilution lines in order to avoid these traps, and where-ever possible combine with other techniques.

As a last approach to the structure of bicontinous microemulsions, we would like to mention the description of the distribution of oil and water phases using random interfaces as proposed by Safran, Pieruschka, and Marcelja, based on ideas by Cahn and Berk [115, 117, 118].

4.5 Small-angle scattering of polymer molecules

Polymer molecules may have scattering contrast when they are suspended in a (usually deuterated) solvent, or if some deuterated molecules are introduced in melts or networks. The chain conformation in melts has been conjectured to be ideal, i.e. Gaussian, by Flory, and in particular Cotton et al did the first SANS measurements to show that this was actually true [119].

It is recalled that the most useful quantity characterizing individually suspended or embedded macromolecules is the radius of gyration R_g. In section 3, we were happy to learn that a single distribution function like the Gaussian, eq.(2), gives us the necessary information about the molecules. It has been used by P. Debye to calculate the scattering function of an ideal linear chain [120], which reads:

$$P_{Debye}(q) = \frac{2}{q^4 R_g^4}\left(\exp\left(-q^2R_g^2\right) - 1 + q^2R_g^2\right) \quad (14)$$

As one can see, it contains three pieces of information only: the low-q limit related to chain mass and contrast, the radius of gyration, and the fractal dimension, fixed to two. One may add that a parametrized form of the much more complicated expression for self-avoiding chains has been proposed by Pedersen and Schurtenberger [121, 122].

As an outlook to more complex situations, small-angle scattering may be used to study individual (isotopically labelled) polymer molecule in a network of otherwise identical chains. Moreover, the network can then be macroscopically deformed, and by observing the microscopic chain deformation on a 2D-detector, the transmission of stresses from the macroscopic to the microscopic scale can be investigated [123-126]. Another complex but common situation is the conformation of macromolecules on a surface. If the surface is a nanoparticle, then small-angle scattering can be used to access the polymer structure [81, 127-130]. If on the other hand, the surface is flat, then reflectometry and possibly GISANS are the methods of choice [95, 131-133].

4.6 Small-angle scattering of colloids and aggregates

Small-angle scattering of colloids may be discussed separately for form factors P(q) and structure factors S(q). Ignoring prefactors, the form factors are the intensities measured in dilute suspension, in absence of interactions, where S(q) equals one. The intuitive picture outlined in sections 4.2 can be used to have an idea of the dominant intermediate-q scattering law. More exact expressions may be obtained by Fourier transforming the contrast function describing the geometry of the particles. For the most important (and easiest) case, spheres of radius RS, one finds:

$$P(q) = 9\frac{[sin(qR_S) - qR_S\, cos(qR_S)]^2}{[qR_S]^6} \tag{15}$$

Many expressions, often necessitating a numerical integration, have been proposed for different geometries, like cubes for instance (see J. S Pedersen in ref. [2]). As a little exercise, the reader may do the small-q development of eq.(15) and find the Guinier expression, eq. (10a).

The interaction between spheres (or any objects) are described by S(q). Note that we directly apply two simplifications here, namely monodispersity, and isotropy, which allows us to use the abovementioned structure factors [98, 99, 101, 134] also for liquids or gases of colloids with excluded volume, and possibly electrostatic, or sticky interactions. The scattered intensity is then directly given by the product of P(q) times S(q), with a prefactor related to contrast and concentration. A large literature exists on such structure factor calculations based on MC simulations or integral equation theories [135], cf. the chapter by R. Klein in ref. [2], or the one by L. Belloni in ref. [136], respectively, all of which are outside the scope of this introductory chapter.

We would like to discuss shortly another approach to particle interactions, which is often encountered in experiments, mostly unwantedly, but not always. Sticky interactions, e.g. due to van der Waals (dispersion) forces, may induce the formation of aggregates, i.e. particle clustering. The latter may be described by a structure factor of spheres, which may also be looked at as the form factor of aggregates after multiplication by the sphere form factor. Then the thermodynamic problem turns into a geometrical one, because particles now sit on center-to-center distances of two radii. The structure factor of an aggregate of N monodisperse spheres can be described by the following expression:

$$S(q) = 1 + \frac{2}{N}\sum_{i>j}^{N} \frac{sin\left(q(r_j - r_i)\right)}{q(r_j - r_i)} \tag{16}$$

where the Debye average has been used [120, 137]. The 'secret' of this expression is hidden in the fact that eq.(16) expresses only the isotropically averaged Fourier transform of the center-of-mass positions, but we do not know the latter yet. By building a thermodynamic or geometrical model, one may propose a set of positions [138], and calculate S(q). Its properties are: (i) a high-q limit of one, (ii) a low-q limit of N, i.e. giving the aggregation number from a low-q measurement, (iii) a peak at 2π/2R= π/R, (iv) possibly a power-law domain at low-q characterizing aggregate fractality, and finally, (v) a correlation hole just before the peak. Note that different ways as compared to eq.(16) to calculate such structure factors exist, like not doing the isotropic average, or using the pair distribution function, their suitability depending on the application.

It immediately follows that one can deduce an estimate of R and of N from such measurements. Moreover, one may determine the internal density from the depth of the correlation hole [139]. For polydisperse spheres, and polydisperse aggregates, equivalent relationships can be derived. To finish this discussion, one may also use eq.(16) to proceed with an inverse approach, called reverse Monte Carlo (RMC, see ref. [140]). This allows one to find a distribution of particles in space compatible with the scattered intensity.

Polymer nanocomposites as a typical soft-matter system have been introduced above. The knowledge of aggregate structure is of primary importance, and we have outlined here how one can analyze the corresponding scattering data. Moreover, it is important to characterize not only filler but also chain structure. The latter is possible using the framework of polymer scattering discussed above, with isotopic substitution both to highlight the chain structure, and contrast match the particles. This is called a zero-average contrast (ZAC) measurement.

To summarize, neutrons are thus perfectly suited for a complete dynamical and structural analysis of nanocomposites.

5 Conclusion

We have tried to present a general view on soft matter systems. They are entropy-dominated systems held together by weak interactions, which explore configurational space by Brownian motion. This leads to many fascinating properties of soft matter, like self-assembly of surfactants, forming systems of great morphological variety. They can be understood with the help of a few simple concepts presented in this chapter, like the bending elasticity described by the Helfrich Hamiltonian. Similar approaches have been developed for microemulsions, the statistical mechanics and dynamical properties of polymers, or the random walks or aggregation of colloids. Cold neutrons are perfect probes for such nano-scale structures, and thermally activated transitions or fluctuations, due to their penetration power, their wave-length and energy. In the second part of this chapter, we have tried to outline an intuitive approach to small-angle scattering by typical soft matter structures, and summarized the main tools to extract structural information from scattered intensities.

I thank the organizers of the French-Swedish winter school on neutrons and soft matter, in particular Max Wolff and Fabrice Cousin, for inviting me to Uppsala, as well as for critical reviewing of the present chapter. Some ideas and comments are due to lectures and discussions with S. Egelhaaf (Düsseldorf), A. Ryan (Sheffield), and O. Glatter (Graz).

References

[1] T. Imae, T. Kanaya, M. Furusaka, N. Torikai, Neutrons in Soft Matter, John Wiley and Sons, Hoboken, New Jersey, 2011.
[2] P. Lindner, Neutrons, X-ray and Light Scattering, North Holland, Elsevier, Amsterdam, 2002.
[3] F. Cousin, I. Grillo, J. Jestin, J. Oberdisse, Neutrons et Matière Molle, EDP Sciences, 2010.
[4] F. Cousin, Small angle neutron scattering, in: M. Ceretti, W. Paulus, M.H. Mathon, C. Ritter (Eds.) Jdn 21 - Neutrons and Materials for Energy, 2015.
[5] C.W. Macosko, Rheology: Principles, Measurements and Applications, Wiley/VCH, New York, 1994.
[6] A. Einstein, Annalen Der Physik, 19 (1906) 371-381.
[7] E.J.W. Verwey, J.T.G. Overbeek, Theory of the stability of lyophobic colloids, Amsterdam: Elsevier., 1948.
[8] J.N. Israelachvili, D.J. Mitchell, B.W. Ninham, Journal of the Chemical Society-Faraday Transactions Ii, 72 (1976) 1525-1568.
[9] D.J. Mitchell, B.W. Ninham, Journal of the Chemical Society-Faraday Transactions Ii, 77 (1981) 601-629.
[10] J. Oberdisse, C. Couve, J. Appell, J.F. Berret, C. Ligoure, G. Porte, Langmuir, 12 (1996) 1212-1218.
[11] P. Alexandridis, U. Olsson, B. Lindman, Abstracts of Papers of the American Chemical Society, 213 (1997) 221-POLY.
[12] P. Alexandridis, U. Olsson, B. Lindman, Langmuir, 13 (1997) 23-34.
[13] P. Alexandridis, U. Olsson, B. Lindman, Langmuir, 14 (1998) 2627-2638.
[14] H. Qiu, M. Caffrey, Biomaterials, 21 (2000) 223-234.

[15] A. Salonen, S. Guillot, O. Glatter, Langmuir, 23 (2007) 9151-9154.
[16] S. Salentinig, A. Yaghmur, S. Guillot, O. Glatter, Journal of Colloid and Interface Science, 326 (2008) 211-220.
[17] M. Tomsic, S. Guillot, L. Sagalowicz, M.E. Leser, O. Glatter, Langmuir, 25 (2009) 9525-9534.
[18] S. Guillot, S. Salentinig, A. Chemelli, L. Sagalowicz, M.E. Leser, O. Glatter, Langmuir, 26 (2010) 6222-6229.
[19] J. Oberdisse, T. Hellweg, Advances in Colloid and Interface Science (2017) to be published.
[20] W. Helfrich, Zeitschrift Fur Naturforschung C-a Journal of Biosciences, C 28 (1973) 693-703.
[21] G. Porte, Current Opinion in Colloid & Interface Science, 1 (1996) 345-349.
[22] P. Herve, D. Roux, A.M. Bellocq, F. Nallet, T. Gulikkrzywicki, Journal De Physique Ii, 3 (1993) 1255-1270.
[23] G. Porte, Journal of Physics-Condensed Matter, 4 (1992) 8649-8670.
[24] G. Porte, J. Appell, P. Bassereau, J. Marignan, Journal De Physique, 50 (1989) 1335-1347.
[25] R. Strey, J. Winkler, L. Magid, J. Phys. Chem., 95 (1991) 7502-7507.
[26] Q. Berrod, K. Lagrené, J. Ollivier, J.-M. Zanotti, Inelastic and Quasi-Elastic Neutron Scattering. Application to Soft-Matter, in: F. Cousin, M. Wolff (Eds.) Neutrons and Soft Matter, EDP Sciences, 2018.
[27] T. Hellweg, D. Langevin, Physical Review E, 57 (1998) 6825-6834.
[28] T.N. Zemb, S.T. Hyde, P.J. Derian, I.S. Barnes, B.W. Ninham, J. Phys. Chem., 91 (1987) 3814-3820.
[29] M. Duvail, J.F. Dufreche, L. Arleth, T. Zemb, Physical Chemistry Chemical Physics, 15 (2013) 7133-7141.
[30] I. Danielsson, B. Lindman, Colloids and Surfaces, 3 (1981) 391-392.
[31] Y. Chevalier, T. Zemb, Reports on Progress in Physics, 53 (1990) 279-371.
[32] L.E. Scriven, Nature, 263 (1976) 123-125.
[33] F. Larche, J. Rouviere, P. Delord, B. Brun, J.L. Dussossoy, Journal De Physique Lettres, 41 (1980) L437-L440.
[34] M. Kahlweit, R. Strey, D. Haase, H. Kunieda, T. Schmeling, B. Faulhaber, M. Borkovec, H.F. Eicke, G. Busse, F. Eggers, T. Funck, H. Richmann, L. Magid, O. Soderman, P. Stilbs, J. Winkler, A. Dittrich, W. Jahn, Journal of Colloid and Interface Science, 118 (1987) 436-453.
[35] B. Lindman, K. Shinoda, U. Olsson, D. Anderson, G. Karlstrom, H. Wennerstrom, Colloids and Surfaces, 38 (1989) 205-224.
[36] U. Olsson, K. Shinoda, B. Lindman, J. Phys. Chem., 90 (1986) 4083-4088.
[37] M. Kahlweit, R. Strey, G. Busse, Physical Review E, 47 (1993) 4197-4209.
[38] R. Strey, Colloid Polym. Sci., 272 (1994) 1005-1019.
[39] K. Shinoda, in: K. Shinoda (Ed.) Solvent properties of surfactant solutions, Marcel Dekker, New York, 1967.
[40] K. Shinoda, S.E. Friberg, Emulsions and Solubilization, Wiley-Interscience, New York, 1986.
[41] R. Johannsson, M. Almgren, R. Schomacker, Langmuir, 9 (1993) 1269-1273.
[42] V. Ganesan, G.H. Fredrickson, Journal of Chemical Physics, 113 (2000) 2901-2917.
[43] S. Wellert, H. Imhof, M. Dolle, H.J. Altmann, A. Richardt, T. Hellweg, Colloid Polym. Sci., 286 (2008) 417-426.
[44] S. Vargas-Ruiz, C. Schulreich, A. Kostevic, B. Tiersch, J. Koetz, S. Kakorin, R. von Klitzing, M. Jung, T. Hellweg, S. Wellert, Journal of Colloid and Interface Science, 471 (2016) 118-126.

[45] M. Doi, S.F. Edwards, The Theory of Polymer Dynamics, Oxford University Press, Oxford, 1986.
[46] M. Doi, Introduction to polymer physics, Oxford Science Publications, Clarendon Press, Oxford, 1996.
[47] B. Frick, D. Richter, Science, 267 (1995) 1939-1945.
[48] B. Ewen, D. Richter, Neutron spin echo investigations on the segmental dynamics of polymers in melts, networks and solutions, in: Neutron Spin Echo Spectroscopy Viscoelasticity Rheology, 1997, pp. 1-129.
[49] A. Arbe, J. Colmenero, D. Richter, M. Monkenbusch, L. Willner, B. Farago, Pramana-Journal of Physics, 63 (2004) 33-40.
[50] A.R. Bras, R. Pasquino, T. Koukoulas, G. Tsolou, O. Holderer, A. Radulescu, J. Allgaier, V.G. Mavrantzas, W. Pyckhout-Hintzen, A. Wischnewski, D. Vlassopoulos, D. Richter, Soft Matter, 7 (2011) 11169-11176.
[51] J. Combet, Polyelectrolytes and Small Angle Scattering in: F. Cousin, M. Wolff (Eds.) Neutrons and Soft Matter, EDP Sciences, 2018.
[52] L. Rubatat, Block Copolymer Electrolytes for Fuel Cells and Secondary Batteries, the Small Angle Neutron Scattering Inputs in: F. Cousin, M. Wolff (Eds.) Neutrons and Soft Matter, EDP Sciences, 2018.
[53] S. Mazoyer, L. Cipelletti, L. Ramos, Physical Review Letters, 97 (2006).
[54] G. Brambilla, D. El Masri, M. Pierno, L. Berthier, L. Cipelletti, G. Petekidis, A.B. Schofield, Physical Review Letters, 102 (2009).
[55] E. Michel, M. Filali, R. Aznar, G. Porte, J. Appell, Langmuir, 16 (2000) 8702-8711.
[56] F. Molino, J. Appell, M. Filali, E. Michel, G. Porte, S. Mora, E. Sunyer, Journal of Physics-Condensed Matter, 12 (2000) A491-A498.
[57] E. Michel, J. Appell, F. Molino, J. Kieffer, G. Porte, Journal Of Rheology, 45 (2001) 1465-1477.
[58] J.F. Berret, G. Cristobal, P. Herve, J. Oberdisse, I. Grillo, European Physical Journal E, 9 (2002) 301-311.
[59] P. Herve, M. Destarac, J.F. Berret, J. Lal, J. Oberdisse, I. Grillo, Europhysics Letters, 58 (2002) 912-918.
[60] J.F. Berret, J. Oberdisse, Abstracts of Papers of the American Chemical Society, 226 (2003) U404-U405.
[61] J.F. Berret, P. Herve, O. Aguerre-Chariol, J. Oberdisse, Journal of Physical Chemistry B, 107 (2003) 8111-8118.
[62] J.-F. Berret, J. Oberdisse, Physica B-Condensed Matter, 350 (2004) 204-206.
[63] J.-F. Berret, B. Vigolo, R. Eng, P. Herve, I. Grillo, L. Yang, Macromolecules, 37 (2004) 4922-4930.
[64] K. Yokota, M. Morvan, J.-F. Berret, J. Oberdisse, Europhys. Lett., 69 (2005) 284-290.
[65] J.-F. Berret, K. Yokota, M. Morvan, R. Schweins, J. Phys. Chem. B, 110 (2006) 19140-19146.
[66] J. Fresnais, J.-F. Berret, L. Qi, J.-P. Chapel, J.-C. Castaing, O. Sandre, B. Frka-Petesic, R. Perzynski, J. Oberdisse, F. Cousin, Phys. Rev. E, 78 (2008) 040401.
[67] I. Hoffmann, S. Prevost, M. Medebach, S. Rogers, N.J. Wagner, M. Gradzielski, Tenside Surfactants Detergents, 48 (2011) 488-494.
[68] L. Chiappisi, I. Hoffmann, M. Gradzielski, Soft Matter, 9 (2013) 3896-3909.
[69] L. Chiappisi, H. Yalcinkaya, V.K. Gopalakrishnan, M. Gradzielski, T. Zemb, Colloid Polym. Sci., 293 (2015) 3131-3143.
[70] H. Frielinghaus, D. Byelov, J. Allgaier, D. Richter, B. Jakobs, T. Sottmann, R. Strey, Applied Physics a-Materials Science & Processing, 74 (2002) S408-S410.
[71] S. Maccarrone, J. Allgaier, H. Frielinghaus, D. Richter, Langmuir, 30 (2014) 1500-1505.

[72] M. Abkarian, G. Massiera, L. Berry, M. Roques, C. Braun-Breton, Blood, 117 (2011) 4118-4124.
[73] A. Banc, C. Charbonneau, M. Dahesh, M.S. Appavou, Z. Fu, M.H. Morel, L. Ramos, Soft Matter, 12 (2016) 5340-5352.
[74] M. Dahesh, A. Banc, A. Duri, M.H. Morel, L. Ramos, Food Hydrocolloids, 52 (2016) 1-10.
[75] B. Wedel, M. Zeiser, T. Hellweg, Zeitschrift Fur Physikalische Chemie-International Journal of Research in Physical Chemistry & Chemical Physics, 226 (2012) 737-748.
[76] T. Hellweg, Journal of Polymer Science Part B-Polymer Physics, 51 (2013) 1073-1083.
[77] Y. Hertle, T. Hellweg, Journal of Materials Chemistry B, 1 (2013) 5874-5885.
[78] X.L. Wang, M. In, C. Blanc, M. Nobili, A. Stocco, Soft Matter, 11 (2015) 7376-7384.
[79] M. Tatou, Nanocomposites modèles silice-latex: Etude des propriétés rhéologiques et de la structure des charges et des chaînes par Diffusion de Neutrons aux Petits Angles, in: LCVN UMR 5587, University of Montpellier II, Montpellier, 2010.
[80] M. Tatou, A.C. Genix, A. Imaz, J. Forcada, A. Banc, R. Schweins, I. Grillo, J. Oberdisse, Macromolecules, 44 (2011) 9029-9039.
[81] A.C. Genix, M. Tatou, A. Imaz, J. Forcada, R. Schweins, I. Grillo, J. Oberdisse, Macromolecules, 45 (2012) 1663–1675.
[82] A. Banc, A.C. Genix, M. Chirat, C. Dupas, S. Caillol, M. Sztucki, J. Oberdisse, Macromolecules, 47 (2014) 3219–3230.
[83] A. Banc, A.C. Genix, C. Dupas, M. Sztucki, R. Schweins, M.S. Appavou, J. Oberdisse, Macromolecules, 48 (2015) 6596–6605.
[84] G.P. Baeza, A.C. Genix, C. Degrandcourt, L. Petitjean, J. Gummel, R. Schweins, M. Couty, J. Oberdisse, Macromolecules, 46 (2013) 6388–6394.
[85] G.P. Baeza, A.C. Genix, C. Degrandcourt, J. Gummel, A. Mujtaba, K. Saalwächter, T. Thurn-Albrecht, M. Couty, J. Oberdisse, ACS Macro Letters, 3 (2014) 448-452.
[86] G.P. Baeza, A.C. Genix, C. Degrandcourt, L. Petitjean, J. Gummel, M. Couty, J. Oberdisse, Macromolecules, 46 (2013) 317-329.
[87] G.P. Baeza, A.C. Genix, C. Degrandcourt, J. Gummel, M. Couty, J. Oberdisse, Soft Matter, 10 (2014) 6686–6695.
[88] G.P. Baeza, A.-C. Genix, N. Paupy-Peyronnet, C. Degrandcourt, M. Couty, J. Oberdisse, Faraday Discussions, 186 (2016) 295-309
[89] G.P. Baeza, J. Oberdisse, A. Alegria, M. Couty, A.C. Genix, Phys. Chem. Chem. Phys., 17 (2015) 1660-1666.
[90] G.P. Baeza, J. Oberdisse, A. Alegria, K. Saalwaechter, M. Couty, A.-C. Genix, Polymer, 73 (2015) 131-138.
[91] A.C. Genix, G.P. Baeza, J. Oberdisse, European Polymer Journal, 85 (2016) 605-619.
[92] A.-C. Genix, J. Oberdisse, Current Opinion in Colloid & Interface Science, 20 (2015) 293-303.
[93] T. Ederth, Neutrons for scattering: What they are, where to get them, and how to deal with them, in: F. Cousin, M. Wolff (Eds.) Neutrons and Soft Matter, EDP Science, Les Ulis, 2018.
[94] C. Loupiac, A. Helary, S. Desert, F. Ott, Neutron imaging and tomography: applications in food science in: F. Cousin, M. Wolff (Eds.) Neutrons and Soft Matter, EDP Sciences, 2018.
[95] F. Cousin, A. Chennevière, Neutron reflectivity for Soft Matter, in: F. Cousin, M. Wolff (Eds.) Neutrons and Soft Matter, EDP Sciences, 2018.
[96] M. Wolff, Grazing incidence scattering, in: F. Cousin, M. Wolff (Eds.) Neutrons and Soft Matter, EDP Sciences, 2018.
[97] J. Oberdisse, G. Porte, Physical Review E, 56 (1997) 1965-1975.
[98] J.K. Percus, G.J. Yevick, Physical Review, 110 (1958) 1-13.

[99] M.S. Wertheim, Physical Review Letters, 10 (1963) 321-&.
[100] J.B. Hayter, J. Penfold, Molecular Physics, 42 (1981) 109-118.
[101] J.P. Hansen, J.B. Hayter, Molecular Physics, 46 (1982) 651-656.
[102] M. Gradzielski, D. Langevin, L. Magid, R. Strey, J. Phys. Chem., 99 (1995) 13232-13238.
[103] W. Jahn, R. Strey, J. Phys. Chem., 92 (1988) 2294-2301.
[104] L. Auvray, J.P. Cotton, R. Ober, C. Taupin, J. Phys. Chem., 88 (1984) 4586-4589.
[105] V.R. Kirste, G. Porod, Kolloid-Zeitschrift and Zeitschrift Fur Polymere, 184 (1962) 1.
[106] L. Auvray, P. Auroy, Scattering by interfaces: variations on Porod's law, in: P. Lindner, T. Zemb (Eds.) Neutron, X-ray, and Light Scattering, 1st edition, North Holland, Amsterdam, 1991.
[107] Y. Talmon, S. Prager, Nature, 267 (1977) 333-335.
[108] Y. Talmon, S. Prager, Journal of Chemical Physics, 76 (1982) 1535-1538.
[109] Y. Talmon, S. Prager, Journal of Chemical Physics, 69 (1978) 2984-2991.
[110] B.W. Ninham, I.S. Barnes, S.T. Hyde, P.J. Derian, T.N. Zemb, Europhysics Letters, 4 (1987) 561-568.
[111] T.N. Zemb, Colloids and Surfaces a-Physicochemical and Engineering Aspects, 129 (1997) 435-454.
[112] P. Debye, A.M. Bueche, Journal of Applied Physics, 20 (1949) 518-525.
[113] P. Debye, H.R. Anderson, H. Brumberger, Journal of Applied Physics, 28 (1957) 679-683.
[114] M. Teubner, R. Strey, Journal of Chemical Physics, 87 (1987) 3195-3200.
[115] P. Pieruschka, S.A. Safran, Europhysics Letters, 22 (1993) 625-630.
[116] N. Freiberger, C. Moitzi, L. de Campo, O. Glatter, Journal of Colloid and Interface Science, 312 (2007) 59-67.
[117] P. Pieruschka, S.A. Safran, Journal of Physics-Condensed Matter, 6 (1994) A357-A362.
[118] P. Pieruschka, S.A. Safran, Europhysics Letters, 31 (1995) 207-212.
[119] J.P. Cotton, D. Decker, H. Benoit, B. Farnoux, J. Higgins, G. Jannink, R. Ober, C. Picot, J.D. Cloizeau, Macromolecules, 7 (1974) 863-872.
[120] P. Debye, Journal of Physical and Colloid Chemistry, 51 (1947) 18-32.
[121] J.S. Pedersen, M. Laso, P. Schurtenberger, Physical Review E, 54 (1996) R5917-R5920.
[122] J.S. Pedersen, P. Schurtenberger, Macromolecules, 29 (1996) 7602-7612.
[123] J. Bastide, J. Herz, F. Boue, Journal De Physique, 46 (1985) 1967-1979.
[124] J. Bastide, F. Boue, Physica A, 140 (1986) 251-260.
[125] F. Boue, B. Farnoux, J. Bastide, A. Lapp, J. Herz, C. Picot, Europhysics Letters, 1 (1986) 637-645.
[126] F. Boue, J. Bastide, M. Buzier, A. Lapp, J. Herz, T.A. Vilgis, Colloid Polym. Sci., 269 (1991) 195-216.
[127] J. Oberdisse, Current Opinion in Colloid & Interface Science, 12 (2007) 3-8.
[128] G. Carrot, A. El Harrak, J. Oberdisse, J. Jestin, F. Boue, Soft Matter, 2 (2006) 1043-1047.
[129] J.S. Pedersen, M.C. Gerstenberg, Macromolecules, 29 (1996) 1363-1365.
[130] J.S. Pedersen, C. Svaneborg, Current Opinion in Colloid & Interface Science, 7 (2002) 158-166.
[131] G. Fragneto, A. Menelle, European Physical Journal Plus, 126 (2011).
[132] G. Fragneto, European Physical Journal-Special Topics, 213 (2012) 327-342.
[133] J. Penfold, R.K. Thomas, Current Opinion in Colloid & Interface Science, 19 (2014) 198-206.
[134] R.J. Baxter, The Journal of chemical physics, 49 (1968) 2770-2774.

[135] J.P. Hansen, I.R. McDonald, Theory of Simple Liquids, Academic Press, London, 1986.
[136] L. Belloni, La Diffusion de Neutrons aux Petits Angles: mécanique statistique des liquides et traitement des données, EDP Sciences, 2010.
[137] P. Debye, Annalen Der Physik, 46 (1915) 809-823.
[138] J. Oberdisse, Y. Rharbi, F. Boue, Computational and Theoretical Polymer Science, 10 (2000) 207-217.
[139] A.-C. Genix, J. Oberdisse, Soft Matter, 13 (2017), 8144-8155
[140] J. Oberdisse, P. Hine, W. Pyckhout-Hintzen, Soft Matter, 2 (2007) 476-485.

Neutrons for scattering: What they are, where to get them, and how to deal with them

Thomas Ederth*
Division of Molecular Physics, IFM, Linköping University, 581 83 Linköping, Sweden

Abstract. In neutron scattering studies of soft matter, a diverse array of methods and instruments are used, providing information on structure and dynamics on various length and energy scales. However, much of the infrastructure needed for neutron scattering is common for many instruments. After a brief historical retrospect of neutron scattering, this chapter introduces the basic infrastructure needed to conduct scattering experiments. This includes equipment that is used to produce, spectrally adjust and purify, and to deliver neutrons to the instruments where scattering experiments are conducted. The basics of the interaction of neutrons with matter is also introduced, as a preparation for the final sections on the different means at hand for neutron detection.

1 Introduction

1.1 Neutrons as probes for the structure of matter

Radiation from most of the electromagnetic spectrum is used to probe structure and excitations in matter, as befits the required length or energy scales. Neutron radiation have several advantages for such purposes: neutrons have (or, rather, as we will see, can be made to have) wavelengths comparable to interatomic and intermolecular separations (0.1-10 Å), energies of the same magnitude as molecular excitations (1-100 meV), such as molecular vibrations or rotations, and lattice vibrations. Neutrons interact weakly with matter, which has several benefits: they penetrate deeply, so that buried interfaces and bulk samples can be studied, theoretical treatment can be simplified (for example by using the Born approximation and ignoring multiple scattering), and this permits elaborate sample environments and containers to be used. Neutron contrast varies in a non-systematic manner between nearby elements and isotopes, giving good contrast between elements of similar atomic weight, and enabling isotopic labelling. Neutrons also have a magnetic moment, and although studies of magnetism is rare in soft matter, several techniques for studying non-magnetic matter utilizes the magnetic properties of neutrons, e.g. spin-echo techniques and magnetic contrast layers for reflectometry. However, on the down side, we find that neutron sources have low brilliance, resulting in lengthy experiments, low signal-to-noise ratios, or requiring large samples; access is complicated and inconvenient in comparison to lab sources, and could involve several months' of waiting and perhaps lengthy travel. In addition, disregarding the cost of the large neutron facilities themselves, many experiments are still

*e-mail: ted@ifm.liu.se

expensive due to the need for isotopes, large single crystals, large sample volumes, etc. As a consequence, exploitation of the unique properties and possibilities offered by neutrons is regularly done in parallel with extensive investigations with other, complementary, methods, such as scattering (static or dynamic light scattering, X-rays, electrons), scanning probe methods, nuclear magnetic resonance, and many other types of spectroscopy. Thus, neutron scattering methods are unlikely to solve many problems in soft matter science on their own. However, it is often the case, that neutrons can provide unique, and otherwise unavailable information, for solving a given problem. The tutorials in this volume provide tools to understand what neutron scattering is about, and how experiments are conducted, in order to help the reader identify those experiments where neutrons can make a considerable difference.

1.2 Discovery of the neutron

In retrospect, it is easy to overestimate and misrepresent the drama preceding important scientific discoveries, but the discovery of the neutron in 1932 clearly was one that was both anticipated and much needed. The existence of the neutron was predicted by Rutherford twelve years before the actual discovery, on the basis of experimental evidence. Theoretically, the description of atomic nuclei in terms of merely protons and electrons at the time, was associated with several difficulites and inconsistencies, which were resolved only when neutrons were introduced. The course of events behind this discovery is very briefly outlined in the following, and readers are referred to other sources for more detailed accounts [1]. In the early years of the 20th century, it was known that electrons were negatively charged, and that atoms were overall charge-neutral. In 1904, Thomson proposed what was to be known as the "plum pudding model" of the atom, where "...the atoms of the elements consist of a number of negatively electrified corpuscles enclosed in a sphere of uniform positive electrification..." [2]. The Geiger-Marsden experiment in 1909 [3], demonstrated the nuclear nature of atoms by observing the distribution of alpha particles deflected upon passing through a thin gold foil. Based on these observations, Rutherford presented an atomic model in 1911 [4], describing the atom as a small, positively charged, massive nucleus surrounded by a much larger cloud of negatively charged electrons. In 1920, Rutherford [5] suggested a more elaborate model in which the nucleus consisted of positive protons and neutrally-charged particles, e.g. a proton and an electron bound in some way ("nuclear electrons"). Rutherford called these particles neutrons. In 1932, Bothe and Becker [6] found that energetic alpha particles falling on certain light elements produced an unusually penetrating radiation, assumed to be gamma radiation since it was unaffected by electric charge. However, upon interaction with paraffin, Joliot-Curie and Joliot [7] found that this radiation lead to ejection of protons of very high energy. In the same year Chadwick [8, 9] performed similar experiments on beryllium, showing that the new radiation consisted of uncharged particles with about the same mass as the proton - these particles were indeed the neutrons that Rutherford had predicted about a decade earlier, and Chadwick was awarded the 1935 Nobel prize in physics for this discovery.

1.3 The early years of neutron scattering

Within four years of the discovery of the neutron, it had been recognized that neutrons have wave properties, that neutrons could be "thermalized" by collisions in paraffin, and that this radiation could be diffracted by crystalline solids, as demonstrated on powders by Halban and Preiswerk[10] and on single crystals by Mitchell and Powers [11]. The same year, the magnetic moment of the neutron was predicted, and Bloch [12] outlined many of the applications of magnetic neutron scattering that were to follow in the subsequent decades. The early experiments used "white" radiation radium-beryllium sources that were insufficient for quantitative scattering experiments, which had to await the advent of nuclear reactors in the late 1940s. In 1948 Wollan and Shull [13] published a number of powder diffraction patterns, obtained at the first full-scale nuclear reactor, the "Clinton pile" at Oak Ridge. A perhaps more significant milestone for the development of neutron scattering was the publication of scattering length data for about 60 elements and isotopes in 1951 [14]. Since nuclear scattering lengths cannot be calculated from theory, the experimental quantification of these interactions enabled a more systematic investigation of the opportunities offered by neutrons. In particular, neutrons were used early on to investigate systems where X-ray scattering (which was at that time well established) was not appropriate, for example hydrogen-containing materials, for distinguishing elements with similar X-ray form factors, or indeed using the magnetic properties of the neutron. Brockhouse and Stewart pioneered the field of inelastic neutron scattering by determining a phonon dispersion curve with a prototype triple-axis spectrometer in 1955 [15]. (Brockhouse and Shull had to wait until 1994 before they were awarded a Nobel Prize for their contributions to neutron scattering.) The development of methods pertinent to soft matter occurred much later, with SANS determinations of protein [16] and polymer structure [17], chain conformation in bulk polymers [18], and the first reflectometry experiments yet another decade later [19]. Early applications of scattering will be introduced further on in this volume, in subsequent chapters relating to specific methods. For a more detailed account of the development of neutron scattering, the reader is referred to the anthology by Bacon, showing the diversity of the field in its early years [20].

2 Basic properties of neutrons

The neutron is a fermion, and thus has spin 1/2, and is electrically neutral. The rest mass m_n is 1.6749 $\times 10^{-27}$ kg, or 1.00866 u, which is 1.00138 times the proton rest mass. The neutron magnetic moment is -1.042×10^{-9} B, or $-1.913\mu_N$, where μ_N is the nuclear magneton (negative since it is directed opposite to the spin). Free neutrons are unstable, and decay into a proton, an electron and antineutrino with a half-life of approximately 10 min: $\mathrm{n} \rightarrow \mathrm{p} + \mathrm{e}^- + \bar{\nu}_e$. This decay is insignificant for most scattering experiment, since, as we shall see, the time needed for a free neutron to travel from the source to the detector, is of the order of milliseconds.

The charge neutrality means that the interaction with matter is weak, since there is no Coulomb interaction with electrons or nuclei, and the interaction with matter is governed by nuclear and magnetic interactions, as will be explained later. The neutron is not an elementary particle (but a nucleon) since it is composed of three quarks, but the internal degrees of freedom of the neutron are not relevant when used as a probe for structure or excitations of soft matter, since the energies involved in the excitation of the internal structure are far too large. The standard model of particle physics predicts a small electrical dipole moment of the neutron, though this has not been experimentally verified.

Neutrons are quantum objects whose behaviour can be understood via either their wave nature or particle properties. An integral part of the wave-particle dualism is that the wave and particle natures can be manifested at different occasions in the same experiment, though not simultaneously. As a nucleon, we are perhaps more familiar with the particle description, but the neutron is also associated with a velocity-dependent de Broglie wavelength, given by $\lambda = h/p$, where h is Planck's constant and $p = m_n v$ is the momentum.

3 Neutron production

Neutrons are strongly bound in atomic nuclei, the binding energy per nucleon varies between 5 and 10 MeV for all but the lightest elements, and only nuclear reactions can split free neutrons from nuclei. The neutrons used in much of the early experimental work in the 1930s were obtained by bombardment of beryllium with α-particles, and while there are many types of neutron generators in various sizes available [21], the neutron sources that are used for condensed matter investigations are large facilites producing neutrons via fission and spallation. The earliest fission reactors were constructed for military purposes in the late 1940s, specifically for nuclear weapon production, but the excess neutron fluxes that became available were quickly taken advantage of for the development of neutron scattering instruments. Reactors dominated among research neutron sources up the early 1970s, when the development of accelerator-based, pulsed spallation sources started. One important driving force for this was the efficiency enhancements expected with pulsed sources, where fewer neutrons are used more efficiently. For high-power sources, heat removal is a major obstacle, and in spallation sources, less heat is produced per fast neutron. A considerable difference in efficiency is seen upon comparing the 58 MW reactor at ILL with the 160 kW pulsed source at ISIS, where the latter runs at about 0.3 % of the ILL power. The ESS spallation source under construction in Sweden is designed for 5 MW (although it will initially be running at 2 MW).

Looking at the development of neutron facilities, there is currently a clear tendency in favour of spallation sources. There are certainly several technical and scientific reasons for this, but one should not ignore the political dimension. An increasing difficulty in opening new reactor sources is associated with mounting public opinion against nuclear reactors, and fission reactors are also burdened by nuclear proliferation concerns and more elaborate management of high-level radioactive waste. This has the effect that many of the reactor-based neutron sources that are being phased out as they reach the end of their technical lifetime, are not replaced, leading to an overall decline in the availability of neutrons. There are concerns that the current and planned pulsed sources will not have a total capacity that meets the needs from the scientific community, with ensuing decline in both competence and scientific activity in the field.

3.1 Reactor sources

Reactors for neutron research produce neutrons by fission of ^{235}U. After the uptake of a slow neutron, the unstable ^{236}U atom splits into two fragmets (of variable mass), and in addition generating 2-3 neutrons, β and γ radiation and on average ca 180 MeV kinetic energy. The emitted 'prompt' neutrons (emitted within 10^{-14} s) can induce fission in other nuclei, but usually do this after having reduced their speed via collisions, to kinetic energies where uptake by U is more probable. This is achieved by a moderating medium (see section 4), such as water. The reactor at ILL uses the cooling water as the primary moderator, but this is surrounded by a heavy water moderator, reflecting neutrons back into the core again. Many fission fragments also contribute to neutron production via 'delayed' neutrons emerging from their further radioactive decay; these neutrons together with the prompt neutrons sustain the chain reaction that maintains neutron production, if there is sufficient amount of available fissile material, and the delayed neutrons, emitted orders of magnitude later than the prompt neutrons, are essential for controlling the reactor.

Reactors steadily produce neutrons with a broad energy distribution, and are thus said to operate in *continuous mode*, delivering a steady stream of neutrons to the instruments. If neutron pulses are desired, they can be obtained via choppers (see section 6.2). The neutron spectrum produced by reactors have a Maxwellian energy distribution. The large flux of cold neutrons is suitable for large objects and slow dynamics, and reactor sources have the advantage that resolution can easily be tailored to specific experimental requirements.

Reactor sources are, as a rule, run as continuous sources with the chain reaction maintained at a steady state. The pulsed reactor IBR-2 in Dubna is an exception, and its main difference from other reactors consists in mechanical reactivity modulation by rotating neutron reflectors, ensuring that criticality is reached only when the reflectors coincide near the reactor core, generating power pulses with a power of 1850 MW at a rate of 5 or 10 Hz, with a pulse half-width of 340 μs for thermal neutrons.

3.2 Spallation sources

Spallation is a process where high-energy particles, usually protons at GeV energies, hit a target of neutron-rich material, such as heavy metals like tungsten, mercury, tantalum or uranium. The nuclei are highly excited by the collisions, and disintegrate under rapid evaporation of neutrons. These neutrons proceed to excite nearby nuclei, promptly emitting further neutrons, until the process decays. In spallation, the wavelength of the generated neutrons is shorter than the dimensions of the nuclei, resulting in internal excitation of nuclides (intra-nuclear cascade), which further increases the number of emitted neutrons. The emitted neutrons cover a wide energy spectrum with a peak around 2 MeV, but with a continuous distribution toward higher energies, rich in hot neutrons, with a small fraction emerging with nearly the incident proton energy, as a result of near-elastic collisions. With the exception of the latter contribution, the spectrum is not too different from a prompt fission spectrum, but in practice most neutrons will have collided several times before leaving the target, and the extractable spectrum is degraded. The spallation process is extremely short, and neutron emission decays within 10^{-15} s, and is determined by the pulse length of the proton source. Pulse lengths of the order of μs result in short and large-amplitude neutron pulses with very good wavelength resolution for thermal neutrons, and short-pulse spallation facilities operate at frequencies in the range 10-60 Hz. To achieve high proton beam intensities, a linear accelerator is used to pump protons into a synchrotron, where a large number of proton pulses are injected, and accumulated, before being extracted to the target. For these short-pulse[1] sources the induced thermo-mechanical stress in the target is considerable, and several advantages follows from extending pulse lengths to the ms range [22]: With long pulses, the total proton beam energy per pulse can be higher, resulting in an increased neutron flux, but at the expense of reduced resolution (introducing uncertainty in the flight time). This favours applications such as many small-angle scattering or reflectometry experiments in soft matter, which are not dependent on high wavelength resolution. An additional aspect is the considerable costs involved in these facilities. Long-pulse sources do not need a storage ring, and by using only a linear accelerator, a given proton beam power delivered to the target can be achieved at substantially lower cost by using only long pulses. At a lower repetition rate, these will also enable operation of large bandwidth instruments, considerably extending the range of accessible wavelengths, see the comments below on time structure. Major spallation sources in operation today are short-pulse facilities (ISIS, SNS, J-PARC). The planned second target station at SNS, and the ESS facility currently under construction near Lund in southern Sweden, are both long-pulse stations. ESS is designed for 1.3 GeV protons and a 5 MW average beam power, from 2 ms pulses at 16.6 Hz, and using a rotating Tungsten target to handle the released thermal energy more effectively.

Thus, spallation produces neutrons of different energies, and since these have different velocities, they arrive at the detector at different times. Hence, if the arrival time of each neutron on the detector is recorded beside the scattering angle, the corresponding wavelength, and hence the scattering vector q can be determined for each neutron. This is the *time-of-flight* mode, and the widespread use of this mode at spallation sources is the reason why instruments there are all end-of-tube instruments, which is usually not the case at reactor sources.

Due to their pulsed nature, spallation sources are naturally suited for time-of-flight instruments, and from the neutron energy distribution, particulary good for measurements at large energies. The

[1]"Short" and "long" could be viewed in relation to the moderation time. Neutrons need of the order of 100 μs for thermal equilibration, and pulse lengths substantially shorter than this moderation time will result in similar moderated pulse shapes.

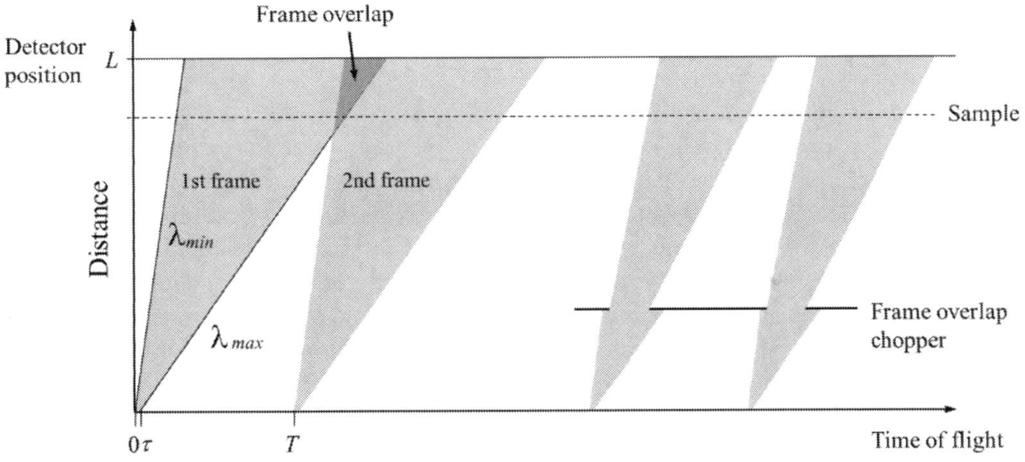

Figure 1. A time-of-flight diagram for a pulsed source generating a pulse of length τ every T seconds. The diagram is shown without choppers (left) and with inserted pulse-defining choppers (right) whose purpose is to separate neutrons from subsequent pulses on the detector.

Swiss spallation source SINQ is unusual in that it is not a pulsed source; the neutron target is bombarded with protons from a cyclotron, delivering a quasi-continuous proton beam, making SINQ a continuous neutron source.

A very important property of pulsed sources is the time structure, that is, the duration and frequency of the pulses, from which time-of-flight diagrams can be constructed. Figure 1 illustrates how the source-detector distance L and the pulse rate $(1/T)$ will define the bandwidth of the instrument. The width of the pulse, τ can be ignored for short-pulse sources, but might noticeably affect the resolution on a long-pulse source. From the diagram it is clear that by increasing L, the resolution can be increased, but this must be accompanied by a lower pulse rate, to keep subsequent pulses separated on the detector, alternatively, overlap of two pulses ("frames") can be prevented with pulse-defining choppers, or with a high-pass filter, removing undesired slow neutrons, but at the same time reducing the bandwidth. Frame overlap must be avoided since the slow neutrons from the previous frame cannot be distinguished from the fast neutrons from the next frame. The leading and trailing edges of the frame are defined by the shortest (λ_{min}) and the longest (λ_{max}) wavelengths, and define the accessible q-range at a given scattering angle.

3.3 Continuous versus time-of-flight operation

Whether neutrons are produced continuously (as is usually the case in reactors) or in pulses (most spallation sources) has several consequences for the type and design of the scattering instrumentation used at the facilities. At continuous sources the beam is usually monochromatized for scattering experiments. This can be achieved by inserting a reflecting monochromator (see section 6.1) into the beam to deflect a specific wavelength to a diffractometer. Other wavelengths can then be extracted from the resulting beam downstream, or a velocity selector can be used at the guide end. The experiments are angle-dispersive, and the scattering vector q is varied by varying the angles of the scattering experiment under the monochromatic beam, with data collected in $\theta/2\theta$-scans. At pulsed sources it would be very wasteful to use only neutrons of a particular wavelength in each pulse, and it makes more sense to use all neutrons, but to keep track of the energy (wavelength) of each neutron. In the latter case, the instruments are static, that is, kept at a fixed angle, and operated in time-of-flight (ToF) mode. This difference has several implications for the design and function of the instruments and the data collection systems, and also for sample environments. This does not visibly affect a SANS instrument much, but for reflectometry it often makes a difference whether the scattering plane is horizontal or vertical, or if angles are fixed or scanned during data collection; not all these options are compatible with, for example, experiments at air-liquid interfaces, to give an obvious example. The fundamentally very different operating principles of reactors and spallation sources have far-reaching implications on many aspects of scattering experiments, from generation, moderation and spectral distribution, to instrument layout and geometry, resolution and data treatment, and are not easily summarized (there is a more elaborate discussion on this topic in section 3.6 of ref. [23]). This pertains also to comparisons of the neutron flux from different types of sources. The time-averaged neutron flux is used to measure the neutron intensity of reactor sources, but how does this compare to the time-averaged or the peak flux of a pulsed source? For similar integrated neutron fluxes, pulsed sources are overall more effective for conducting scattering experiments than reactor sources, and hence the time-averaged flux of a reactor source is usually compared to the peak flux of a pulsed source.

However, the differences between pulsed and continuous sources are also leveled out by development of instrumentation and measurement modes, thereby blurring the distinction between the two types of sources. Some examples of this are reactor instruments (such as the ILL D17 reflectometer) which are equipped with both monochromators and choppers (see section 6.2) to allow selection between monochromatic and time-of-flight operation on the same instrument. The continuous spallation source at the PSI uses many "reactor type" angle-dispersive scattering instruments, but have also chopper instruments for white-beam ToF mode.

4 Moderators and neutron temperature

The energy spectrum of neutrons generated in either a reactor or in a spallation source is inappropriate for most scattering experiments since the maxima in their energy distributions lie in the MeV range, which is far too high. A 1 MeV neutron has a speed of 14000 km/s, and the corresponding de Broglie wavelength is 30 fm, which is clearly too short to be useful for structural characterization of molecular matter. Hence, the energy distribution needs to be shifted into a range that is more useful for scattering purposes, see Table 1. This is accomplished via *moderation*, a process where neutrons are allowed to thermally equilibrate with a volume of material to reduce the average kinetic energy via inelastic collisions. This is most efficiently done with a light material with a large collision cross section but low absorption, such as light or heavy water, graphite or beryllium. Hydrogenous materials are common, since H has a very large scattering cross-section (82 barn; cross-sections are measured in the unit barn (b), where 1 b = 10^{-28} m^2), and a mass similar to the neutron mass, resulting in efficient inelastic dissipation of the kinetic energy. At thermal equilibrium, the neutron velocities are Maxwell-Boltzmann distributed, and translated into wavelength, the neutron beam flux $\phi(\lambda)$ (neutrons per second per unit area), in the wavelength interval $[\lambda, \lambda + d\lambda]$, is then

$$\phi(\lambda) \propto \frac{1}{\lambda^5} e^{-h^2/2m_n kT\lambda^2}$$

This distribution is shown for three different temperatures in Figure 2, and the peak in this distribution occurs at

$$\lambda = \frac{h}{\sqrt{5 m_n k_B T}}$$

Table 1. Neutron temperatures, the corresponding wavelengths and energies, and examples of moderator materials to achieve these temperatures.

	Wavelength (Å)	Energy (meV)	Example of moderator
Fast neutrons	< 0.5		
Hot neutrons	0.5 - 1	80-300	Graphite at 2000 K
Thermal neutrons	1-5	3-80	D_2O at 300 K
Cold neutrons	3-30	0.1-10	Liquid H_2 at 25 K
Ultracold neutrons	> 500		

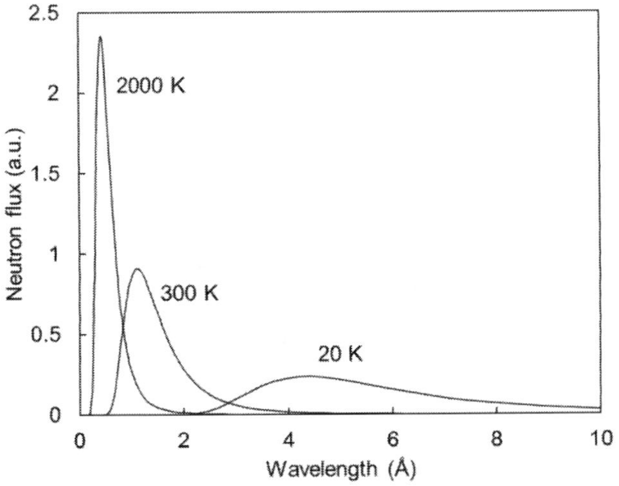

Figure 2. The Maxwellian wavelength distribution of neutrons in thermal equilibrium, at three different temperatures. The three distributions have the same area. At 300 K, the mean neutron wavelength is 1.8 Å, and the mean energy 26 meV. For comparison, we note that photons with λ = 1.8 Å have an energy of 7 keV, and the temperature is 8×10^5 K.

In reality, the energy dependence of the collision cross-section of most materials gives lower-energy neutrons a larger effective cross-section for a given nuclide than fast neutrons, which slightly modifies the energy distribution so that hotter neutrons are more abundant than predicted.

Depending on the incident neutron energy and the moderator material, 10-100 collisions are needed for equilibration, after which the neutrons diffuse isotropically in the moderator. The moderator has one or several exits, allowing the neutrons to enter as a beam into a neutron guide. The isotropic diffusion in the moderator allows the neutron guides to be aligned so that they are not directly facing the primary source, which reduces gamma radiation and uncollided high-energy neutron background in the beam.

Moderators are an intrinsic part of the design of a reactor, since fission neutrons must be moderated to be efficiently captured by ^{235}U, often using heavy water, light water, or graphite, but there is no corresponding such function of the moderator for spallation sources. This means that there is more freedom in designing moderators to suit the needs of the instruments at a spallation source. On the other hand, spallation source moderators are preferably kept small to keep the pulse width narrow, resulting in under-moderation and an appreciable fraction of high-energy (epithermal neutrons). The smaller size also results in less efficient coupling to the neutron guides, while there are no such size constraints at reactor sources. As an example, the moderators are about 0.5 l at the ISIS spallation source Target Station 1.

Thermal neutrons, with wavelengths in the range 1-5 Å, which is of the order of interatomic and molecular distances, have energies around 30 - 700 cm^{-1}, or about 3-80 meV, where many molecular and lattice excitation energies are found. However, for some experiments, especially reflectometry,

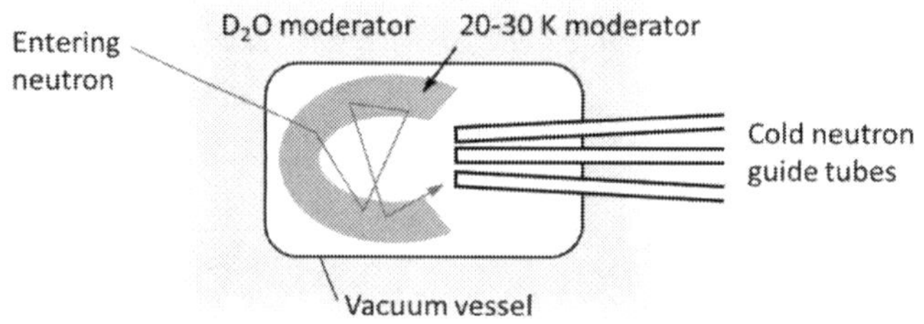

Figure 3. Schematic view of a cold source moderator. Neutrons enter from the surrounding 300 K moderator, typically containing D_2O. Neutrons with high energy have a greater probability (lower cross section) to pass through the cold moderator material, for example liquid hydrogen, into the cavity. Via repeated collisions the kinetic energy is reduced, and as the energy decreases, the cold moderator will be increasingly acting as a reflector, retaining neutrons in the cavity, until they enter one of the guide tubes.

small-angle scattering and biological applications, it is sometimes desirable to have neutrons of longer wavelengths, and these are produced in a cold source. This is commonly a hydrogen-containing material, such as hydrogen or methane kept at temperatures near 20-30 K, often surrounded by a Be reflector to bring diverted neutrons into the moderator, see Figure 3. On the other hand, neutrons with shorter wavelengths are obtained from a hot source. At the ILL this is a graphite block in the heavy water moderator which is heated to ca 2000 K by the collisions of hot neutrons with the graphite.

In pulsed sources, there is a constraint in that the pulse length needs to be kept short, and hence there is an upper limit to the time neutrons can be allowed to diffuse in the moderator before they enter the neutron guides. This is managed via *decoupling* the moderator with a neutron-absorbing material inserted between the moderator and the reflector material. Cadmium absorbs neutrons strongly, but as with most materials the absorption cross-section is wavelength-dependent, and it is almost transparent for energies above 0.3 eV. This means that fast neutrons from the spallation pulse can enter the moderator, but neutrons that have lost too much energy (spent too long time) in the reflector or in the moderator, are removed. The width of the moderator pulse can also be controlled via *poisoning*, where a small amount of neutron-absorbing material, such as gadolinium, is added to the moderator. This has a similar effect as decoupling; it narrows the pulse at low energies where the absorption cross-section is largest, and effectively clears the moderator of neutrons that have spent too long time in the moderator.

4.1 Neutron guides

Using guide tubes [24] to channel neutrons away from the neutron source allows a larger number of instruments to use the same source, since floor space in the immediate vicinity of the source does not limit the number of instruments, but allows instruments to be physically located further away, in guide halls. A location further from the neutron source reduces the radiation background, and placing instruments in a guide hall separated from a reactor source also means that users can access instrument under less strict radiation protection and safety regulations, which benefits both the users and the safety of the neutron facility.

Guide tubes are typically made of glass with neutron-reflective coatings lining the inside walls, with rectangular cross sections (width 1-5 cm, height 5-20 cm). Neutrons travelling near-parallel to the guide walls are kept within the tube by total reflection from the walls. Most solids have neutron refractive indices which are lower than 1 for cold and thermal neutrons, and thus have a critical angle, θ_c, below which neutrons are totally externally reflected (see section 7.2), and hence can be kept inside the guide tube with very low losses. The inside is coated with Ni or with Ni-Ti heterostructures with hundreds of layers, or with supermirrors, [25] multilayer mirrors with gradually varying thicknesses of the layers, resulting in several times higher critical angles than plain Ni coatings (there are also polarized mirrors, using e.g. Co/Ti or Si/Fe, and alternating ^{58}Ni and ^{62}Ni isotope layers). A multilayer coating introduces a Bragg reflection at a certain angle, and the gradient in the lattice spacing of supermirrors effectively produces a range of effective Bragg angles, and therefore a reflectivity which extends to several times the critical angle values expected for other coatings. This allows for curved guide tubes, and guides are usually given a slight curvature in the horizontal plane, because this has some benefits. A small curvature eliminates gamma radiation from the neutron source, since gamma radiation travels straight through the walls of the guides (and into surrounding shielding). Further, since neutron reflectivity is strongly wavelength dependent, the curvature can also be used to eliminate fast neutrons, and the radius can be used to tune the lower cutoff wavelength. This also makes guide tubes more useful the longer the wavelength, and thus much better for use with cold neutrons than for hot neutrons. Where there is not room enough for longer curved guide tubes, multichannel guides with several parallel supermirror plates inserted into the guide are used. These "beam benders" have a geometry like a curved Söller collimator (see next section). Recent developments in guide tubes include elliptic focusing guides, [26] which considerably improve flux at small sample areas and additionally reduce background scattering, since they allow tuning of the beam footprint to fit the sample size and hence minimize illumination of areas outside the sample. However, this will come at the cost of reduced resolution, since beam divergence at the sample increases.

The distance that neutrons can be transmitted in guide tubes is limited by gravity and reflection losses at the walls. For reactors and other continuous sources, there is in principle no limit to the length of a guide tube, except that gravity might have to be taken into account. For pulsed sources, the time structure of the source (the desired bandwidth) sets limits to the distance between the source and the instrument, to avoid mixing of neutrons from subsequent pulses. Guide lengths rarely exceed 100 m, but several instruments at the ESS facility will be placed over 150 m from the spallation target. Neutron guides (and beam paths in general) are evacuated, to reduce losses due to scattering against air molecules, which reduces the flux of a neutron beam by approximately 5 % per m at normal air pressure – and considerably more if the air is humid, due to the large hydrogen cross section.

5 Collimation

Collimators remove neutrons from the beam which do not travel in the desired direction, that is, they reduce beam divergence and increase the angular resolution ($\Delta\theta/\theta$). This is achieved by placing slits or apertures (pinholes) in the beam path that absorb neutrons with too large angular deviation from the desired path, either in the primary beam, or before the detector. In principle, two pinholes would be sufficient, but in practice often more than two are used, to eliminate stray neutrons. Long collimation systems or small apertures obviously give the best collimation, but at the expense of reduced beam intensity, and vice versa. Again, this is a tradeoff between resolution and intensity that must be done for each experiment.

A Söller collimator is a system of parallel plates made of a neutron-absorbing material. Neutrons that hit the plates are absorbed, and the separation and the length of the plates determine the acceptance angle, see Figure 4. This results in collimation in one plane, but with slits or apertures, rectangular or circular collimation is obtained. Two slits before the sample, each limiting the extent of the beam both in the vertical and horizontal directions, one immediately before the sample, the other one or a few meters upstream, ensure that only the sample is illuminated, and that this illumination is a near-parallel beam. Beside maintaining the angular resolution, this also reduces scattering from surrounding equipment and sample holders. Similarly, slits between the sample and detector reduce

https://doi.org/10.1051/epjconf/201818801002

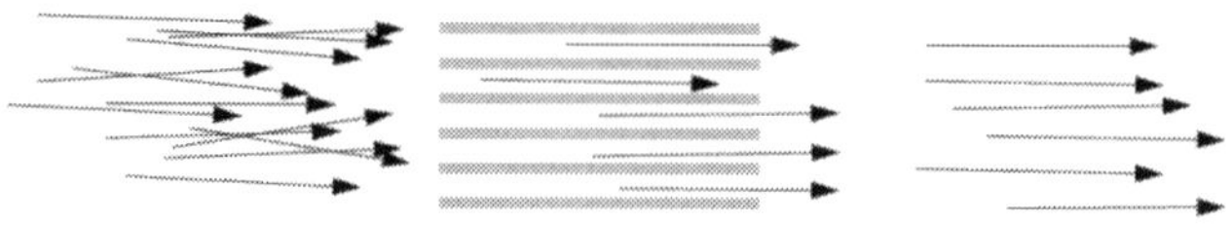

Figure 4. A Söller collimator is a stack of absorbing parallel plates, where the length L and separation between the plates d define the divergence α of the passing netrons, via $\alpha = d/L$.

background noise on the detector by minimizing the detection of neutrons scattered from sample holders. Slits are usually motorized since their openings are regularly varied with the incidence angles, and adjusted for sample size.

A radial collimator has absorbing plates arranged in a fan-shaped structure, these are used to reduce background on large-area detectors, and ensure that only neutrons from those parts of the sample which are both illuminated by the beam and which have a free line-of-sight through the collimator are counted.

6 Monochromators

The majority of the instruments at reactor sources use a fixed wavelength in angle-dispersive configurations for scattering experiments, and use monochromators for defining the wavelength or to analyze the energy of inelastically scattered neutrons. A conceptually simple way of preparing a monochromatic beam is to use the Bragg reflection from a single crystal of some suitable material, inserted into the guide. The Bragg angle θ_B is given by

$$n\lambda = 2d \sin \theta_B, \qquad n = 1, 2, 3, ... \tag{1}$$

where n is the order of the reflection and d the lattice spacing. However, the practical implementation of this principle is somewhat involved. The large size of the neutron guides means that the crystals need to be very large to cover the whole beam cross section. For all experiments there will also be a balance between the flux and the desired wavelength resolution, $\Delta\lambda/\lambda$. Permitting a lower resolution (larger $\Delta\lambda/\lambda$) means that the flux can be increased. Depending on the experiment, the required $\Delta\lambda/\lambda$ can vary from 10^{-4} to 10^{-1}, which means that monochromators will have very different designs in order to fulfil these requirements. For many experiments in soft matter science, the resolution requirements are relatively moderate, and for small-angle scattering $\Delta\lambda/\lambda \approx 10\%$ is not uncommon, and sometimes even 20% can be accepted, with associated benefits in higher available flux.

6.1 Monochromator crystals

Commonly used monochromator crystal materials are silicon, germanium, copper, pyrolytic graphite, or beryllium. A single crystal could be expected to yield a monochromatic beam with $\Delta\lambda/\lambda \approx 1\%$. As mentioned, flux and resolution are important design parameters, and where lower resolution can be accepted, there are several ways of modifying crystals to increase the flux. A single crystal gives high resolution, but low flux. A simple way of increasing the flux is to increase the thickness of the crystal, or to place crystals after each other in the beam, though this compromises the angular resolution. If a single crystal is mechanically strained under heat, dislocations form in the crystal, and the local alignment of the lattice planes is slightly distorted. Introducing such *mosaicity* results in a spread of Bragg angles over the crystal volume, with deviations of the order of 30 arc minutes, and widens the

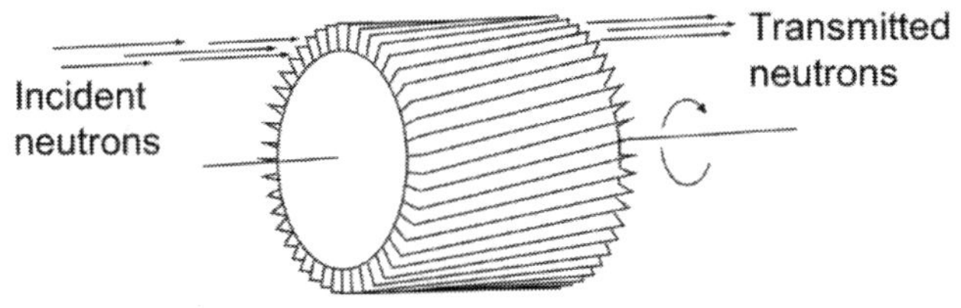

Figure 5. A velocity selector. The rotation speed and the helical pitch of the lamellae along the beam direction allows neutrons within a narrow velocity distribution to pass the selector.

reflected wavelength spread to approximately 5 %, but if this is tolerated it also increases the flux considerably. Alternatively, several crystals can be mounted at slightly varying angles, reducing the need for a large single crystal and allowing instead focussing of the beam. A single crystal that is slightly bent will also have a focussing effect, and at the same time some effective mosaicity since the actual Bragg angle will vary over the curvature.

6.2 Choppers

A *chopper* is a rotating disk made of neutron-absorbing material with at least one aperture, and is used as a shutter which allows beam transmission at certain time intervals. At spallation sources choppers are used to exclude certain wavelengths, and the rotation is synchronized with the pulsed beam (pulse-defining, or frame overlap choppers, see Figure 1). Neutrons of different energies arrive at the chopper at different times, and only those neutrons that arrive when the aperture is aligned with the beam pass through the aperture. In practice, the chopper is often assembled by two disks whose relative angles can be adjusted to allow the size of the aperture (the "angular window") to be adjusted. Two (or more) synchronized choppers separated along the rotation axis have a similar function as a velocity selector, but with greater flexibility since adjustment of both rotation speed and the relative angle of the two disks can be used to control the transmitted bandwidth. Choppers are also used to create pulses for time-of-flight experiments at reactor sources, though this is a wasteful mode of operation, in that only a small fraction of the white-beam neutrons are used.

6.3 Velocity selectors

A velocity selector can be thought of as a twisted and rotating Söller collimator, see Figure 5. Parallel plates arranged in a helical shape on the surface of a cylinder, with neutron-absorbing walls, permit neutrons to pass between the slits if they have a speed that is determined by the helical pitch and rotation speed of the cylinder. Velocity selectors offer lower wavelength resolution than Bragg reflectors, typically 10 % (with ensuing increase in flux), and are most frequently used at end of neutron guide tubes at reactor sources, for e.g. SANS instruments.

6.4 Filters

Filters remove neutrons of unwanted energies from a beam either by deflecting them out of the beam, or by absorbing the neutrons. A pertinent example is higher order Bragg diffractions from a beam monochromator crystal. According to Bragg's law (Equation 1), a diffracted beam will also contain wavelengths of higher order ($n > 1$, giving $\lambda/2, \lambda/3$ etc.), and these need to be filtered away to obtain a monochromatic beam. Filters which diffract the unwanted wavelengths out of the beam direction are often graphite of liquid nitrogen-cooled beryllium, and should be selected so that $\lambda > 2d_f$, where d_f is the lattice spacing of the filter crystal, so that the desired wavelength is not itself Bragg reflected. Suitable filters should have high coherent cross sections, but low incoherent and absorption cross sections. Filters are also cooled to minimize scattering of the transmitted beam by lattice vibrations (phonons) in the crystals.

Filters relying on (resonant) absorption are usually strongly absorbing in certain spectral regions. While Cd is widely used for shielding, and is very effective for thermal neutrons, but has limited absorption for fast neutrons. Certain isotopes with strong nuclear resonances are effective bandstop filters, and are used as thin foils, but most lie in regions of the energy spectrum that are of less interest to work in soft matter.

7 Interactions of neutrons with matter

In the absence of charge and associated Coulomb interactions, neutrons interact weakly with atoms, and do so via either of two mechanisms: *magnetic scattering* or *nuclear scattering*. In magnetic scattering, the magnetic moment of the neutron interacts with the magnetic moment on the atom, originating from unpaired electrons. These are distributed over the atom – over dimensions comparable to the neutron wavelength – resulting in an angle-dependent magnetic form factor, which can be manipulated by reorienting the electron spins via an external magnetic field, and which can also be calculated from electron densities. In nuclear scattering, the neutrons interact directly with the nucleus of an atom via the strong nuclear interaction, and this scattering contribution cannot be predicted from theory. For a small number of strongly absorbing nuclei, such as ^{113}Cd, ^{149}Sm, or ^{157}Gd, there is also *resonance scattering* (as opposed to the two types of *potential scattering* mentioned above) when the neutron and the nucleus forms a compound nucleus; this phenomenon will not be discussed here.

The probability of a neutron to interact with a nucleus is quantified via the cross section σ, which is measured in barns (b). The cross section consists of a term representing the probability of scattering, σ_{scatt} and a term representing the probability of absorption by the nucleus, σ_{abs}, so that

$$\sigma = \sigma_{scatt} + \sigma_{abs}$$

The small neutron cross section for most materials has both advantages and disadvantages; among the latter, the foremost is that weak interaction also means that neutrons are difficult to detect. This will be discussed later on, in section 8. On a positive note, the weak interaction simplifies theoretical treatment of neutron scattering, which is often carried out in the so-called *Born approximation*, or the *kinematic scattering approximation*, in which it is assumed that the beam traverses the sample unaffected by it. Hence, refraction, multiple scattering and beam extinction by scattering of the primary beam are all neglected.

7.1 Scattering lengths

The strength in the interaction of a neutron with a nucleus is given by the scattering length b, which is a phenomenological quantity, and which must be determined empirically, since it cannot be calculated.

The scattering lengths for the elements, or for specific isotopes of these, do not vary in a systematic manner with nucleus composition; they have been plotted for some elements in Figure 6, where also the scattering lengths of individual isotopes are indicated for H and Ni. For non-absorbing nuclei, the scattering length is independent of the neutron wavelength. The differences in scattering length between hydrogen (^{1}H) and deuterium (^{2}H = D) is essential for many applications in soft matter research. Nickel has a very rich variety in scattering length between different isotopes, which is exploited for neutron mirrors and many other aspects of neutron research in the solid state. The seemingly arbitrary variation in scattering length between nearby elements and isotopes is an advantage over X-ray scattering, which can provide similar structural information, but with X-rays the scattering

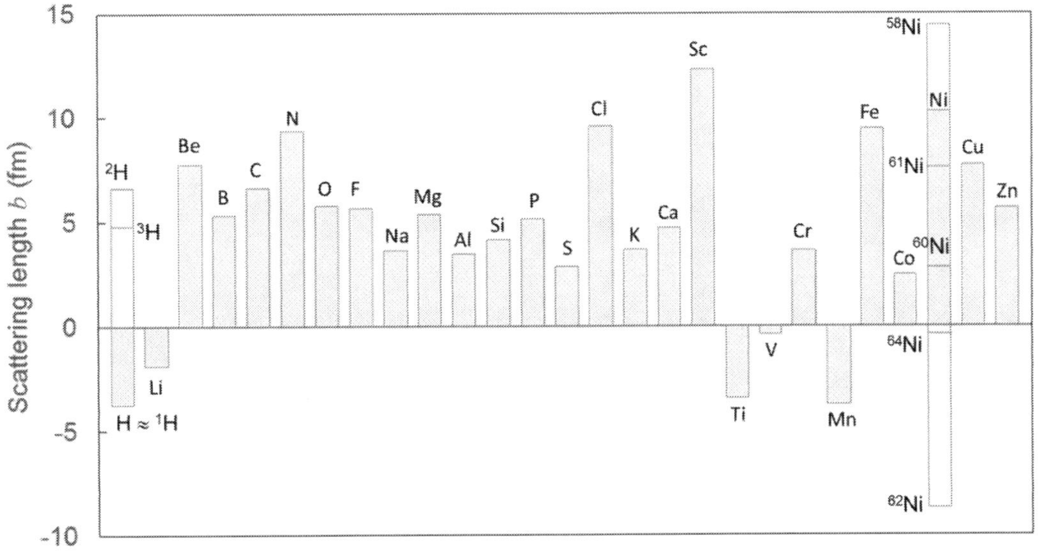

Figure 6. Neutron (bound, coherent) scattering lengths for some elements. For hydrogen and nickel, the scattering length of the elements in their natural compositions are indicated with the grey bars, but also the scattering length of specific isotopes are indicated with the transparent bars.

length is proportional to the density of electrons, meaning that nearby elements can be very difficult to distinguish due to the low contrast between them. The scattering length can be either positive or negative, and of significant importance for the use of neutrons to probe the structure of matter is the fact that hydrogen and deuterium have different signs of their scattering lengths, $b_H = -3.74$ fm and $b_D = 6.67$ fm, respectively, and that mixing these isotopes can be used to modulate the contrast in a system.

A free nucleus will not have the same scattering length as a bound nucleus, such as a nucleus restricted by a chemical bond or a lattice. The free (b_{free}) scattering length is related to the bound scattering length (b) via

$$b_{free} = \frac{M}{M + m_n} b$$

where M is the nuclear mass and m_n the neutron mass. Nuclei are rarely completely free, and in most cases bound scattering lengths are tabulated and used. From the relation above it is clear that free and bound scattering lengths are substantially different only for low mass elements, such as hydrogen, since the ratio $M/(M + m_n)$ approaches unity as M becomes large.

The interaction of a wave with some medium – the potential that the wave will meet – depends on the local average scattering length of the atoms, but also on the number density of the scattering nuclei, and the product of the two is the scattering length density (SLD), and is given by

$$\rho = \sum_i b_i n_i \tag{2}$$

where b_i is the scattering length of nuclear species i, and n_i its number density. Scattering length densities of some materials are shown in Table 2.

Table 2. Neutron scattering length densities (SLDs) for some materials. The right column includes solids often used as substrates in reflectometry.

	SLD (10^{-6} Å^{-2})		SLD (10^{-6} Å^{-2})
H_2O	-0.56	Si	2.07
D_2O	6.38	SiO_2	3.39
Octane	-0.53	Quartz	4.2
d-octane	6.43	Sapphire	5.4

The scattered intensity from inhomogeneities in the scattering length (densities), arising from, say molecules X in a solvent of Y, is proportional to the square of the difference in SLD, sometimes called the *contrast factor* $\Delta\rho^2$:

$$\Delta\rho^2 = (\rho_X - \rho_Y)^2$$

7.2 The refractive index

The optical properties of visible light, and its interaction with matter, are usually described in terms of angle of incidence θ and refractive index n, but in neutron scattering it is customary to use scattering length density ρ and momentum transfer q, since these are easily related to the reciprocal quantities that are natural for scattering studies. The mean scattering length density ρ is related to the refractive index n as:

$$n = 1 - \frac{\lambda^2}{\pi}\rho + i\lambda C \tag{3}$$

The constant C in the imaginary term is a function of the absorption cross section of the material, but with the exception of a few strongly absorbing substances, such as cadmium, samarium, or gadolinium, the absorption is very small and this term is usually neglected. Magnetic materials have a contribution to the refractive index also from the neutron spin, which can be used in conjunction with polarised neutrons to study magnetism, though this is not considered here. The weak interaction of neutrons with matter means that the refractive index is usually very close to unity, with the quantity $(n-1)$ of the order of 10^{-6} for most materials. Note also that a positive scattering length implies that $n < 1$.

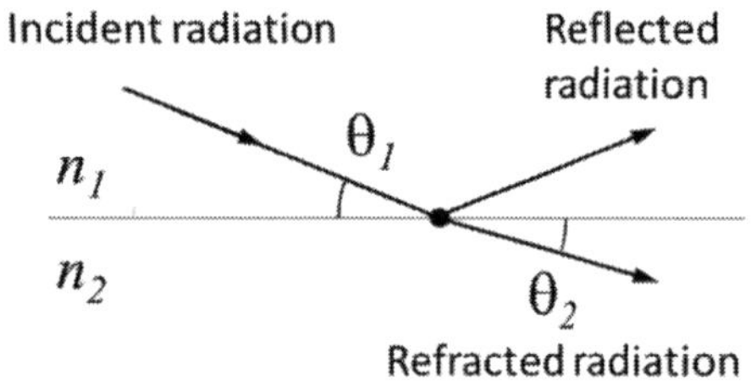

Figure 7. The geometry for the reflection and refraction of a beam incident on a surface at an angle θ_1. The reflected beam leaves the surface at an angle θ_1, while the latter is refracted at and angle θ_2.

At an interface separating two media with refractive indices n_1 and n_2, the ratio of the refractive indices will determine the angle (θ_2) at which radiation incident at an angle θ_1 will be refracted. With the conventions according to Figure 7, Snell's law gives:

$$n_1 \cos\theta_1 = n_2 \cos\theta_2 \tag{4}$$

If $n_2 > n_1$, then $\theta_2 > \theta_1$ and there exists a real angle of refraction, θ_2, for all incidence angles θ_1. However, since most elements have positive scattering lengths, then according to equation (3), the

 https://doi.org/10.1051/epjconf/201818801002

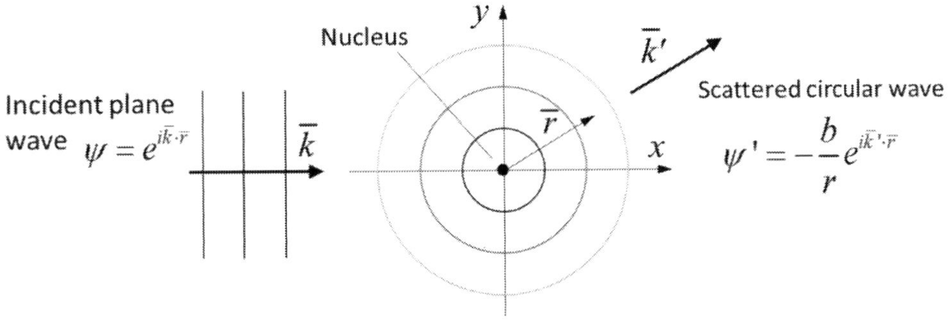

Figure 8. A wave ψ with unit amplitude travelling in the positive x direction with wavenumber $k = |\bar{k}|$ encounters a point scatterer (meaning that the size of the object is much less than the wavelength). The wave is scattered spherically, with an amplitude b/r, where r is the distance from the scatterer (placed at $r = 0$). The positive constant b is the scattering length of the nucleus, and is a measure of the strength of the interaction. If the scattering is elastic, then $|\bar{k}| = |\bar{k}'|$. (After Pynn [27].)

neutron refractive index is < 1, and whenever material 1 is air or vacuum, $n_2 < n_1$, and there is a real angle of refraction only for incidence angles larger than a critical angle θ_c, at which the angle of refraction is 0, and:

$$\cos\theta_c = \cos\theta_1 = \frac{n_2}{n_1} \tag{5}$$

Where $\theta_1 < \theta_c$, total external reflection occurs, and only an evanescent wave penetrates into material 2. With the ratio n_2/n_1 deviating very little from unity, the angles at which total external reflection is observed are consequently very small, typically $< 0.1°$.

As long as inhomogeneities in the scattering length density are small, it is the average scattering length density in a medium that is important, and it is thus possible to mix, for example, hydrogenated and deuterated substances to vary the average scattering length density in a structure or a solvent to obtain a certain refractive index. Isotopic substitution is indeed a common method of *contrast variation*, i.e. changing the refractive index of a substance, and due to the large differences in scattering length between hydrogen and deuterium, this is especially so with hydrogenated compounds. The scattering length densities for H_2O and D_2O are such that most materials have SLDs that fall between these two forms of water, and this applies also to many other hydrogenated solvents and substances. In particular the H_2O/D_2O ratio can be selected to make specific parts of a sample "invisible" to the neutrons. This *contrast matching* is widely used to highlight different parts of a system, and by conducting experiments on the same system but in different contrasts, also details of very complex structures can be revealed, as will be discussed in more detail in the subsequent chapters.

There are also certain "null matrix" alloys that result in zero coherent scattering lengths. Ti (b_c = −3.44 fm) and Zr (b_c = +7.16 fm), or Mn (b_c = −3.75 fm) and Cu (b_c = +7.72 fm) are use to form such alloys that are useful for engineering purposes.

7.3 Point scatterers

Consider the scattering from a nucleus, placed at $\bar{r} = 0$, which we initially define as a spinless isotope with real scattering length b. Since the atomic nuclei are much smaller than the wavelengths used for neutron scattering, the nuclei act as point scatterers, resulting in isotropic scattering. For an incident

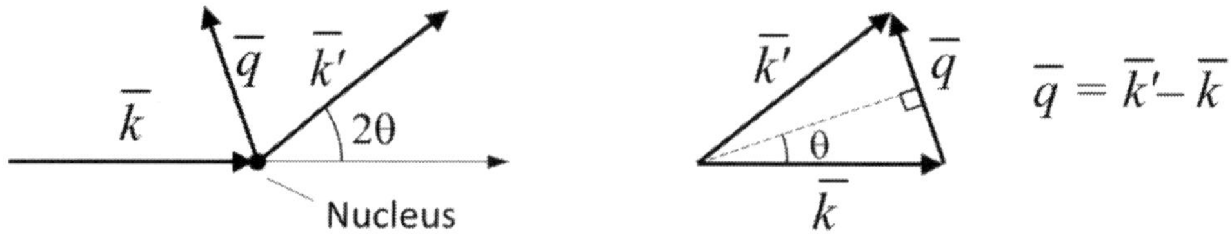

Figure 9. The relation between the momentum transfer $\bar{q}$ and the incident and scattered wavevectors, $\bar{k}$ and $\bar{k}'$.

plane wave ψ, the scattered wave ψ' is then spherical (see Figure 8):

$$\psi = e^{i\bar{k}\cdot\bar{r}} \quad \text{and} \quad \psi' = -\frac{b}{r}e^{i\bar{k}'\cdot\bar{r}}$$

The only (mathematically) possible potential giving rise to spherical (isotropic) scattering is a delta function, and we thus approximate the potential using the Fermi pseudopotential:

$$V(\bar{r}) = \frac{2\pi\hbar^2}{m} b\delta(\bar{r})$$

If we consider the potential from a group of i particles, each with scattering length b_i and at position $\bar{r}_i$, the resulting potential is

$$V(\bar{r}) = \sum_i \frac{2\pi\hbar^2}{m} b_i\delta(\bar{r}_i)$$

The scattering amplitude from this set of particles can be obtained via a Fourier transformation of this potential, yielding

$$A(\bar{q}) = \sum_i b_i e^{i\bar{q}\bar{r}_i} \tag{6}$$

where q is the scattering vector, or the momentum transfer, i.e. the difference between the scattered and the incident wavevectors, as shown in Figure 9. The magnitude of the scattering vector can also be expressed in angle of incidence θ and the wavelength λ (keeping in mind that $k = |\bar{k}| = |\bar{k}'| = 2\pi/\lambda$):

$$q = |\bar{q}| = |\bar{k}' - \bar{k}| = \sqrt{\bar{k}'^2 + \bar{k}^2 - 2\,\bar{k}'\cdot\bar{k}} = \sqrt{k'^2 + k^2 - 2|\bar{k}'||\bar{k}|\cos 2\theta} =$$

$$= \sqrt{2k^2(1-\cos 2\theta)} = \sqrt{2k^2(2\sin^2\theta)} = 2k\sin\theta = \frac{4\pi\sin\theta}{\lambda} \tag{7}$$

In an experiment, we measure the flux J' of neutrons scattered from the sample onto a detector subtending the solid angle $d\Omega = \sin\theta\, d\theta\, d\phi$ in the direction (θ, ϕ), see Figure 10. Here, 2θ is the scattering (deflection) angle[2], and ϕ is the azimuthal angle, i.e. the direction defined by the angle rotating around the incident wave direction. If the incident flux is J, then the ratio J'/J represents the fraction of neutrons scattered into the solid angle $d\Omega$, and it defines the *differential scattering cross section*:

$$\frac{d\sigma}{d\Omega} \equiv \frac{J'}{J}$$

[2]This is the convention used in, for example, reflectometry, where θ is the incidence angle. In small-angle scattering, the scattering angle (the angle formed between $\bar{k}$ and $\bar{k}'$) is θ, and the equations involving the scattering angle adjusted accordingly.

 https://doi.org/10.1051/epjconf/201818801002

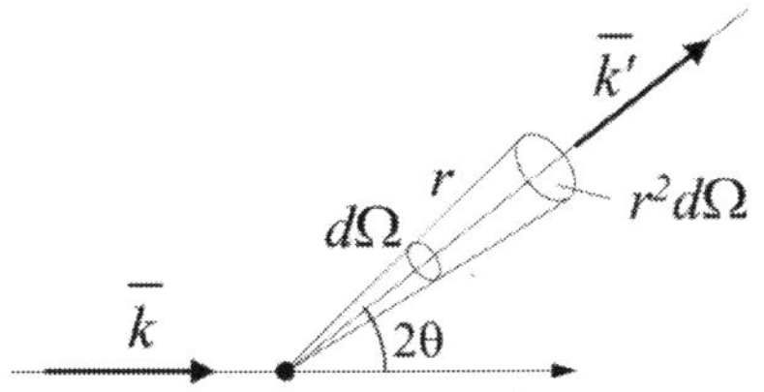

Figure 10. Geometry of a scattering experiment. The azimuthal angle ϕ rotates in the plane perpendicular to the incident wavevector $\bar{k}$.

If J is a plane wave it represents the transmitted energy per unit area per second, while the spherical wave J' is the transmitted energy per solid angle per second. The ratio J'/J then has units of area per solid angle. Integrating $d\sigma/d\Omega$ over all solid angles yields the total scattering cross section:

$$\sigma = \int \frac{d\sigma}{d\Omega} d\Omega$$

Expressed in the waves ψ and ψ', the incident flux is $v|\psi|^2$ and the total scattered flux into a sphere of radius r is $4\pi r^2 v|\psi'|^2$, where v is the neutron speed, and with σ defined as the scattered flux/incident flux, this gives

$$\sigma = \frac{4\pi r^2 v|\psi'|^2}{v\psi^2} = \frac{4\pi r^2 v|-\frac{b}{r}e^{ik'r}|^2}{v|e^{ikx}|^2} = 4\pi b^2$$

For elastic scattering where $|\bar{k}| = |\bar{k}'|$ we have

$$d\sigma(\bar{q}) = \frac{J' r^2 d\Omega}{J} = |A(\bar{q})|^2 \, d\Omega$$

and within the Born approximation, this becomes

$$\frac{d\sigma}{d\Omega}(\bar{q}) = |A(\bar{q})|^2 \tag{8}$$

7.4 Coherent and incoherent scattering

If we insert (6) into equation (8), and take into account that the nuclei b_i can be different isotopes, as well as have different nuclear spins, this results in a variation of scattering lengths, and to obtain the scattering amplitude from an array of N particles we need to average over the distribution of scattering lengths:

$$\frac{d\sigma}{d\Omega}(\bar{q}) = |A(\bar{q})|^2 = \left\langle \sum_i^N b_i e^{i\bar{q}\bar{r}_i} \sum_j^N b_j e^{-i\bar{q}\bar{r}_j} \right\rangle = \left\langle \sum_{i,j}^N b_i b_j e^{i\bar{q}\bar{r}_{i,j}} \right\rangle = \sum_{i,j}^N \left\langle b_i b_j e^{i\bar{q}\bar{r}_{i,j}} \right\rangle \tag{9}$$

The scattering lengths b_i and b_j depend on the isotope and the spin state of each nucleus, and since they are independent of each other, the product of their expectation values is equivalent to the expectation value of their product. For the averages of the scattering lengths we obtain:

$$i \neq j : \quad \langle b_i b_j \rangle = \langle b_i \rangle \langle b_j \rangle = \langle b_i \rangle^2 = \langle b \rangle^2$$

$$i = j : \quad \langle b_i b_j \rangle = \langle b_i^2 \rangle = \langle b^2 \rangle$$

To rewrite the result for $i = j$ we use the relation

$$\left\langle (b - \langle b \rangle)^2 \right\rangle = \left\langle b^2 - 2b\langle b \rangle + \langle b \rangle^2 \right\rangle = \langle b^2 \rangle - \langle b \rangle^2 \Rightarrow \langle b^2 \rangle = \langle b \rangle^2 + \left\langle (b - \langle b \rangle)^2 \right\rangle \tag{10}$$

Inserting this into equation (9), we obtain (noting that $\bar{r}_{i,j} = 0$ for $i = j$):

$$\sum_{i,j}^{N} \left\langle b_i b_j e^{i\bar{q}\bar{r}_{i,j}} \right\rangle = \sum_{i \neq j}^{N} \langle b \rangle^2 \left\langle e^{i\bar{q}\bar{r}_{i,j}} \right\rangle + \sum_{i=j}^{N} \langle b \rangle^2 + \sum_{i=j}^{N} \left\langle (b - \langle b \rangle)^2 \right\rangle \tag{11}$$

Now, the first two terms on the right are in $\langle b \rangle^2$ and can be combined into a sum for all $\{i, j\}$, and after again applying relation (10) to the last term we obtain

$$\frac{d\sigma}{d\Omega}(\bar{q}) = \langle b \rangle^2 \sum_{i,j=1}^{N} \left\langle e^{i\bar{q}\bar{r}_{i,j}} \right\rangle + N\left(\langle b^2 \rangle - \langle b \rangle^2\right) \tag{12}$$

From (12) we see that the scattering cross section has two contributions, the first term on the right still contains the phase factors $e^{i\bar{q}\bar{r}}$, and corresponds to scattering from nuclei with the same value of the average $\langle b \rangle$. This is a *coherent* superposition of pairs of scatterers, and takes into account all interference effects, and hence contains the structural information. This is the scattering we would obtain if all the scattering lengths were exactly $\langle b \rangle = b_c$, the coherent scattering length. The second term on the right does not contain phase information and hence does not result from interference, but from the random distribution of the actual scattering lengths in the system. This is the *incoherent* contribution, i.e. scattering from single atoms that superimpose in an incoherent manner. For a single isotope with zero nuclear spin, there is thus no incoherent contribution, but for other atoms, the total scattering cross section σ_{scatt} has two contributions, so that $\sigma_{scatt} = \sigma_c + \sigma_i$, and can now write the coherent and incoherent contributions to the scattering cross section as

$$\sigma_c = 4\pi \langle b \rangle^2 = 4\pi b_c^2$$

$$\sigma_i = 4\pi \left(\langle b^2 \rangle - \langle b \rangle^2\right) = 4\pi b_i^2$$

from which it is clear that the magnitude of σ_i is determined by the mean square deviation of the scattering lengths. These relations hold under the assumption that there is no correlation between the scattering lengths and the positions of the nuclei, and are valid for spin and isotope incoherence. The situation where (more or less) random variations in scattering potential arise as a result of, for example, mixing of different components in a system, is referred to as *diffuse* scattering, even though this is also incoherent if the variations are genuinely random.

The coherent and incoherent contributions can be considered from a different perspective. The combination of a nucleus with non-zero spin I and the neutron with spin $1/2$ can have two spin states, either parallel ($I + 1/2$) or antiparallel ($I - 1/2$), and these will result in different scattering lengths; b_+ and b_-, respectively, and with different degeneracies. For b_+ there are $2I + 2$ possible states, and for b_- there are $2I$ states, and thus $4I + 2$ states in total. In calculating the total scattering length, the contributions from b_+ and b_- must be weighted accordingly, with the weighting factors

$$w_+ = \frac{2I+2}{4I+2} = \frac{I+1}{2I+1} \quad \text{and} \quad w_+ = \frac{2I}{4I+2} = \frac{I}{2I+1}$$

Values of b_+ and b_- are available for a limited number of isotopes, but for a particular isotope with mass number A, the means of b and b^2 are obtained via

$$\langle b \rangle_A = w_+ b_+ + w_- b_- = \frac{(I+1)b_+ + I b_-}{2I+1}$$

$$\langle b^2 \rangle_A = w_+ b_+^2 + w_- b_-^2 = \frac{(I+1)b_+^2 + I b_-^2}{2I+1}$$

If these means are again averaged over the isotopes for an element, with proper weight for their relative abundance, $\langle b\rangle$ and $\langle b^2\rangle$ can be calculated, and we have, as above, $b_c = \langle b\rangle$, and

$$b_i = \sqrt{\langle b^2\rangle - \langle b\rangle^2} = \sqrt{w_+ w_-}(b_+ - b_-)$$

If these relations are inverted, we can calculate b_+ and b_- from b_c and b_i:

$$b_+ = b_c + \sqrt{\frac{I}{I+1}}b_i \quad \text{and} \quad b_- = b_c - \sqrt{\frac{I+1}{I}}b_i$$

Calculating the coherent and incoherent contributions from assemblies of elements (molecules, solutions, etc.) involves a similar weighting procedure. Consider the compound $X_m Y_n$ with m X atoms and n Y atoms. For these, we use the weighting factors (number fractions) $w_X = m/(m+n)$ and $w_Y = n/(m+n)$, from which the coherent scattering cross section (per atom) is

$$\sigma_c(X_m Y_n) = 4\pi\langle b\rangle^2 = 4\pi(w_X b_X + w_Y b_Y)^2$$

For example, for water, this yields σ_c (H_2O) $= 4\pi[w_H b_H + w_O b_O]^2$ fm^2 $= 4\pi[(2/3)(-3.739) + (1/3)5.803]^2 = 0.039$ barn per atom, on average, or $3 \cdot 0.039 = 0.117$ barn per water molecule. A similar calculation for D_2O gives 15.3 barn per molecule. Incoherent contributions are calculated similarly; the spin and isotope incoherence contributions for the compound $X_m Y_n$ are obtained from

$$\sigma_{i,spin}(X_m Y_n) = w_X \sigma_i(X) + w_Y \sigma_i(Y)$$

$$\sigma_{i,isotope}(X_m Y_n) = 4\pi\left(\langle b^2\rangle - \langle b\rangle^2\right), \quad \text{where } \langle b\rangle = w_X b_X + w_Y b_Y \text{ and } \langle b^2\rangle = w_X b_X^2 + w_Y b_Y^2$$

After simplification, these two contributions yield the resulting incoherent cross section per atom:

$$\sigma_i(X_m Y_n) = w_X \sigma_i(X) + w_Y \sigma_i(Y) + 4\pi w_X w_Y (b_X - b_Y)^2$$

Again, for illustration, inserting the numbers for H_2O and D_2O gives 168.1 and 4.17 barn per molecule, respectively, showing the huge differences in the coherent and incoherent contributions for H_2O and D_2O.

It is of interest that for example Vanadium has a very small coherent contribution to the scattering cross section, so that the entire contribution to σ_{scatt} is incoherent. This is not wavelength dependent, and Vanadium is thus used for calibration purposes, to determine the flux profile as a function of the wavelength.

For structural characterization, the incoherent scattering, which does not contain any structural information, adds a constant background, independent of q, and which is often determined emprically and then subtracted from the data. In studies of dynamics of atoms or molecules with quasielastic or inelastic scattering, the incoherent scattering is exploited. Since it depends on the correlation between the positions of one nucleus at different times (self-correlation), it provides information about dynamics or diffusion, and is particularly useful for H due to its high incoherent contribution.

7.5 Absorption

Neutron absorption cross sections are small for most nuclei – often much smaller than the scattering cross section – so that absorption can in most cases be ignored. The most significant absorbing isotopes are ^{3}He, ^{6}Li, ^{10}B, ^{113}Cd, and ^{157}Gd. These few strongly absorbing isotopes are used in various forms in neutron detectors and for neutron shielding.

For most isotopes the absorption is proportional to the wavelength, so that $\sigma_{abs}(\lambda) \propto \lambda$, though this does not apply to some absorbing isotopes, such as ^{113}Cd, and ^{157}Gd, which are strong resonant absorbers in the thermal neutron energy range, with sharp absorption peaks of very high cross section. If these materials are needed, but not for their absorption properties so that absorption is to be avoided, there are often isotopes which have low absorption, for example ^{11}B and ^{160}Gd. Common neutron absorbing materials are made of Cd sheets, B_4C, or borated aluminium.

Using Beer's law for transmission, $T = I/I_0 = exp(-Nl\sigma_{abs})$, where N is the number density of atoms, we obtain the thickness l of a Cd sheet (using $\sigma_{abs} = 2520$ b) that is needed to absorb 99.9% of the neutrons in a beam to ca 0.6 mm.

In contrast to X-rays, metals tend to be rather transparent to neutrons, with low absorption cross sections, but also with low scattering cross sections; aluminum and zirconium are two good examples, which also makes them useful as sample containers or window materials. Note that even though absorption of neutrons is small for many materials, they can still be activated, why some care must be taken in the handling of these materials immediately after exposure to neutrons.

8 Detection

The weak interaction of neutrons with matter poses particular challenges for neutron detection. In a detector, we want as many neutrons as possible to be counted, unless it is a beam monitor, in which case we want a low, but controllable effect of the measurement. Detectors may be grouped according to the working principles, or depending on their function and purpose. Neutrons have to be detected indirectly, via generation of secondary charged particles, and this can be done in slightly different ways: *i)* Absorption with prompt emission. Most detectors for scattering purposes fall into this category, and it includes prompt emission of detectable radiation from nuclides such as ^{3}He, ^{6}Li, ^{10}B and ^{235}U. *ii)* Elastic scattering (proton recoil), where collisions cause ionization, which is easy to detect, or *iii)* via activation, that is, reaction with absorbing materials, which emit radiation at a later time. This is clearly not suitable for detectors in scattering experiments.

8.1 Gas-filled detectors

Gas-filled detectors are used for detection of thermal neutrons via various nuclear reactions (or for detecting fast neutrons via recoil interaction). These have a gas enclosed in a thin-walled metal tube cathode (typically made from aluminium), and are filled with isotopes such as ^{3}He, or ^{10}B in the form of BF_3, and have an anode wire in the center of the tube, and with a high voltage applied between the anode and the cathode. An incident neutron causes a nuclear reaction, and the high kinetic energy of the reaction products ionizes gas atoms, which in turn ionize other gas atoms in an avalanche-like multiplication process. The induced charges result in a measureable electrical pulse. An important parameter for detectors is their gamma-ray sensitivity, and their capacity to distinguish gamma rays from neutrons, since gamma-ray generation can be considerable in the neutron beam.

In a ^{3}He detector, an incident neutron produces ^{3}H and ^{1}H; the proton ionizes the surrounding gas atoms to create the charges that are detected by the counter.

Boron-filled detectors, in the form of gaseous BF_3 are also used for detection of thermal neutrons, but are less sensitive than ^{3}He counters. However, the energy of the neutron reaction with BF_3 is higher than for ^{3}He, allowing better discrimination against gamma radiation.

These counters are used in a voltage range where the current pulse is proportional to the energy deposited in the gas – not the incident neutron energy, but to the number of captured neutrons – and hence they are called *proportional counters*. It is also common to add additional gases to improve performance; a heavy gas (such as Ar) reduces the range of the fission products, and ensures that more of their kinetic energy is deposited in the gas (and not in the container walls). This will also improve the charge collection time but also increase sensitivity to gamma rays. Polyatomic gases, such as methane or carbon dioxide in ^{3}He counters, act as quenchers in that they dampen the avalanche process by taking up kinetic energy in rotational modes, and improve the resolution.

Counters based on ^{3}He have been widespread for long, but supply of ^{3}He has been limited in recent years; ^{3}He is obtained as a decay product of tritium, which in itself is a byproduct of nuclear reactors or from weapons programs. Increased demand of radiation detectors for security, and decreasing supply, lead to the "Helium-3 crisis" in 2009, with soaring prices and reduced availability. Since then, replacement technologies have been a very hot topic in detector development.

Boron-lined gas-filled proportional counters react similarly to BF_3 gas-filled proportional detectors. Detection is by means of a ^{10}B coating rather than boron or ^{3}He gas, enabling a higher neutron sensitivity.

Recently, boron carbide (B_4C) has been proposed as an alternative liner material for gas-filled counters, where aluminium blades are coated with a thin film containing high amounts of ^{10}B.[28]

A fission chamber is a type of gas-filled detector where some fissionable material (typically ^{235}U) is coated as a thin film inside the chamber, and this layer is exposed to some detector gas (alternatively, it is filled with UF_6 gas). After a neutron has induced a fission event, fission fragments travel in nearly opposite directions, and the ionization from the fragment that enters the gas is sensed by the detector. The detection efficiency is rather low, but these detectors are thus useful as beam monitors, where the intensity, or the white beam spectrum, of the beam incident on a sample is measured, and it is desired that the attenuation should be low. Monitors are also made from low-pressure ^{3}He cells.

Proportional counters are made position-sensitive by using a resistive center wire, instead of a good conductor. In this case, the charge generated by an incident neutron can be measured at both ends of the wire, and the charge mesured at either end is proportional to the distance from the charging event, and hence the location of the incident neutron can be determined. By stacking a large number of such position-sensitive counters, a two-dimensional position-sensitive neutron detector is obtained.

8.2 Scintillation detectors

Scintillator-based detectors use absorbing isotopes in a salt mixed with some scintillator material, for example ^{6}LiF or $^{10}B_2O_3$ used with the scintillating ZnS, or $LiCaAlF_6$ mixed with a scintillating dopant like Cesium or Europium. An incident neutron causing a reaction in the absorbing isotope will emit radiation which is converted to light by the scintillator material, and a photodetector/photomultiplier records the generated light. It is essential that the scintillator material is transparent to the emitted radiation; if emitted photons are re-absorbed, this will reduce the efficiency. Scintillators are generally both faster and have better spatial resolution than gas detectors, but have much higher sensitivity to gamma rays. The more compact design, however, gives them several advantages over gas detectors in arrays or two-dimensional sensors. An *Anger camera* is a two-dimensional scintillation detector with one single slab of scintillator material, placed in front of an array of photomultipliers. An incident neutron will generate light on several detectors, and by analysing the photodetector signals, the location of this event can be determined to within a few millimeters.

Acknowledgements

I thank the organizers of the 2016 French-Swedish Winter School on Neutrons and Soft Matter and J. Oberdisse for critical reading of this chapter, and B. Nagy for discussions and comments.

References

[1] A. Pais, *Inward bound: Of Matter and Forces in the Physical World* (Clarendon, Oxford, 1988)
[2] J.J. Thomson, Philosophical Magazine Series 6 **7**, 237 (1904)
[3] H. Geiger, E. Marsden, Proceedings of the Royal Society of London. Series A **82**, 495 (1909)
[4] E. Rutherford, Philosophical Magazine Series 6 **21**, 669 (1911)
[5] E. Rutherford, Proceedings of the Royal Society of London. Series A **97**, 374 (1920)
[6] H. Becker, W. Bothe, Zeitschrift für Physik **76**, 421 (1932)
[7] I. Joliot-Curie, F. Joliot, Comptes Rendus **194**, 876 (1932)
[8] J. Chadwick, Nature **129**, 312 (1932)
[9] J. Chadwick, Proceedings of the Royal Society of London. Series A **136**, 692 (1932)
[10] H. von Halban, P. Preiswerk, Comptes Rendus de l'Académie des Sciences (Paris) **203**, 73 (1936)
[11] D.P. Mitchell, P.N. Powers, Physical Review **50**, 486 (1936)
[12] F. Bloch, Physical Review **50**, 259 (1936)
[13] E.O. Wollan, C.G. Shull, Physical Review **73**, 830 (1948)
[14] C.G. Shull, E.O. Wollan, Physical Review **81**, 527 (1951)
[15] B.N. Brockhouse, A.T. Stewart, Physical Review **100**, 756 (1955)
[16] D.M. Engelman, P.B. Moore, Proceedings of the National Academy of Sciences **69**, 1997 (1972)
[17] R.G. Kirste, W.A. Kruse, J. Schelten, Die Makromolekulare Chemie **162**, 299 (1972)
[18] J.P. Cotton, D. Decker, H. Benoit, B. Farnoux, J. Higgins, G. Jannink, R. Ober, C. Picot, J. des Cloizeaux, Macromolecules **7**, 863 (1974)
[19] J.B. Hayter, R.R. Highfield, B.J. Pullman, R.K. Thomas, A.I. McMullen, J. Penfold, Journal of the Chemical Society, Faraday Transactions 1: Physical Chemistry in Condensed Phases **77**, 1437 (1981)
[20] G.E. Bacon, *Fifty years of neutron diffraction : the advent of neutron scattering* (Hilger, Bristol, 1987)
[21] J. Reijonen, Proceedings of the 2005 Particle Accelerator Conference pp. 49–53 (2005)
[22] F. Mezei, Physica B: Condensed Matter **234**, 1227 (1997)
[23] B. Willis, C. Carlile, *Experimental neutron scattering* (Oxford University Press, Oxford, 2009)
[24] H. Maier-Leibnitz, T. Springer, Reactor Science and Technology **17**, 217 (1963)
[25] F. Mezei, Communications on Physics **1**, 81 (1976)
[26] J. Stahn, U. Filges, T. Panzner, The European Physical Journal - Applied Physics **58**, 11001 (2012)
[27] R. Pynn, *Neutron Scattering—A Non-destructive Microscope for Seeing Inside Matter* (Springer US, Boston, MA, 2009), pp. 15–36
[28] C. Höglund, J. Birch, K. Andersen, T. Bigault, J.C. Buffet, J. Correa, P.v. Esch, B. Guerard, R. Hall-Wilton, J. Jensen et al., Journal of Applied Physics **111**, 104908 (2012)

Neutron imaging and tomography: Applications in food science

Camille Loupiac[1,2,*], *Simone* Scussat[1], *Arnaud* Hélary[2], *Sylvain* Desert[2], and *Frédéric* Ott[2]

[1]Equipe PCAV, UMRA 02.102 PAM, AgroSup Dijon, Université de Bourgogne-Franche Comté, 21000 Dijon, France
[2]Laboratoire Léon Brillouin, UMR 12 CEA – CNRS, CEA Saclay, 91191 Gif-sur-Yvette, France

Abstract. Neutron imaging and tomography allow observing the structure of large objects (from few to hundred centimeters) at a resolution around 100 µm. This non-destructive method is based on the detection of the transmission of a neutron beam through an object in 2D (imaging) or 3D after rotating the sample (tomography). The components of the object attenuate the neutron beam differently, depending on their composition, and produce accurate pictures of objects which give information about their structure. Neutron imaging has been used for quality control purposes in industries (aircraft, motor engineering…), but also find application from materials sciences (fuel cells, lithium batteries) through earth science (petrology, geology) to cultural heritage. More recently studies have been performed on food science topics.

1 Introduction

Neutron imaging and tomography are non – invasive and non – destructive methods which allow observing the structure of large objects (from few to hundred centimeters) at a resolution around 100 µm. Neutron imaging has been used for quality control purposes in industries that require precision machining such as aircraft or motor engineering [1-3]. Real-time imaging also allows following fast movement (few seconds) within objects, such as oil distribution or water uptake [4-7]. For example, today, one important application of neutron radiography is in testing the performance of fuel cells (in operando) by imaging water or hydrogen flow in the cells (in situ) [8-9]. Neutron imaging yields not only information on the quantity of the adsorbed water but also on its spatial distribution. Due to its high sensitivity to small amounts of hydrogenous compounds in a matrix, neutron radiography can be used to measured humidity transport, and is powerful for a number of applications- including studies of water propagation in heat exchangers and concrete structures, internal structure within geological samples, water uptake by plants, oil and fuel distribution in car engines and fuel injectors [1, 2, 3]. Neutron tomography has also been used in the cultural heritage studies and art world or in archeology to authenticate paintings and examine artifacts made of metal or stone [10-12]. Wood density is one of the most important wood characteristics for assaying wood quality as it provides an indication of the physical and mechanical properties of wood, such as strength, stiffness, hardness and hygroscopic properties [13,14].

So, neutron imaging and tomography have been more and more used these last years to study and qualify the structure of materials in complex environments and more particularly their changes as

* Corresponding author: camille.loupiac@agrosupdijon.fr

function of various parameters such as constraints, relative humidity, temperature, processing, or after storage etc. The size of the observed samples can varied from few millimeters to hundred centimeters. The resolution is around hundred micrometers. In the following text, we are going in a first part to introduce the principle of neutron imaging and the main characteristics of the existing instruments. Then, we will present a short overview of the results of the few studies performed on food science topics (food processing like drying and cooking, 3D microstructure of food matrices ….) or related fields (defects in packaging, hydration of root and plant leaves in agronomy…). We will show that food matrices can also be studied by neutron imaging and tomography during processing and over different conditions (T, P…).

2 Principle and instruments

2.1 Principle

Neutron imaging is based on the detection of the transmission of a neutron beam through an object. When this beam passes through the object components attenuate the neutron beam differently, depending on their composition, and produce accurate pictures of objects which give information about their structure. Like X-ray imaging, neutron imaging is made possible by the fact that different elements have different attenuation coefficients (Table 1). Hydrogen for example has a higher attenuation coefficient than lead or copper, which means that the signal detected for hydrogen will not be as strong as the signal detected for lead or copper. Moreover, the unique ability of neutron imaging to provide information on materials and structure, sometimes inside objects opaque to X-rays makes neutron radiography and tomography quite powerful. The contrast uniqueness in neutron imaging is based on the fact that neutrons interact with the nucleus inside the atom instead of electrons (as occurs with X-rays). This leads to a widely different contrast mechanism as compared to X-rays. Many organic materials are quite opaque to neutrons, whereas many metals are relatively transparent; hence they can be easily penetrated by neutrons. The interaction between the neutron beam and a material is generally described by the attenuation coefficient, which is a measure of the extent to which a material attenuates incident radiation. The attenuation coefficient (see Table 1) depends strongly on the elemental composition of the material and the actual neutron energy.

Table 1. Attenuation coefficients with thermal neutrons (cm^{-1})[from 3].

1a	2a	3b	4b	5b	6b	7b	8			1b	2b	3a	4a	5a	6a	7a	0
H 3.44																	He 0.02
Li 3.30	Be 0.79											B 101.60	C 0.56	N 0.43	O 0.17	F 0.20	Ne 0.10
Na 0.09	Mg 0.15											Al 0.10	Si 0.11	P 0.12	S 0.06	Cl [illegible]	Ar 0.03
K 0.06	Ca 0.08	Sc 2.00	Ti 0.60	V 0.72	Cr 0.54	Mn 1.21	Fe 1.19	Co 3.92	Ni 2.05	Cu 1.07	Zn 0.35	Ga 0.49	Ge 0.47	As 0.67	Se 0.73	Br 0.24	Kr 0.61
Rb 0.08	Sr 0.14	Y 0.27	Zr 0.29	Nb 0.40	Mo 0.52	Tc 1.76	Ru 0.58	Rh 10.88	Pd 0.78	Ag 4.04	Cd 115.11	In 7.58	Sn 0.21	Sb 0.30	Te 0.25	I 0.23	Xe 0.43
Cs 0.29	Ba 0.07	La 0.52	Hf 4.99	Ta 1.49	W 1.47	Re 6.85	Os 2.24	Ir 30.46	Pt 1.46	Au 6.23	Hg 16.21	Tl 0.47	Pb 0.38	Bi 0.27	Po	At	Rn
Fr	Ra 0.34	Ac	Rf	Ha													

Lanthanides	Ce 0.14	Pr 0.41	Nd 1.87	Pm 5.72	Sm 171.47	Eu 94.58	Gd 1479.04	Tb 0.93	Dy 32.42	Ho 2.25	Er 5.48	Tm 3.53	Yb 1.40	Lu 2.75
Actinides	Th 0.59	Pa 8.46	U 0.82	Np 9.80	Pu 50.20	Am 2.86	Cm	Bk	Cf	Es	Fm	Md	No	Lr neut.

In accordance with the attenuation law, the intensity of the neutron beam behind the samples can be calculated as follows:

$$I = I_0 \times e^{-(\Sigma d)} \quad (1)$$

Where, I_0 is the incident intensity of the neutron beam (grey levels); I is the weakened intensity of the neutron beam (grey levels); Σ is the attenuation coefficient (cm^{-1}); d is the sample thickness (cm). Based on this equation, the attenuation coefficient Σ can be calculated by:

$$\Sigma = ln\ (I_0/I)\ /d \quad (2)$$

The attenuation coefficient is determined by the nuclear density N and the interaction probability σ for the respective elements constituting the sample. However, the effective attenuation coefficient Σeff consists of the attenuation coefficients of the involved elements:

$$\Sigma eff = \sum_{n}^{i=1} N_i \sigma_i \quad (cm^{-1})$$

(3)

Where N is the nuclear density (nuclei/cm^3) and σ is the total cross section of interaction with the neutron of an element (cm^2)

$$If\ Ni = \frac{\rho i}{Mi} L\ (nuclei/cm^3)$$

(4)

Where ρ is the density (g/cm^3), L is the Avogadro's constant and M is the atomic weight, the density of a sample can be evaluated through the following equation:

$$\rho = \frac{\Sigma M}{L\sigma} (g/cm^3) \quad (5)$$

For example, this equation (5) has been used to measure wood density and to compare the obtained results with the standard measurements realized by X-ray densitometry [13]. Density obtain with the two techniques are quite similar, and neutron brings several advantages, in particular its high sensitivity to hydrogen which allows more recently the visualization and quantification of liquid water in softwoods [14].

Transmission-based imaging using radiation could benefit to a great extent from the technical progress made in computing and data processing, allowing for the development and optimization of methods, measuring equipment, and research facilities. The increased brightness of modern neutron sources as well as pulsed nature of some of them, allow the improvement of the resolution of neutron tomography and radiography. The performance of detection devices (spatial and timing resolution) has to meet the challenges posed by those new imaging experiments in order to achieve the desired high resolution (around 10 μm), reduced the data acquisition times (few msec) and fully utilize the capabilities of modern neutron sources. Indeed, the improvement of spatial resolution in neutron imaging opens new fields of applications [4, 15-18].

2.2 Instruments

The main components of a neutron imaging instrument are presented in Figure 1 [16].

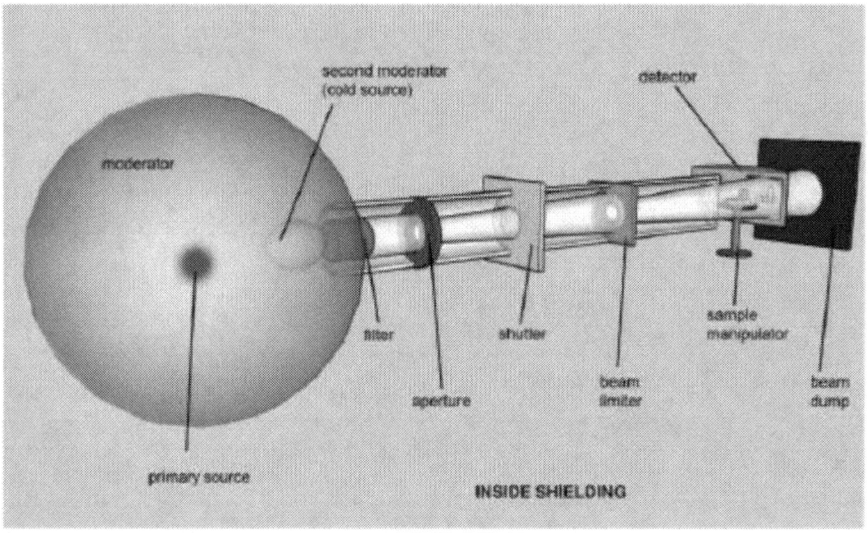

Figure 1. Schematic overview of a beamline design of a neutron imaging facility with the main components [from 16].

2.2.1 Source: Cold versus thermal

The cold neutron source allows different investigations compare to thermal ones. First the macroscopic cross-sections for materials are larger for cold neutrons due to the energy dependence of the absorption cross-section. The practical effect is that a higher contrast is achieved with cold neutrons due to energy dependence of the absorption cross-section.

2.2.2 Beamline and collimation

The initial neutrons from the source have to be tuned accordingly to get a well collimated beam of either thermal or cold neutrons. First of all, a moderator slows down the neutrons before a part of them can form the beam for imaging. This has to be done with apertures, filters, beam limiters and energy selectors. Most neutron imaging instruments are currently based on the simple pinhole geometry, in which the best achievable spatial resolution is limited by the L/D ratio, where L is the distance between the "source" pinhole with a diameter D and the detector position. The beam size at the detector position is defined by the beam divergence and the distance L. In order to achieve higher spatial resolution, it is necessary to keep L/D high (by increasing L/decreasing D) and to position the sample as close as possible to the detector such that the image of the sample projected onto the detector can be considered to be formed by a parallel-beam that is perpendicular to the detection plane. The divergence of the beam (and therefore available beam size) at the detector position is thus also determined by the coating of the neutron guide. Most facilities possess two measuring positions at varying distances to the source and hence varying collimator lengths.

Kardjilov et al. [19] studied by simulation the impact of the neutron guide geometry on the image quality. Two focusing neutron guides (linearly tapered or cone-beam and elliptic) and a straight neutron guide with circular pinhole (for comparison) each of total length 3m were simulated using Monte Carlo simulation code MCSTAS. The standard (straight guide with circular pinhole) configuration provides a beam that is just 1/9 the area of the beams produced by the focusing guides. Both experiments and simulations demonstrate superior spectral and geometry homogeneity of the cone-beam setup compared to the conventional pinhole geometry. The cone-beam geometry enables large beam cross section at relative short distances from the focal point. Use of the elliptic guide allowed enlargement of the beam dimensions by a factor of 3 at a distance of 5 m from the end of the guide to the detector at CONRAD (Berlin), thus enabling tomographic investigations of large samples. Results of Monte Carlo simulations as function of the different neutron guide configurations are presented in Figure 2.

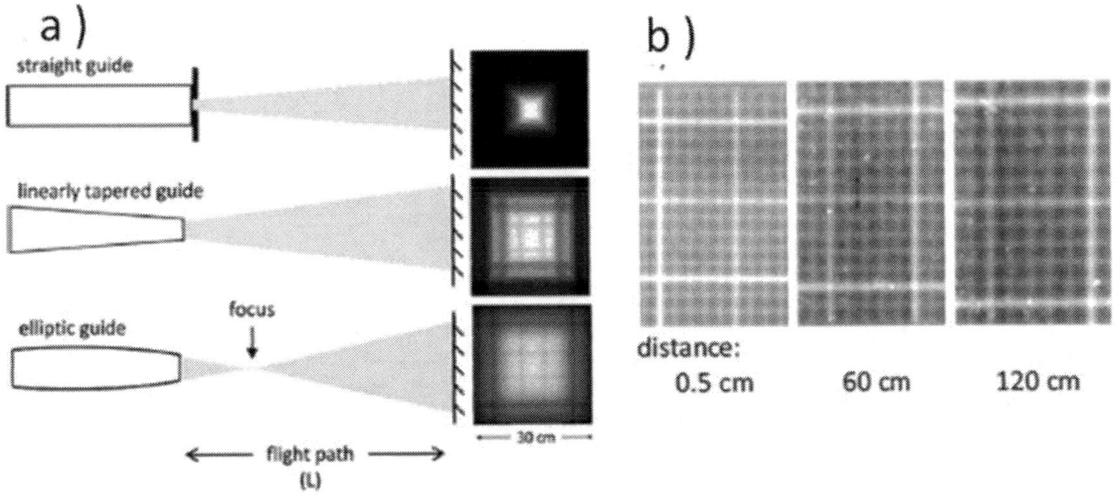

Figure 2. Impact of neutron guide geometry on image quality [From 19].
a) Monte Carlo simulations of the intensity distribution 5m behind the guide exit in a detector plane of 30 x 30 cm^2 of different neutron guide: -straight guide with a pinhole of 1 cm; - linearly tapered guide (coating m = 3.0, length: 3 m, entrance: 3x3 cm^2, exit: 1x1 cm^2); - elliptic guide (coating m= 3.0, length: 3m, entrance: 3x3 cm^2, exit: 0.89x0.89 cm^2, F1= 7.5 m, F2= 0.2 m).
b) Imaging experiments showing the magnification of a grid with 1 mm periodicity measured at different distance from the detector in the cone-beam geometry.

2.2.3 Detection

The key component of a neutron imaging setup remains the imaging detector, a neutron sensitive two-dimensional area placed perpendicular to the beam in order to observe the transmitted neutrons behind the object of investigations. A typical setup of the detection part of a neutron imaging instrument is shown in Figure 3.

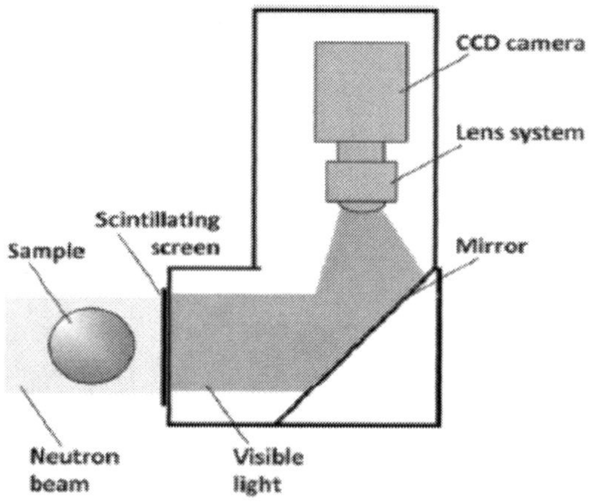

Figure 3. Detector setup for neutron imaging based on a CCD camera [from 19].

Among the detector systems used, the most common and highly flexible digital ones is based on cooled CCD cameras which are optimized for very high light sensitivity and long exposure times. The primary neutron sensitive detector is a scintillating screen (based on Lithium-6 or Gadolinium converters) which is observed by the CCD via a suitable lens system. To avoid any direct exposure of the sensitive electronics of the camera the emitted scintillation light is mirrored by 90° or 180°.

2.2.4 Scintillators

Neutron detection is a more complicated task than the registration of ionizing particles. The neutron detection process is not possible in a direct manner because neutrons do not carry a charge which is needed for ionisation. Therefore a nuclear excitation reaction is needed, which can be absorption, collision or even fission. To enable a sensitive response, conversion materials with a high interaction cross-section are commonly used: Gd, B-10, Li-6, He-3 or U-235. The neutron capture results in gamma-, beta-radiation or the emission of charged particles, which initiate light emission in suitable materials, electronic excitation or other measurable signals (Koerner et al., 2000 ;Lehmann et al., 2004). A common used screen is made of LiZnS(Ag). LiZnS has relatively low neutron capture from Li-6 but offers high output (photons/neutron) from photon emission of ZnS. Gadox (Gd_2O_2S) exhibits much higher neutron capture from the very high cross section of gadolinium, with 90% absorption in a layer of 10 µm, but also a significantly lower light output. The free-mean-path length of conversion electrons (Resolution = 12 µm) is an order of magnitude less than the range of free-mean path length of triton particle in Li-6 (Resolution= 130 µm). The disadvantage of gadox scintillating screens in neutron imaging is their sensitivity to X-rays and the low light due to low energy produced.

The first step towards developing a high resolution detector system is the improvement of scintillating screen. The thickness of the scintillating screen significantly influences the resolution. Thinner scintillating layers improve the spatial resolution but decrease the efficiency, which leads to much longer exposure times. A step forward was the production of thin LiZnS(Ag) scintillating screens going down to 50 µm layer thickness. The response of different scintillator screens has been studied by Kardjilov et al. [18] and the optimization of scintillators thickness and composition is resumed in Table 2.

Table 2. Spatial resolution of different scintillator screens: impact of thickness and composition [18].

Scintillator	Thickness (µm)	Exposure Time (s)	Resolution (µm)

Lithium	400	1x20	192
"	200	1x20	137
"	100	1x20	108
"	50	1x20	72
Gadox	30	3x20	76
"	20	3x20	57
"	10	3x20	50
"	5	3x20	48

As reported in this table, a slightly better resolution is obtained with the 50 µm lithium scintillator compared to the 30 µm gadolinium one. This is due to the high level of noise observe with the latter. However, 50 µm is the lowest thicknesses which can be obtained with lithium. Another key component of the detection system allowing to increase the quality of the resolution and which has been studied by Kardjilov et al. [19] is the optical lens associated to the camera. Indeed as can be seen in Table 3, by using lenses from 105 mm to 200 mm with same scintillator screens (for example 5 µm) the resolution obtained goes from 48 µm to 25 µm.

Table 3. Conjunction of different optical lenses and various gadox scintillator thicknesses [19].

Scintillator / lens	Thickness (µm)	Exposure time (s)	Resolution (µm)
Gadox /105 mm lens	30	3 x 20	76
"	5	3 x 20	48
Gadox / 200 mm lens	30	3 x 30	61
"	5	3 x 30	25

Other ways to improve the resolution of the detection has been reported by Boillat et al. [15] and were typically designed for applications like fuel cell imaging. Indeed in this specific case, the resolution requirement has to be obtained in different directions (observation of the membrane of the cell orientated in position parallel and / or perpendicular to the neutron beam), such making anisotropic setups attractive. In a first step, the resolution was increased by using the detector in a tilted position as illustrated in Figure 4. The effect of the angle (α with 90° as value for the normal non tilted configuration) on the resolution was studied and an improvement of the resolution is observed. The resolution of 8.7 µm obtained with a tilting angle of 3° is probably still the highest resolution reported for such neutron imaging setup.

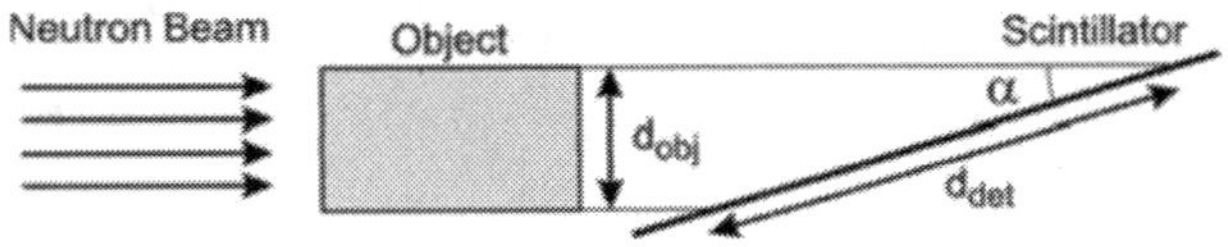

Figure 4. Principle of detector tilting (α). A feature of size of the object (d_{obj}) is projected on a larger distance d_{det} in the detector plane [15].

However a strong limitation for the resolution comes from the distance (Ld) between the sample and the detector. Also the tilting of detector may result in an increase in the Ld value, around 60 mm in a first setup which was decrease to this study to 25 mm by using an optical adapter which allowed placement of the scintillator in a tilted position without rotating the detector box.

Finally, by using a large source diameter and by increasing the L/D ratio specifically in one direction using beam limiters (Figure 5), the negative effect on the neutron flux can be mitigated.

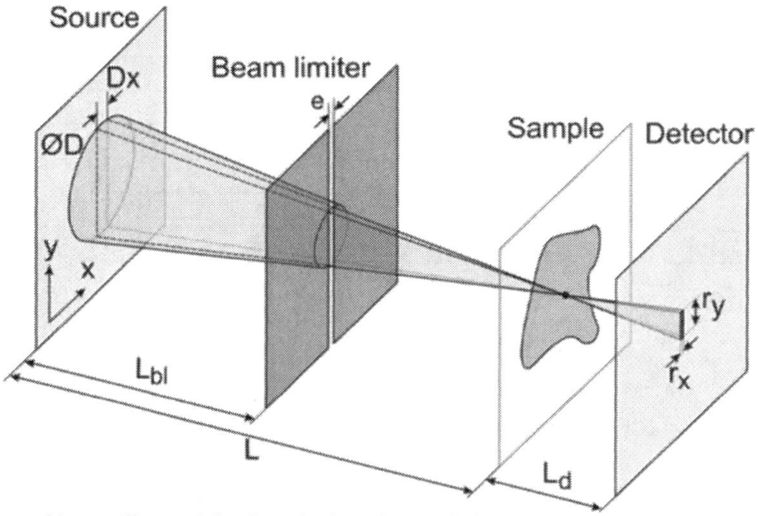

Figure 5. Geometry used by Boillat et al. [15] to obtain anisotropic beam collimation perfectly dedicated to through and in plane measurements.

Moreover, the requirement of spatial resolution for in plane imaging for example for fuel cell water migration can be perfectly attended with this configuration. Table 4, resumes the different parameters used in this study to perform such experiments and to increase the resolution.

Table 4. Typical setup parameters for in-plane imaging of fuel cells.

Parameter	**Anisotropic**	**Isotropic**
sample-detector distance	25 mm	25 mm
L/Dx	1200	1200
L/Dy	89	1200
Tilting angle	10°	10°
Neutron flux	1.13×10^7	0.84×10^6
Horizontal resolution	20	20
Vertical resolution	200	50
Exposure time	10 s	2 min

To measure the obtained resolution as function of the tilting angles, radiograms of a test object consisting of a gadolinium pattern on a silicon substrate were acquired and are presented in Figure 6. The test object is a grid of perpendicular lines, but the feature used for the measurement of the resolution is a single line with a width of 150 µm. As observed on Figure 6A, some distortion of the object are introduced which has been attributed to the slightly bended surface of the detector. As shown in the same figure, these distortions can be corrected in a post processing step.

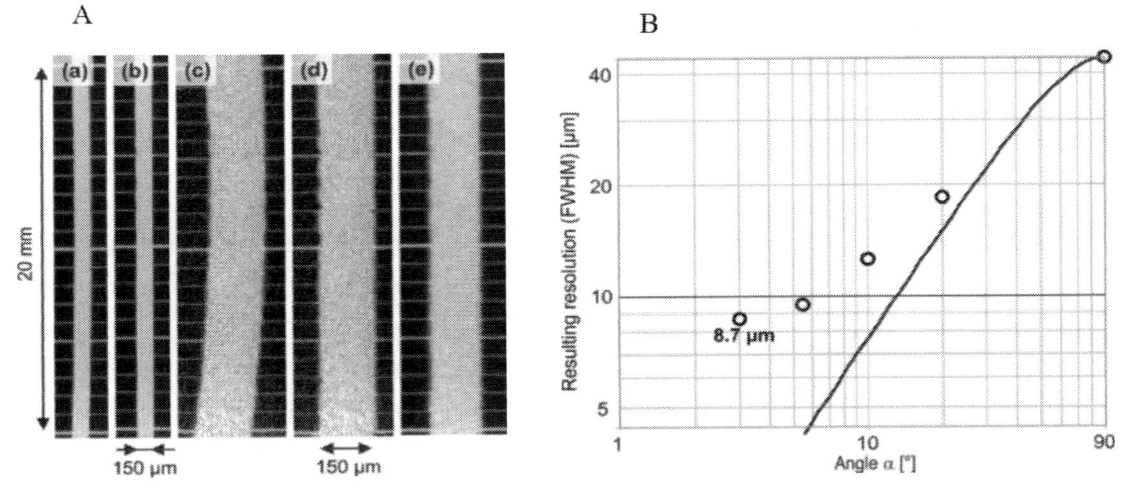

Figure 6. Radiograms of a test object consisting of a gadolinium pattern on a silicon substrate [15].
A- Images of 150 µm wide test object. a) α= 10°, b) α= 10° corrected, c) α= 3, d) α= 3° corrected, and e) same image as (b) magnified for comparison with (d).
B- Measured horizontal resolution as function of the tilting angle α. The solid line represents the expected resolution as expected by the following equation: $r_{eff} = r_{det} \times \sin \alpha$

The resolution of 8.7 µm obtained with a tilting angle of 3° is the highest resolution reported for neutron imaging.

This tilting geometry has been used by Boillat et al. [15] in the determination of liquid water in operating polymer electrolyte fuel cells (PEFCs). Figure 7 presented the evolution of the water distribution measured in the cathode side of an operating cell (70°C, 2 bar, 1 A/cm^2).

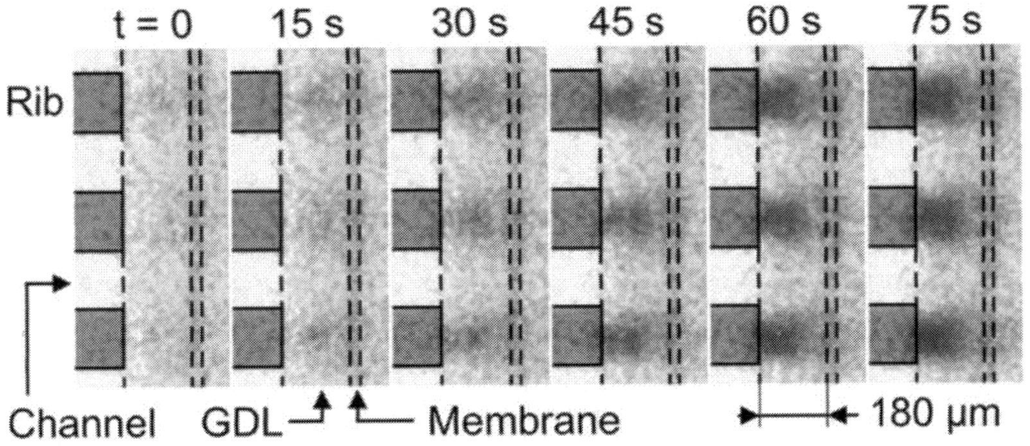

Figure 7. Evolution of the water distribution measured in the cathode (GDL) of an operating PEFC. Cathode gas flow is dry. Anode relative humidity is changed from 40 to 100% at t=0. Due to the magnifying effect of the tilted detector, the horizontal and vertical scales are not identical. The cell had an active area of 0.5 cm^2 with a depth of 5 mm in the beam direction [15].

The combination of spatial and temporal resolutions is clearly sufficient to observe the behavior of the water distribution during such a transient state. In particular, it can be observed that liquid water first appears in the depth of the gas diffusion layer before condensing in the layer just around the membrane.

The layer just next to the membrane corresponding to the microporous layer exhibits a significantly lower content of liquid water.

2.2.5 New detector: Microchannel plates

Neutron counting detectors with MicroChannel Plates (MCP) provide some unique capabilities for high resolution applications [20-22]. MCPs have been widely used for electron/ion/photon detection, however neutron detection with 10B and/or Gd- enriched MCPs is quite new technology. Standard MCPs are basically a glass disk (~ 1 mm thick) with densely packed cylindrical pores, each microscopic in dimension, typically ranging from 5 to 10 μm diameter. Each hole operates as an independent electron multiplier, providing high-resolution 2-D imaging with a spatial resolution which is channel pitched limited, typically 7 to 12 μm between adjacent pores. The process of neutron detection by the MCP consists of 3 stages: absorption of incoming neutron within the MCP glass, escape of the reaction products into adjacent pore, release of charged particles form the pore walls which initiate the electron avalanche in the pore [23]. Detectors with MCPs were measured to have efficiency levels as high as 70 % for cold neutrons, sub-15 μm spatial resolution and a 1 μs timing resolution. Many different types of readout approaches exist for MCP detectors, one of them is the Medipix2/Timepix readout. One of the disadvantages of this system is the small active area (14 x 14 mm^2 per one chip).

2.2.6 Resolution: Definition and measurement

There are 3 competing demands in neutron imaging to optimize for each detection system and its practical use: spatial resolution, time resolution and image quality. As previously explained, the spatial resolution is not only defined by the detector but also by the beam properties. However with a beam collimation ratio L/D better than 500, the resulting image will be only influenced by the detector performance, when a small distance between the object and the detector plane can be realized. The time resolution is mainly given by the detector efficiency, meaning how fast an image can be produced in the sensitive area. This is the exposure time, texp, where neutrons are applied and converted into detectable signal. The time to download the image information to a permanent storage facility (development or readout time, tread) is influencing the time resolution too. The real frame rate is then defined as FR = 1/ (texp +tread). The image quality can be described in the first order by the signal-to-noise ratio of the image (S/N), which increases usually with longer exposure time. Therefore, a wide dynamic range is to prefer, typically 16 bits and more. The noise level is given by 3 contributions, the statistics level of the detection process itself (shot noise), that from the readout process (readout noise) and from the background effects as dark current or random effects from background radiation (dark noise). Spatial resolution is also related to the image quality because an image with high noise level is inherently blurred too.

The spatial resolving power of an image-forming system is the ability to separate image of two neighboring objects points. Various statements of the resolution have been given in the literature, including the pixel size of the detector, the thickness of the scintillator, or observability of small holes [24-26]. The Modulation Transfer Function (MTF) is a common metric used to quantify the spatial resolution. Methodology used to measure it, is described by the ISO 12233 standard [24]. The initial task in MTF measurement is the identification of suitable edges for the analysis. The edge must be oriented with a minimum tilt angle and this is made by selecting a rectangular region of interest (ROI) on a specific target (example gives in Figure 8a). For example, the region over which the calculations are done is defined as shown in Figure 8b. The Edge Spread Function (ESF) represents the pixel response in terms of gray levels along a line(s) perpendicular to the edge measured on an opaque object with a straight line (see Figure 8b and 8c).

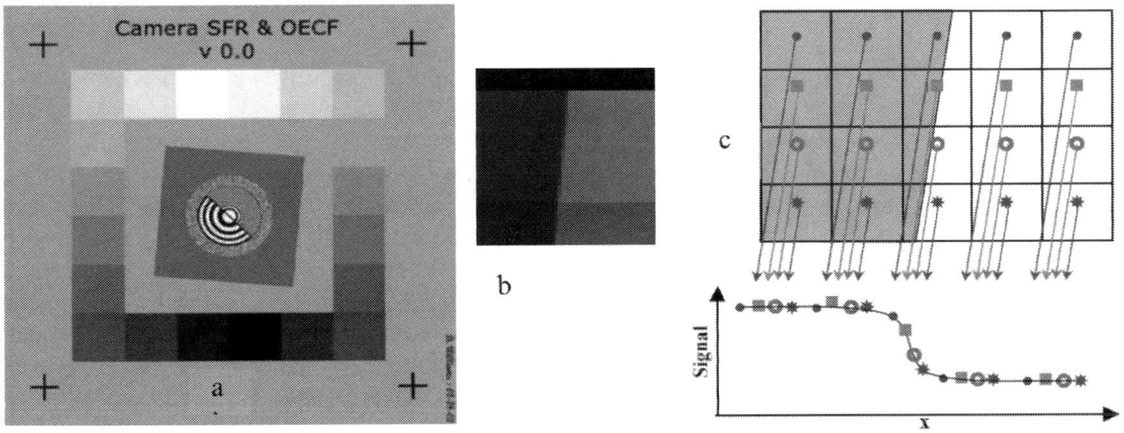

Figure 8. Resolution a) Edge prototype object, b) Selected Region of interest, c) Projection of data along the edge [24].

Edge-gradient analysis is a line spread function (LSF) transformation technique that derives the LSF. The term can be summarized as "the lost of contrast caused by the imaging system at any given spatial frequency". Rather than a simple spatial resolution limit, the imaging capabilities are fully described across the entire spatial frequency range (usually measured in the line pairs per millimeters, lp/mm). An MTF curve clearly depicts that the finer detail, the lower contrast, highlighting the inter-related nature of resolution and contrast [25,26].

Grunzweig et al [27] described their works on the fabrication and the application of a novel neutron imaging test device made of gadolinium and specifically designed for a real time evaluation of the spatial resolution of a neutron imaging detector. The developed test device consists of two geometrical patterns, a Siemens star and a square of grid lines (see Figure 9). The diameter of the Siemens star is 20 mm and consisted of 128 spokes. The period of the inner spoke is 40 μm (25 line pairs mm-1) and the outer one is 500 μm (2 line pairs mm-1). Concentric radial markers are at periods of 50, 100, 200, 300, 400 and 500 μm. The second test pattern is a square of grid lines. Its size is 24 x 24 mm2. The distance between two adjacent grid lines is 1 mm. The grid lines have linewidths of 50, 100, and 150 μm.

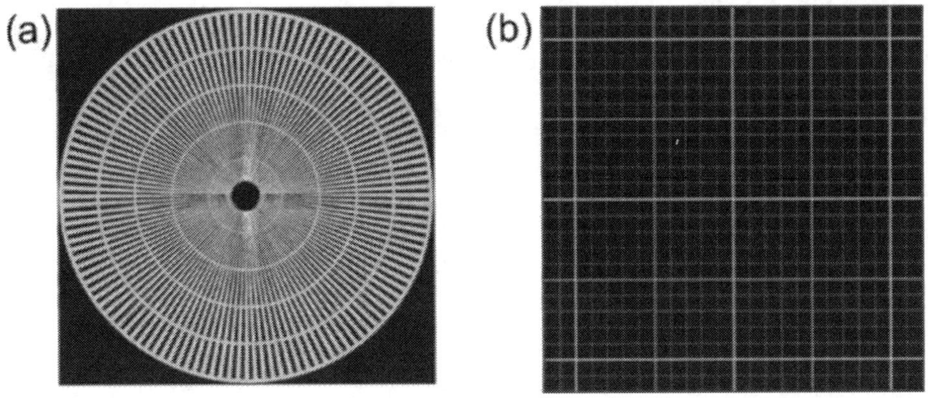

Figure 9. Geometrical patterns of the test device dedicated to the characterization of the detector performance. (a) Geometry of the Siemens star which is ideal for identifying the resolution limit in all directions. (b) Geometry of the square of grid lines which is designed to provide a convenient means for measuring distortions [27].

3 Applications in food science

Neutron imaging and tomography are excellent tools for evaluating water diffusion and uptake inside composite structure. These imaging methods were able to locate small quantities of water. For example, in food science, extrusion, drying and heating processing are technology commonly used. These processes are often applied in food and pharmaceutical industries for affecting product microstructure, product chemistry or the macroscopic shape of products. The texture of manufactured foods is related to their mechanical properties which in turn depend on their structure and the way it has been generated during processing. This is also the case for solid food foams, among which cereal products are particularly relevant. Microstructural characteristics such as cell size heterogeneity, macroporosity can be determined by neutron imaging. Moreover stability over storage at different relative humidity and how this storage can affect the porosity are key parameters that still need to be better characterized and understood. The problems observed at the macro-scale have prompted interest in understanding the processes going on a smaller scale, the pore scale for example. The intention is to gain an understanding of the fundamental pore-scale processes and thereby to be able to understand and describe the governing macro-scale processes. These topics, processing-structure-hydration can be covered by imaging measurements and the following part of this review will present some recent studies performed on food matrices.

3.1 Imaging during processing

3.1.1 Drying fish

Among food, the dry products allow to have a shelf life longer than liquid products, at room temperature and often under a reduced volume facilitating their packaging and transport. The drying of food is a process still studied today. The main aim is to decrease the development of microorganisms or to stop some chemical degradation by reducing the availability of the water. To reduce water content as low as possible in food matrices, two main drying processes are used mainly the spray-drying and for high value products the freeze drying. Such processes with different steps involving low or high temperatures and vacuum can deteriorate the structure and the organoleptic properties of the final products. Moreover, after drying, the ageing of the powders as function of the storage conditions need to be control as water sorption by the powders or the low moisture products can increase the mobility of the ingredients and of the water. This increase of mobility and water availability over storage can again provoke some physical change in the matrices and be detrimental for the final properties (texture, solubility, oxidation, flavor…) of the products. In this context, there is still a strong need of better understanding and characterization of the water repartition and mobility in food products during drying steps but also during storage at specific relative humidity. Neutron imaging and tomography have been used to analyze drying process of fish [28]. Different type of fish (mackerel and squid) involving different steps before drying (as previous salting steps for example), or different weights (small, middle and large) have been dried. Water losing during process was analyzed through the neutron beam. The analysis of the neutron image recorded on these samples showed that the loss of water was not homogeneously distributed among the fish. To have a better characterization on the spatial distribution of water during drying, tomography was also performed on mackerel dried by conventional process (soft breeze) and freeze dried. Some results are presented in Figure 10. In the freeze dried fish the repartition of remaining water seems to be more homogeneous along the overall fish than in the classically dried sample. Whatever the drying process, it was shown that in some part of head or organs, water content was kept high.

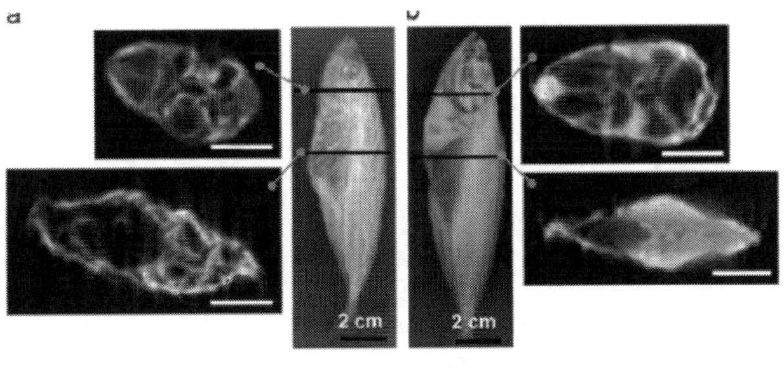

Figure 10. Neutron images of fish (center) and reconstructed tomography slices (one at the level of the head of the fish and another at the level of the shoulder) recorded on mackerel treated by freeze drying (a- left) and by conventional drying (soft breeze) (b- right) [28].

3.1.2 Cooking of meat

Cooking meat modifies its structure resulting in the improvement of its taste, aroma and texture. In fact, most of the biochemical reactions which are responsive of these changes are linked to the proteins which constitute meat: myosin, heamprotein and collagen. The changes in the meat microstructure and the concomitant juice release have been recently observed by neutron imaging [29]. The analysis of the attenuation cartography (as presented in Figure 11) realized on images collected during cooking in a semi-professional oven (placed in front of the neutron detector) highlights the fact that the main microscopic shrinkage of the meat happens after 70°C (after myosin and collagen denaturation). More important changes of attenuation have been observed at higher temperature (around 80°C) and are related to juice release following the protein denaturation.

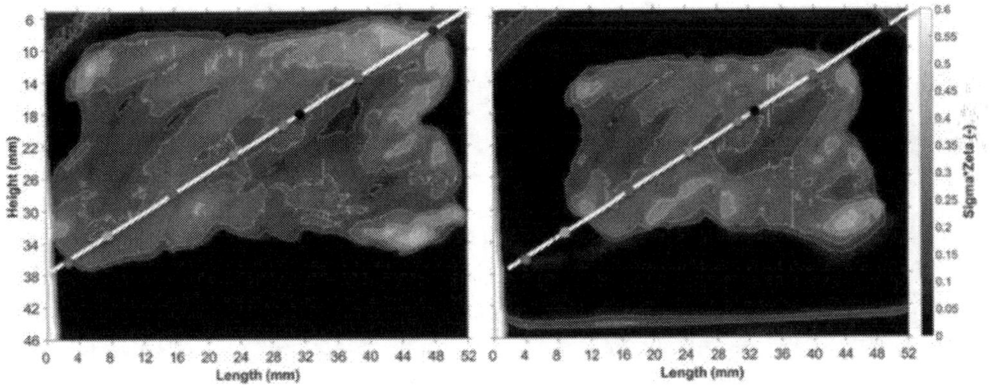

Figure 11. Cartographies of beef meat slices: raw (left) and cooked at 106 °C (right) [29].

In this study, the process of cooking in the oven has been recapitulated in the neutron beam. The main goal was to better determine the key temperature to observe morphological changes due to contraction of the meat fibres caused by cooking and migration of water (juice) inside the sample and towards the outside.

3.2 Imaging structure and defects of food and packaging

3.2.1 Water loss in fruit

During storage and shelf life, fruit loose water which reduces their weight and causes the decrease of the organoleptic quality and acceptance by the consumer. A multiphysics model was applied to study the dehydration of apple tissue accompanied by large mechanical deformation [30]. The model incorporated a water transport model and a nonlinear viscoelastic deformation model. To visualize and quantify the water transport in apple tissue neutron imaging and tomography were performed on cylindrical samples of apples (diameter 11 mm and height 20 mm). Dehydration of these cylindrical apple tissues has also been studied by gravimetrical measurements and by numerical modelisation. As shown in Figure 12, the experimental mass loss results (neutron and gravimetrical) and the predicted results are in agreement.

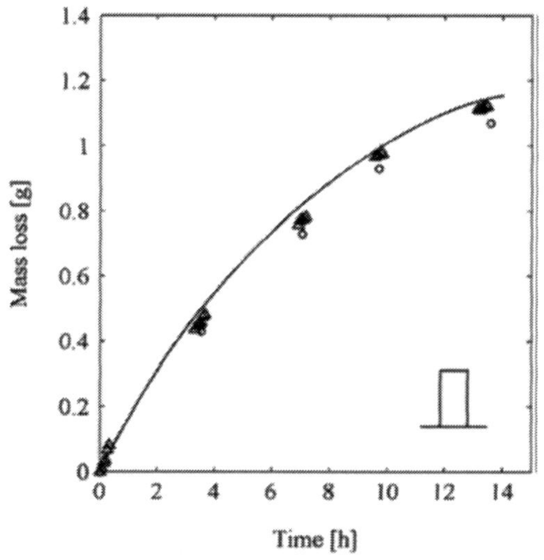

Figure 12. Water loss from apple tissue overtime: balance (o), radiography (Δ) and model (solid line) [30].

The accuracy of the model was checked more in detail by comparing the predicted local water distribution with those from the neutron tomography. In addition to moisture content, also shrinkage predicted by numerical model was compared with experiments by neutron to further validate the model as shown in Figure 13.

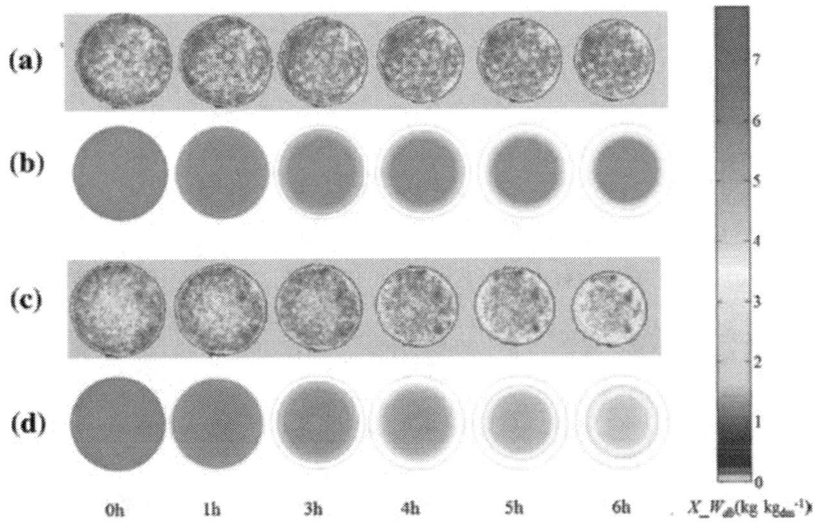

Figure 13. Horizontal slices of water content through the apple tissue as function of time. (a) Reconstructed neutron tomography image at the middle layer of the sample (y = 10mm). (b) Numerical model at the middle layer. (c) Reconstructed neutron tomography iamge near the top layer (y = 17.5 mm). (d) Numerical model near the top layer (17.5 mm). (X_Wdb: dry base water content)[30].

Neutron tomography revealed the heterogeneity of the water distribution of apple tissue. The total water loss and local water content obtained by experiments and simulations are comparable, indicating the numerical model developed in this study is a viable alternative to neutron.

3.2.2 Structural analysis of dried corn kernels

Neutron tomography was used for the structural analysis of dried corn kernels. The analysis of the image (Figure 14) obtained by neutron tomography showed that this technique was able to identify well-known internal structure of corn kernel. Many structures visible in the corn kernel reconstructions were identifiable as well-known anatomical features of corn kernels as shown in Figure 14.

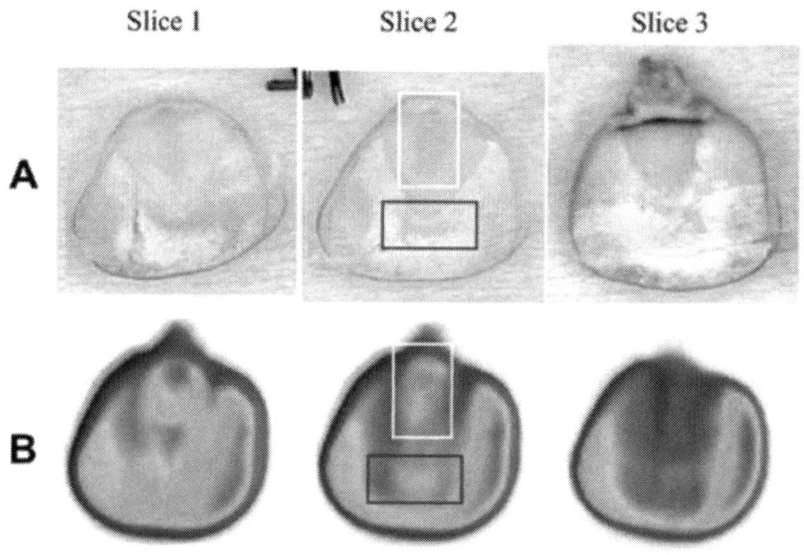

Figure 14. Reconstructed (B) vs. physical kernel slices (A) [31].

The main goal of this study [31] was to determine if these un-destructive experiments can discriminate different type of treatment of kernels. Indeed, kernels which were infected, or not, by aflatoxin producing fungus Aspergillus flavus were analyzed. Two varieties of kernel were used, one susceptible to fungus infection and one resistant. Infected kernels were found to have a lower neutron attenuation in some specific part of their anatomical structure (scutellum and embryo regions) possibly caused by lower hydrogen concentrations due to fungal degradation.

3.2.3 Defect in corks stoppers

Cork is used in a variety of products going from construction materials to gaskets, but its most important use is as a stopper for premium wines. The first known use of cork as a closure dates back to the fifth century BC, when it was used with Greek amphora. Nevertheless, the rise of cork started at the fifteenth century with the beginning of glass wine bottles. The principle requirements for cork stoppers are the homogeneity of the cork and the lack of cavities / and or cracks. Cork commonly used for wine stoppers comes from the bark of the oak tree Quercus suber L. The quality grading of cork stoppers is based on visual analysis of transverse and tangential sections of cork planks, taking into account the three main types of defects: pores (lenticular channels), physiological anomalies (nails, clay), pathogenic anomalies (insect galleries). Whatever the technique used, cork stoppers are grading in different classes. Generally there are 7 or 8 qualities [32]. Class 0 represents the best quality while 6 or 7 (depending on the classification used) is the poorest. As presented in Figure 15, the defects of cork stoppers from two classes (0 and 4) were analyzed by neutron imaging and tomography [33].

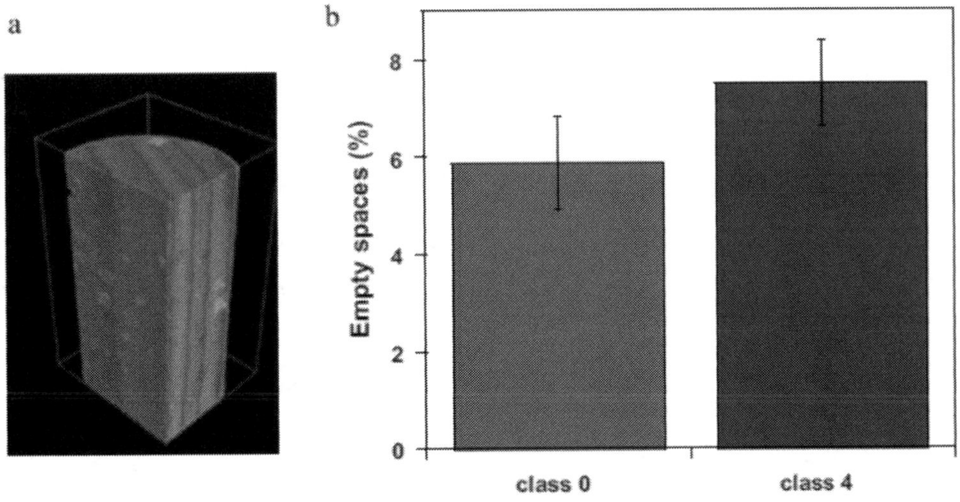

Figure 15. Analysis of defects in cork stoppers by neutron tomography. a) Visualization in 3 dimensions of a natural cork stopper (grade 0) obtained by neutron tomography; b) Histogram representing the differences of empty space content between natural corks from grade 0 and grade 4 [33].

Surface analysis of defects obtained from photography leads to more surface defects in class 4 than in class 0. Neutron radiography and tomography are powerful methods that really show the defects inside the material. Moreover, tomography also allows observing defects distribution along the whole stopper and possible interconnectivity.

4 Conclusion

Neutron imaging and tomography are more and more used for high spatial (resolution of few micrometers on large objects) and temporal (few second to obtain a 2D image and few hours to obtain 3D reconstruction) characterization of the microstructure of food, packaging or in agronomy. The main advantages of the neutron arise from the fact that they are selective (mainly to hydrogen), they can penetrate deep level of aluminium, so studies can be performed in complex surrounding sample, and they are non-destructive. Recently neutron imaging (and tomography) has been used to study the impact of process (temperature, pressure, drying…) in –situ and determine the change of structure of the food as function of this process. Thanks to the high selectivity of neutron, moreover through their high sensitivity to hydrogen (compared to deuterium), the migration of water in food or in vegetable, or leave and plants by neutron imaging experiments is more and more studied. These experiments allow to better understand the crucial role of water, hydration on microstructural properties of food matrices and moreover their link to organoleptic properties (such as texture, color etc..).

References

1. B. Schillinger, E. Lehmann, P. Vontobel, Physica B, **59**, 276 (2000)
2. N. Kardjilov, A. Hilger, I. Manke, M. Strobl, W. Treimer, J. Banhart, Nuclear Instruments and Methods In Physics Research Section A, **542**, 16 (2005)
3. I.S. Anderson, R. McGreevy, H. Bilheux, Neutron Imaging and Applications: A Reference for the imaging Community, **(Springer, New York, 2009)**

4. M. Dierick, J. Vlassenbroeck, B. Masschaele, V. Cnudde, L. Van Hoorebecke, A. Hillenbach, Nuclear Instruments and Methods In Physics Research Section A, **542**, 296 (2005)
5. B. Schillinger, H. Abele, J. Brunner, G. Frei, R. Gähler, A. Gildemeister, A. Hillenbach, E. Lehmann, P. Vontobel, Nuclear Instruments and Methods in Physics Research A, **542**, 142 (2005)
6. N. Takenaka, T. Kadowaki, Y. Kawabata, I.C. Lim, C.M. Sim, Nuclear Instruments and Methods in Physics Research A, **542**, 129 (2005)
7. M. Kang, H.Z. Bilheux, S. Voisin, C.L. Cheng, E. Perfect, J. Horita, J.M. Warren, Nuclear Instruments and Methods In Physics Research Section A, **708**, 24 (2013)
8. S. Deabate, G. Gebel, P. Huguet, A. Morin, G. Pourcelly, Energy Environ. Sci., **5**, 8824 (2012)
9. M. Klages, S. Enz, H. Markotter, I. Manke, N. Kardjilov, J. Scholta, Journal of Power Sources, **239**, 596 (2013)
10. K., Ryzewski, S. Herringer, H. Bilheux, L. Walker, B. Sheldon, S. Voisin, J-C Bilheux, V. Finoocchiaro, Physics Procedia, **43**, 343 (2013)
11. D. Mannes, F. Schmid, J. Frey, K. Schmidt-Ott, E. Lehmann, Physics Procedia, **69**, 653 (2015)
12. J. Teixeira, R. Magli, C. Loupiac, European Journal of Mineralogy, **27 (3)**, 289 (2015)
13. D. Mannes, E. Lehmann, P. Cherubini, P. Niemz, Trees, **21**, 605 (2007)
14. M. Sedighi-Gilani, M. Griffa, D. Mannes, E. Lehmann, J. Carmeliet, International Journal of Heat and Mass Transfer, **55**, 6211 (2012)
15. P. Boillat, G. frei, E.H. Lehmann, G.G. Scherer, A. Wokaum, Electrochem. Solid-State Letters, **13**, B25 (2010)
16. E. H. Lehmann, A. tremsin, C. Grunzweig, I. Johnson, P. Boillat, L. Josic, Journal of Instrumentation, **6**, C01050 (2011)
17. E.H. Lehmann, P. Vontobel, G. Frei, G. Kuehne, A. Kaestner, Nuclear Instruments and Methods In Physics Research Section A, **651**, 1 (2011)
18. N. Kardjilov, M. Dawson, A. Hilger, I. Manke, M. Strobl, D. Penumadu, F.H. Kim, F. Garcia-Moreno, J. Banhart, Nuclear Instruments and Methods In Physics Research Section A, **651**, 95 (2011)
19. N. Kardjilov, A. Hilger, M. Dawson, I. Manke, J. Banhart, M. Strobl, P. Boni, Journal of Applied Physics, **108**, 034905 (2010)
20. A.S. Tremsin, J.V. Vallerga, J.B. McPhate, O.H.W. Siegmund, W.B. Feller, L. Crow, R.G. Cooper, Nuclear Instruments and Methods in Physics Research A, **592**, 374 (2008)
21. A.S. Tremsin, J.B. McPhate, J.V. Vallerga, O.H.W. Siegmund, J.S. Hull, W.B Feller, E. Lehmann, Nuclear Instruments and Methods in Physics Research A, **604**, 140 (2009)
22. A.S. Tremsin, J.B. McPhate, J.V. Vallerga, O.H.W. Siegmund, W.B. Feller, E. Lehmann, L.G. Butler, M. Dawson, Nuclear Instruments and Methods In Physics Research Section A, **652**, 400 (2011)
23. A.S. Tremsin, W.B. Feller, R.G. Downing, Nuclear Instruments and Methods in Physics Research A, **539**, 278 (2005)
24. M. Estribeau, P. Magnan, Proceedings of SPIE, **5251**, 243(2004)
25. F. Kharfi, O. Denden, A. Bourenane, T. Bitam, A. Ali, Applied Radiation And Isotopes, **70**, 162 (2012)
26. S.H. Williams, A. Hilger, N. Kardjilov, I. Manke, M. Strobl, P.A. Douissard, T. Martin, H. Riesemeier, J. Banhart, Journal of instrumentation, **7**, P02014 (2012)
27. C. Grünzweig, G. frei, E. Lehman, G. Kühne, C. David, Review of Scientific Instruments, **78**, 053708 (2007)
28. K. Tanoi, Y. Hamada, S. Seyama, T. saito, H. Likura, T. M. Nakanishi, Nuclear Instruments and Methods in Physics Research A, **605**, 179 (2009)
29. S. Scussat, F. Ott, A. Hélary, S. Desert, P. Cayot, C. Loupiac, Food Biophysics, **11**, 207 (2016)
30. W. Aregawi, T. Defraeye, S. Saneinejad, P. Vontobel, E. Lehmann, J. Carmeliet, D. Derome, P. Verboven, B. Nicolai, Int. J. Heat and Mass Transfer, **67**, 173 (2013)

31. 31. T.E. Cleveland IV, D.S. Hussey, Z-Y. Chen, D.L. Jacobson, R.L. Brown, C. Carter-Wientjes, T.E. Cleveland, M. Arif, J. of Cereal Science, **48**, 517 (2008)
32. A. Ferreira, F. Lopes, H. Pereira, Annals of Forest Science, **57(2)** 187 (2000)
33. A. Lagorce-tachon, T. Karbowiak, C. Loupiac, A. Gaudry, F. Ott, C. Alba-Simionesco, R. Gougeon, V. Alcantara, D. Mannes, A. Kaestner, E. Lehmann, J-P. Bellat, Journal of Food Engineering, **149**, 214 (2015)

Polyelectrolytes and small angle scattering

Jérôme Combet*

Institut Charles Sadron, CNRS-Université de Strasbourg, 23 rue du Loess, B.P. 84047, 67034 Strasbourg Cedex 2, France

Abstract. We present an introduction to the application of small angle X-ray and neutron scattering to the study of polyelectrolyte solutions. We aim to give a simple overview of the structural information that can be gained with these techniques. In particular, we show how neutron scattering associated to isotopic labelling enables the determination of the different partial structure functions as well as the dispersion state and the average conformation of polyions.

1 Introduction

Polyelectrolytes (PEs) are polymers with ionisable groups that dissociate in polar solvent (such as water). Due to the long-range electrostatic interactions as well as the presence of dissociated counterions, their solutions show distinct properties than those of neutral polymers [1]. Their remarkable solubility in water is an important characteristic that can be used for instance to modify the rheological properties of aqueous solutions (thickener, gel...). The presence of electrical charges opens up additional possibilities trough the formation of electrostatic complexes with oppositely charged species (colloids, surfactants, PEs, proteins...) leading to new potential applications (flocculent, drug delivery, chemosensors, coatings...). The structural organisation of PEs solutions is a fundamental point to control in order to improve their applications. It concerns the average conformation and the dispersion state of macroions, but also, the distribution of counterions in the solution. In this context, small angle neutron and X-ray scattering (SANS and SAXS) are well-suited techniques since they allow a well-adapted distance scale (from few Å to several hundreds) to be probed.

In this contribution, we aim to present how scattering experiments can help us to understand the structural properties of PEs solutions[1]. We first briefly introduce PEs and recall their main characteristics (Sec. 2). Then, we give the basics of small angle scattering theory. We present the coherent scattering cross section and the partial structure functions related to the polyions and the counterions. We particularly emphasize the importance of the neutron scattering technique and the contrast variation method to separate these different contributions, but also, to extract the intramolecular (form factor) and intermolecular structure functions associated to the polyions (Sec. 3). We present experimental measurements and show the effects of the concentration, contrast factor and linear charge density on the scattering cross sections (Sec. 4). Finally, we give an example

* Corresponding author: jerome.combet@ics-cnrs.unistra.fr

[1] A more detailed and complementary description can be found in [2].

of polyions form factor and intermolecular structure function and discuss their main characteristics (Sec. 5).

2 Polyelectrolytes

PEs are polymers with ionisable groups (Fig. 1). They can be classified according to the polyions properties such as the sign of the ionized group (polycations/polyanions) or their chemical charge fraction (highly charged/weakly charged PEs). Distinctions can also be made on the basis of the weak or strong acid/base character of the monomers (strong PEs/weak PEs), their hydrophilic/hydrophobic properties, the intrinsic rigidity of the chains... In this contribution, only salt-free aqueous solutions of linear, hydrophilic, flexible and strong PEs in the semidilute regime (i.e. above their overlap concentration c_p*) will be considered.

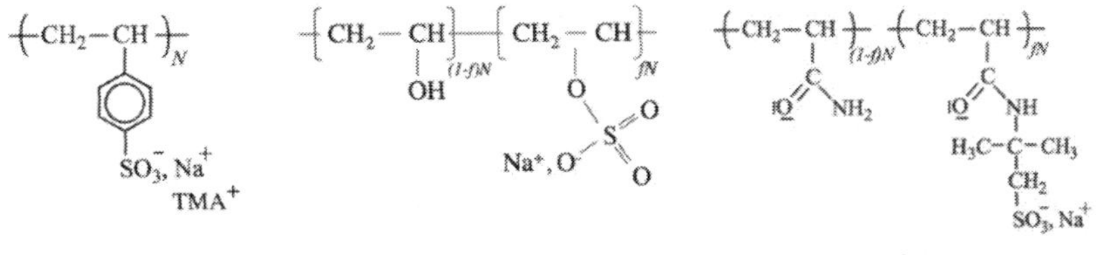

Fig. 1. Sodium/tetramethylammonium poly(styrene sulfonate): NaPSS, TMAPSS; Sodium salt of partially sulfuric acid esterificated poly(vinyl alcohol): NaPVS; Sodium salt of Poly(acrylamide-co-sodium-2-acrylamido-2-methylpropane sulfonate): AMAMPS.

Polyions can be considered as copolymers carrying ionized monomers and neutral ones (non-dissociated or non-ionisable groups of distinct nature, Fig. 2a). The structure of the solutions is highly dependent on the solvent quality (usually H_2O) for the polymer backbones. Since the different type of monomers may have a different hydrophilic/hydrophobic character, the macromolecular organisation results from the balance between the electrostatic repulsions that stretch the polyions, and the hydrophobic interactions that lead to their collapse. In the following, we will consider that the physics of the system is mainly governed by the electrostatic repulsions (hydrophilic PEs).

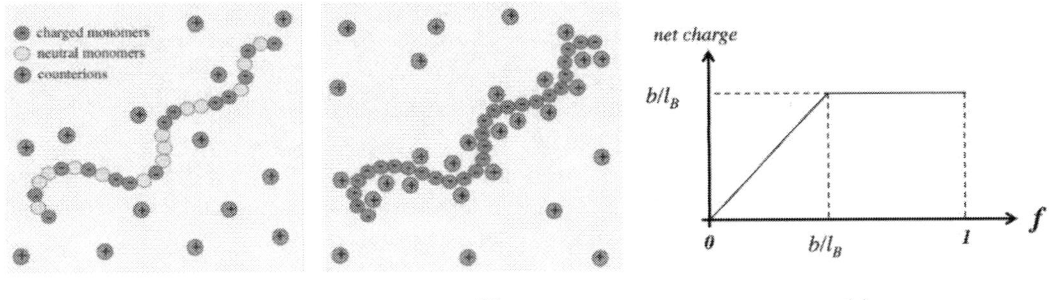

Fig. 2. Schematic representation of a PE solution. The typical size of the monomer is b, the average distance between charge is l, the degree of polymerization is N. There are N_c charged monomers and counterions. The chemical charged fraction f is defined as $f=N_c/N$. (a) weakly charged PE; (b) highly charged PE ($N_c=N$, $f=1$) and condensation phenomenon; (c) net charge fraction versus chemical charge fraction f according to the MO theory.

The properties of PEs solutions naturally depend on the linear charge density, i.e. the fraction of dissociated groups along the polyions. The chemical charge fraction f is defined as the ratio of the number of ionized groups N_c to the degree of polymerization N ($f=N_c/N$, $0 \le f \le 1$). In the case of rigid (or locally rigid) polyions with monomers of size b and an average distance between ionized groups l, f is also equal to b/l.

For high chemical charge fractions ($f \to 1$), the electrostatic potential around the polyions is very huge and part of the counterions is trapped in the vicinity of the chains (Fig. 2b). This condensation phenomenon leads to a charge renormalization and an effective charge fraction $f_{eff} < f$. In the framework of the Manning-Oosawa theory (MO) [3, 4], it occurs when the average distance between charged monomers is equal to the Bjerrum length[2] l_B (monovalent counterions). The variation of the net charge fraction with the chemical charge fraction f is presented in Fig. 2c. For $f < b/l_B$ (weak-coupling regime), it corresponds to the chemical charge fraction f; for $f > b/l_B$ (strong-coupling limit), it is constant and given by the effective charge $f_{eff} = b/l_B$. The counterion condensation threshold corresponds to the boundary between these regimes $f_{threshold} = b/l_B$.

3 Small Angle Scattering

A PE solution is a complex system containing polyions and counterions (solute) as well as solvent molecules (usually water). In a small angle scattering experiment, we measure the total differential scattering cross section per unit volume[3] $\Sigma^{total}(\vec{q},c_p)$ (cm^{-1}), which contains the coherent scattering cross section $\Sigma^{coh}(\vec{q},c_p)$ related to the solute, and a flat "background" $\Sigma^B(c_p)$, which has to be removed [5]. This later contribution combines scattering from the solvent as well as incoherent scattering from the solute in the case of neutron scattering. For highly charged PEs (f=1), it can be written

$$\Sigma^B(c_p) = \Phi_s \Sigma^S + c_p N_a \sigma^{inc} / 4\pi \qquad (1)$$

where Σ^S (cm^{-1}) is the differential scattering cross section of the solvent; Φ_s its volume fraction; c_p, the PE concentration (mol cm^{-3}); σ^{inc} (cm^2), the incoherent cross section of a monomer including its counterions (cm^2) and N_a the Avogadro's number (mol^{-1}). In the following, we will only discuss the coherent scattering cross section related to the solute, $\Sigma^{coh}(\vec{q},c_p)$, that contains all the information about the structural organization of the PE.

3.1 Scattering cross sections and partial structure functions

In a small angle scattering experiment, the spatial resolution (> 5Å) is always larger than the inter-atomic distances. Atoms are no longer considered as the elementary scatterers and can be replaced by monomers or molecules (molecular counterions, solvent) [5].

The coherent scattering cross section per unit volume $\Sigma^{coh}(\vec{q},c_p)$ (in cm^{-1}) can be written

$$\Sigma^{coh}(\vec{q},c_p) = \frac{1}{V}\sum_{i,j} K_i K_j \left\langle \exp\left[i\vec{q}\left(\vec{r}_i - \vec{r}_j\right)\right]\right\rangle, \quad K = a - a_s \frac{v}{v_s} \qquad (2)$$

[2] In pure water at 25°C, l_B=7.14 Å.

[3] We consider that standard data corrections have already been performed.

where K_i and K_j are the contrast lengths (cm) of the elementary scatterers i and j (monomers or counterions) relative to the solvent. a and a_s are the coherent scattering lengths (cm) of the scatterers and the solvent[4], and v and v_s, their partial molar volumes (cm^3/mol). These quantities can be determined from the atomic or chemical composition of the scatterers, assuming additivity of their coherent scattering lengths and their partial molar volumes. Contrast lengths of the monomer (m) K_m and the counterions (c) K_c relative to the solvent (s) can also be written as a function of the scattering length densities ρ_m, ρ_c, ρ_s (cm^{-2})[5]

$$\begin{cases} K_m = a_m - a_s \dfrac{v_m}{v_s} = \dfrac{v_m}{N_a}(\rho_m - \rho_s) \\ K_c = a_c - a_s \dfrac{v_c}{v_s} = \dfrac{v_c}{N_a}(\rho_c - \rho_s) \end{cases} \quad \text{with} \quad \rho_m = \frac{a_m}{v_m} N_a,\ \rho_c = \frac{a_c}{v_c} N_a,\ \rho_s = \frac{a_s}{v_s} N_a \tag{3}$$

The coherent scattering cross section of the solute (Eq. (2)) can be split in three distinct terms by considering specifically monomer-monomer, counterion-counterion and monomer-counterion pairs in the summation (as illustrated in Fig. 3), leading to three partial structure functions S_{mm}, S_{cc} and S_{mc} associated to the monomers (m) and the counterions (c)[6]:

$$\Sigma^{coh}(q) = K_m^{\ 2} S_{mm}(q) + K_c^{\ 2} S_{cc}(q) + 2K_m K_c S_{mc}(q) \tag{4}$$

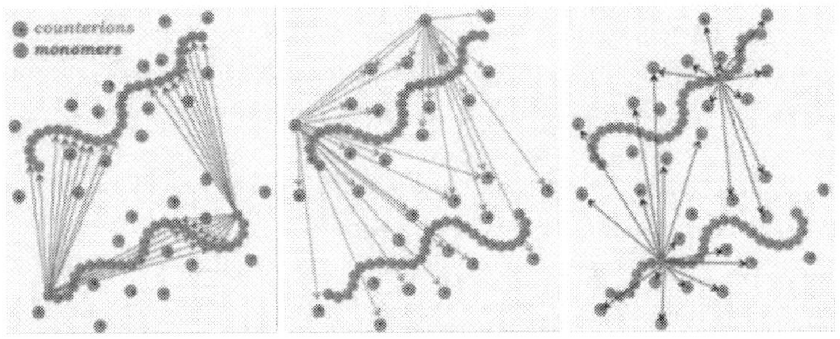

(a) (b) (c)

Fig. 3. Illustration of the three distinct summations according to the nature of the elementary scatterers pairs (arrows): (a) monomer-monomer pairs; (b) counterion-counterion pairs; (c) monomer-counterion pairs. These summations lead to the three partial structure functions S_{mm}, S_{cc}, and S_{mc} respectively.

The partial structure functions (Eq. (4)) are weighted by $K_m^{\ 2}$ and $K_c^{\ 2}$ (cm^2), the contrast factors of the monomers and counterions relative to the solvent.

These partial structure functions contain all the information about the organisation of the monomers (S_{mm}), the counterions (S_{cc}), and the cross structure (S_{mc}). S_{mm} combines intramolecular (form factor) and intermolecular terms related to the macroions. For

---

[4] For neutrons, the scattering lengths of atoms and ions depend on the nuclei properties. For X-rays, they are related to their number of electrons Z_e: a_X (10^{-13} cm) = $2.82\ Z_e$.

[5] Using conventional units, we have: $\rho(10^{10}\,\text{cm}^{-2}) = \dfrac{\sum a(10^{-13}\,\text{cm})}{\sum v(\text{cm}^3\text{mol}^{-1})} 6.022(10^{23}\,\text{mol}^{-1})$.

[6] We will mainly consider the modulus of the scattering vector q considering an isotropic scattering. The c dependence of the scattering functions will only be mentioned if necessary.

counterions, which are not covalently bounded, this distinction could be meaningless. However, for highly charged PEs, they condense around the polyions and partly reproduce the organisation of the polyions. The cloud of condensed ions can therefore be considered as a dynamical large scale "self assembled" system: the notion of form factor is recovered and the separation between intra- and intermolecular terms makes sense. S_{cc} therefore reflects the average conformation and the dispersion state of the cloud of condensed counterions. In this description, the contribution of the free counterions only appears as a flat contribution in the scattering cross section.

In a neutron scattering experiments, it is possible to determine each of the partial structure functions S_{mm}, S_{cc} and S_{mc} by using isotopic labelling and the contrast variation method, or the contrast matching method. This can be achieved using a subtle combination of deuterated macroions or counterions and a suited mixture of H/D solvent molecules. For H_2O/D_2O mixtures, the scattering length density of the solvent is given by

$$\rho_s = \phi_v^H \rho_s^H + (1-\phi_v^H)\rho_s^D \tag{5}$$

where ϕ_v^H is the volume fraction of the H water. A judicious choice of ϕ_v^H allows to match the scattering length density of the solvent ρ_s to ρ_c (K_c=0) or ρ_m (K_m=0) and to directly determine the macroion ($\rho_s = \rho_c$, $\Sigma^{coh}(q) = K_m^{\;2} S_{mm}(q)$) or the counterion ($\rho_s = \rho_m$, $\Sigma^{coh}(q) = K_c^{\;2} S_{cc}(q)$) partial structure functions. This provides a real advantage over other small angle scattering techniques (SAXS or light scattering). In SAXS measurements, we are usually in an intermediate contrast situation ($K_c \neq 0$ and $K_m \neq 0$) and all partial structure functions contribute to the coherent scattering cross section.

All these structure functions combine intra- and intermolecular contributions. Behind this apparent complexity, they are usually well suited to characterize the dispersion state of the macroions (and the condensed counterions). Indeed, the scattering functions usually display a maximum that is characteristic of the organization of the polyions in salt-free solutions.

3.2 Intra- and intermolecular structure functions

When we are concerned with the form factor of macroions and their intermolecular correlations, it is necessary to go beyond the previous scattering cross sections. We will assume hereafter that K_c=0 and therefore, that the contribution of counterions to the scattered intensity is negligible. The coherent scattering cross section becomes $\Sigma^{coh}(q,c_p) = K_m^{\;2} S_{mm}(q,c_p)$, or simply, $\Sigma^{coh}(q,c_p) = K^2 S(q,c_p)$. $S(q,c_p)$ is usually called the *total structure function* and noted $S_T(q,c_p)$.

By differentiating the monomer pairs belonging to the same polyions or to different chains, $S(q,c_p)$ can be split in two distinct parts associated with intra- and intermolecular correlations respectively (Fig. 4):

$$S(q,c_p) = S_T(q,c_p) = c_p g(q,c_p) = S_1(q,c_p) + S_2(q,c_p) = c_p g_1(q,c_p) + c_p^{\;2} g_2(q,c_p) \tag{6}$$

In this expressions, $S_1(q,c_p)$ and $S_2(q,c_p)$ are the intra- and intermolecular structure functions (cm^{-3}). $S_1(q,c_p)$ contains all the information about the conformation of a single molecule, while $S_2(q,c_p)$ is related to their interactions. $g(q,c_p)$ is a pair correlation function in the Van Hove's formalism [6]; $g_1(q,c_p)$, the form factor (without unit); $g_2(q,c_p)$, the intermolecular correlation function (cm^3). c_p is the PE monomer concentration (cm^{-3}). Since electrostatic interactions are repulsive, $S_2(q,c_p)$ and $g_2(q,c_p)$ are negative functions.

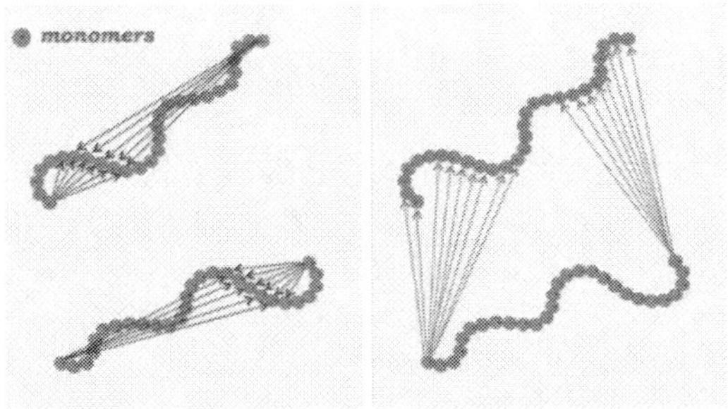

(a) (b)

Fig. 4. Illustration of the two types of summation. Counterions are neglected. (a) monomer pairs along the same polyions: intramolecular structure function $S_1(q)$; (b) monomer pairs on distinct polyions: intermolecular structure functions $S_2(q)$.

To determine the form factor of the polyions $g_1(q,c_p)$, the "classical" extrapolation to zero concentration is not feasible in salt-free PEs solution: the dilution modifies the strength of the electrostatic interactions and therefore, the average conformation of the chain. Using neutron scattering and its unique isotopic H/D substitution possibility, other methods are however possible. The "Zero Average Contrast" technique (ZAC) enables the determination of $g_1(q,c_p)$ in a single measurement. For a mixture of identical H/D chains (total monomer concentration c_p), the coherent scattering cross section can be written

$$\Sigma^{coh}(q,c_p) = K_H{}^2 S_{HH}(q,c_p) + K_D{}^2 S_{DD}(q,c_p) + 2K_H K_D S_{HD}(q,c_p) \tag{7}$$

If H and D macroions have the same degree of polymerization, their form factor $g_1(q,c_p)$ and their intermolecular correlation function $g_2(q,c_p)$ are identical. For a fraction x of hydrogenated chains, the partial structure functions are:

$$\begin{cases} S_{HH}(q,c_p) = xc_p g_1(q,c_p) + x^2 c_p{}^2 g_2(q,c_p) \\ S_{DD}(q,c_p) = (1-x)c_p g_1(q,c_p) + (1-x)^2 c_p{}^2 g_2(q,c_p) \\ S_{HD}(q,c_p) = x(1-x)c_p{}^2 g_2(q,c_p) \end{cases} \tag{8}$$

and Eq. (7) becomes

$$\Sigma^{coh}(q,c_p) = [xK_H{}^2 + (1-x)K_D{}^2]c_p g_1(q,c_p) + [xK_H + (1-x)K_D]^2 c_p{}^2 g_2(q,c_p) \tag{9}$$

The second term, related to the intermolecular correlations can be cancelled providing that the average contrast $[xK_H + (1-x)K_D]$ is equal to zero. We usually choose $x = 0.5$, leading to $K_H = -K_D = K$. The coherent scattering cross section is thus a simple function of the form factor:

$$\Sigma^{coh}(q,c_p) = K^2 c_p g_1(q,c_p) = K^2 S_1(q,c_p)$$
$$g_1(q,c_p) = \frac{\Sigma^{coh}(q,c_p)}{K^2 c_p} \quad (10)$$

The knowledge of $g_1(q,c_p)$ (or $S_1(q,c_p)$) and $g(q,c_p)$ (or $S(q,c_p)$) allows the determination of the intermolecular correlation function $g_2(q,c_p)$ (or $S_2(q,c_p)$):

$$g_2(q,c_p) = S_2(q,c_p)/c_p{}^2 = \left(g(q,c_p) - g_1(q,c_p)\right)/c_p$$
$$S_2(q,c_p) = S(q,c_p) - S_1(q,c_p) \quad (11)$$

4 Coherent scattering cross sections

In this section, we present experimental scattering cross sections $\Sigma^{coh}(q)$ measured on different PE solutions in different situations. We comment their general features in the light of the previous chapters, and in particular, their variations with the concentration, the contrast conditions and the chemical charge fraction. We finally give a simple picture for the structural organisation on the basis of a theoretical model.

4.1 General properties and concentration dependence

SANS and SAXS measurements on salt-free PE solutions usually show a maximum at a position q^* that shifts to larger q-values as the concentration is increased [7-9]. This peak is characteristic of the electrostatic repulsions and disappears when salt is added to the solution (screening of the electrostatic interactions). In the dilute regime ($c_p<c_p^*$, i.e. below the critical overlap concentration), q^* scales as $c_p^{1/3}$. The maximum is the signature of a soft position order ("liquid-like order") between the centres of mass of the chains [10]. In the semidilute regime ($c_p>c_p^*$), macroions are entangled. The maximum is now related to the correlations between polyion segments and q^* scales as $c_p^{1/2}$. For higher concentrations (concentrated solutions), other scaling exponents can also be observed according to the intrinsic thickness of the chains [11, 12].

Typical SAXS measurements ($\Sigma^{coh}(q,c_p)/c_p$) in the semidilute regime are presented in Fig. 5. They were obtained using synchrotron radiation (D2AM beamline, ESRF, Grenoble, France) on salt-free aqueous solutions of poly(styrene sulfonate) polyions (PSS^-) with two different counterions (TMA^+ and Na^+). For these highly charged systems (f=1), a large number of counterions are condensed and trapped along the chains. The theoretical effective charge fraction f_{eff} is close to 0.36, corresponding approximately to one effective charge every three monomers.

The scattering curves for these two systems are qualitatively similar. They show a broad scattering peak at a similar position q^* which is displaced to larger q values when the concentration is increased. General comments can however be mentioned on these two systems.

If we consider the contrast condition for TMAPSS, $\rho_m > \rho_c \approx \rho_s$: the scattering length density of the counterions is close to that of the solvent (Fig. 5c). The contrast length K_c is therefore small compared to that of the monomer K_m. We are in a situation where the coherent scattering cross section measured by X-ray is essentially related to the monomer partial structure function $S_{mm}(q)$. For NaPSS solutions, this is no longer the case

https://doi.org/10.1051/epjconf/201818803001

$\rho_m \approx \rho_c > \rho_s$, $\Sigma^{coh}(q,c_p)$ is a linear combination of three different partial structure functions. As it will be discussed in the next chapter, this explains the difference observed in the shape of the scattering cross sections for the highest q values, and more particularly, the faster reduction in the case of NaPSS [2, 13].

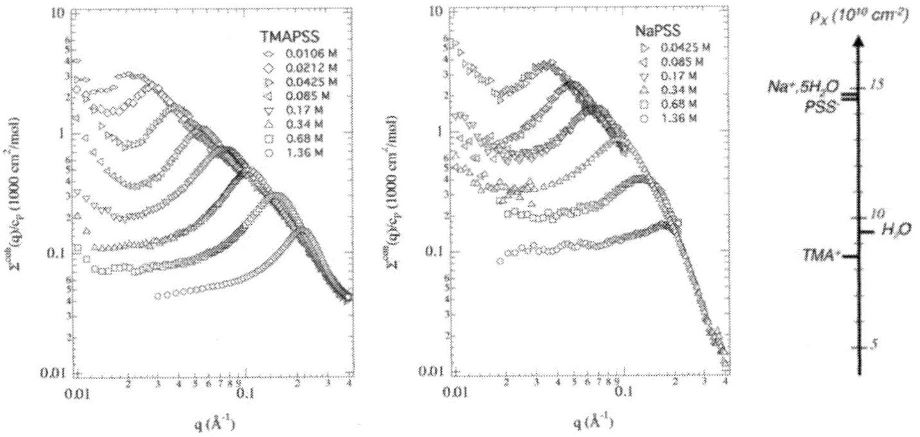

(a) (b) (c)

Fig. 5. $\Sigma^{coh}(q)/c_p$ measured by SAXS on salt-free aqueous solutions of TMAPSS (a) and NaPSS (b) at different concentrations. For these PEs, f=1, and f_{eff}=0.36. (c) X-ray scattering length densities of the different components. For Na^+, we consider the ρ value corresponding to the hydrated ion (Na^+, $5H_2O$).

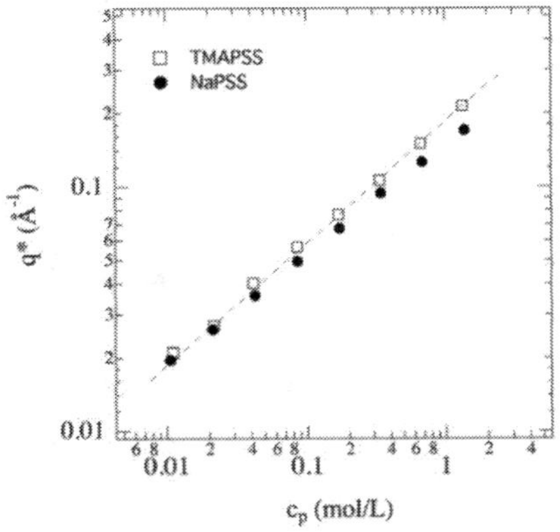

Fig. 6. Variation of q^* with the concentration for TMAPSS and NaPSS. Dashed line corresponds to the $c_p^{1/2}$ power law.

For these two systems, the variation of q^* as a function of the concentration can be described by a power law close to $c_p^{1/2}$ (Fig. 6). This is typical of hydrophilic PEs in the semidilute regime, where the interactions are dominated by the electrostatic repulsions. For hydrophobic PEs, this behaviour is usually lost, the power law exponent is reduced [14] and different regimes in concentration are observed [15]. A close inspection of the q^* variation however reveals smaller values for NaPSS especially at high concentrations. This effect is not due to the hydrophobicity and could be explained by the combination of the different partial structure functions and [2, 13].

4.2 Contrast factors variations: Partial structure functions S_{mm} and S_{cc}

Experimental scattering cross sections are usually related to both components: polyions and counterions. To characterise their respective organization in the solution, it is necessary to determine their partial structure functions S_{mm} and S_{cc}. This can be achieved with neutron scattering by using the contrast matching method. It requires the adjustment of the solvent scattering length density to that of the monomers (S_{cc} measurement) or that of the counterions (S_{mm} measurement) using suited H_2O/D_2O mixtures.

Such measurements have been carried out on semidilute salt-free aqueous solutions of TMAPSS (f=1) and are presented in Fig. 7a (D22 beamline, ILL, Grenoble, France). They were obtained using deuterated polyions (PSS^-_D) and hydrogenated counterions (TMA_H^+) in H_2O and D_2O. Since $\rho_{D_2O} = \rho_{PSS^-_D}$ and $\rho_{H_2O} = \rho_{TMA^+}$ (Fig. 7c and Table 1), measurements in D_2O leads to S_{cc} ($K_m \approx 0$, $\Sigma^{coh}(q) = K_c^{\,2} S_{cc}(q)$), while experiment in H_2O provides S_{mm} ($K_c \approx 0$, $\Sigma^{coh}(q) = K_m^{\,2} S_{mm}(q)$).

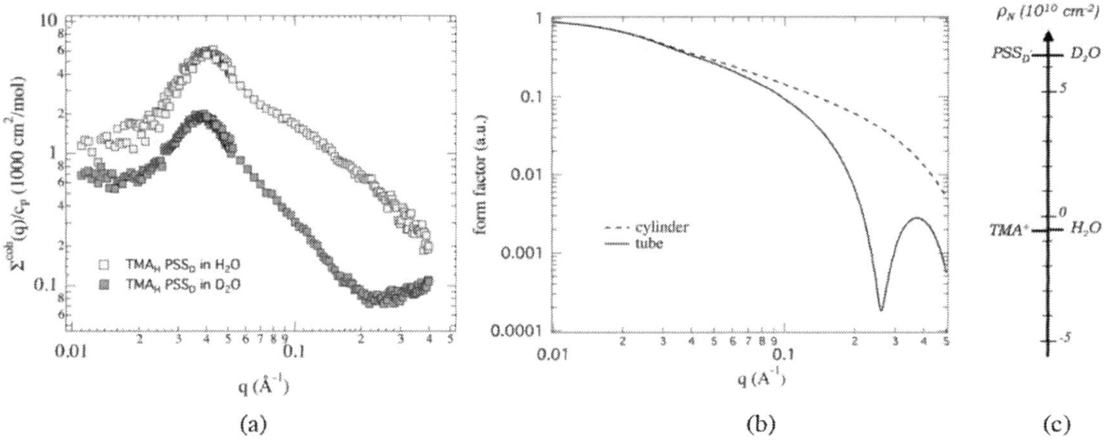

Fig. 7. (a) coherent scattering cross section $\Sigma^{coh}(q)$ measured by SANS on TMA_HPSS_D (f=1, f_{eff}=0.36) at c_p=0.0425 mol/L in H_2O and D_2O. In H_2O, $\Sigma^{coh}(q)=K_m^{\,2}S_{mm}(q)$. In D_2O, $\Sigma^{coh}(q)=K_c^{\,2}S_{cc}(q)$. (b) form factors (a.u.) of a rigid cylinder (radius: 5 Å, length: 200 Å), and for a tube (internal radius: 5 Å, external radius: 13 Å, length: 200 Å). (c) neutron scattering length densities of the different components.

Table 1. SANS parameters for TMA^+_H and PSS^-_D in H_2O and D_2O. a_N: neutron coherent scattering length, K_N neutron contrast length, K_N^2 neutron contrast factor.

	a_N	v	ρ_N	K_N in H_2O	K_N^2 in H_2O	K_N in D_2O	K_N^2 in D_2O
	(10^{-13}cm)	(cm^3/mol)	(cm^{-2})	(10^{-13} cm)	(10^{-26} cm^2)	(10^{-13} cm)	(10^{-26} cm^2)
PSS_D^-	120.17	114	6.35	130.8	17109	-0.7	0.5
TMA_H^+	-8.92	84.2	-0.64	-1.1	1.2	-98.2	9643
H_2O	-1.68	18.07	-0.56				
D_2O	19.15	18.07	6.38				

These measurements show important properties and complement the previous SAXS analysis. The main observation concerns the presence of a maximum in the partial structure function of counterions (S_{cc}). Its position is almost identical to that observed in the monomer structure function S_{mm}[7]. This means that the distribution of counterions is not homogeneous in the solution and reproduces the spatial organization of the macroions [44]. This is a clear manifestation of the counterion condensation process. In the absence of condensation, one would expect a completely flat partial structure function S_{cc}.

Another point concerns the shape of the scattering cross sections above q^*, and more particularly, the presence of a minimum around 0.2 $Å^{-1}$ in S_{cc}. This behaviour originates from the form factor of the scattering objects. While a polyion can be assimilated to a rigid cylinder of a few Å in diameter (at a local scale), the cloud of condensed counterions can be viewed as a tube surrounding this cylinder. For large scattering vectors ($q>q^*$), we can reasonably assume that the main contribution to S_{cc} and S_{mm} comes from the form factor of these different entities[8]. The form factors of a tube and a cylinder are shown in Fig. 7b. A minimum and a first oscillation are clearly observed for the tube whereas they are out of the selected q range for the cylinder. This explains the shape behaviour observed in S_{cc} around 0.2 $Å^{-1}$. Furthermore, in between 0.06 and 0.2 $Å^{-1}$, the decay of the theoretical form factor is much steeper in the case of the tube, a fact that is also observed in the experimental scattering cross sections[9].

We can also learn quantitative information from the height of the maximum. Around q^* (0.04 $Å^{-1}$), the form factors are very close to each other. If we made the assumption that the scattering is dominated but the contribution of the form factor, the difference in intensities only originates from the contrast length and the concentration of monomers and condensed counterions. The latter can therefore be estimated, leading to an effective charge fraction of 0.4, very close to the theoretical one (0.36).

In conclusion, the shape of the partial structure functions ($q>q^*$) can be *qualitatively* understood with the help of the form factor of the scattering entities. We can note that an exact calculation of the high q behaviour of S_{cc} (above q^*) is quite complex to determine since counterions do not form a wall of constant scattering length density. It should be better approximated using the real counterion density profile determined from the Poisson-Boltzmann equation within a cylindrical cell model [16-18].

4.3 Effect of the linear charge density

To vary the chemical charge fraction, it is necessary to synthesize a random copolymer with ionisable and neutral monomers. Pioneer works were proposed by El Brahmi et al. using acrylamide-acrylic acid copolymers [19] and by Nishida et al. using esterificated poly(vinyl alcohols) (NaPVS) [20]. In the latter study the position of the PE peak was measured as a function of the concentration and the chemical charge fraction above and below the MO counterion condensation threshold (SAXS experiments).

Beside the classical $c_p^{1/2}$ evolution, the most interesting point was the increase of q^* with the chemical charge fraction f according to the power law $q^* \propto f^{\approx 0.3}$ up to f_{crit} (Fig. 8). For higher charge fractions, q^* was found to be almost constant. The crossover value f_{crit} being close to the theoretical condensation threshold ($f_{threshold}$=0.36), this phenomenon was interpreted as the consequence of the onset of the counterion condensation process leading

---

[7] For higher concentrations small differences are observed [2, 13], q^* appears at slighltly lower values for S_{cc}. This effect is related the form factor of the scattering entities.

[8] This assumption is justified in the light of the $S_1(q)$ and $S_2(q)$ structure functions presented in Sec. 5.2.

[9] This is consistent with the SAXS observations on TMAPSS and NaPSS (Sec. 4.3).

to a charge renormalization. In other words, despite the increase in the chemical charge fraction, both the effective charge fraction and the peak position remain constant.

Another study was performed by Essafi et al. by using poly(acrylamide-co-sodium-2-acrylamido-2-methylpropane sulfonate) (AMAMPS) [21]. In this work, different charge fractions above $f_{threshold}$ were investigated (SANS, SAXS experiments). They could evidence the invariance of q^* for highly charged macroions. A close inspection of the macroion partial structure function (S_{mm}, SANS measurements) showed a similar organisation in the strong-coupling regime. The authors also tried to determine the effective charge through osmotic pressure measurements [22] and showed that is was in striking agreement with MO predictions.

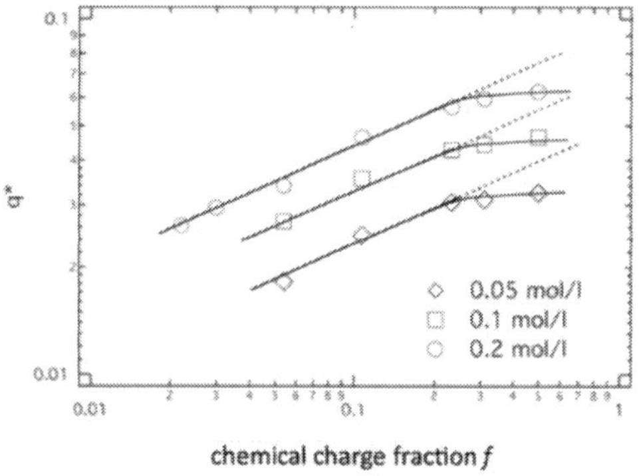

Fig. 8. Variation of q^* with the chemical charge fraction f for NaPVA at three different concentrations in semidilute regime (or close to c_p^*, SAXS measurements, from Ref. [20]). The lines correspond to a power law $q^* \propto f^{\alpha}$ with α=0.3.

All these studies highlight an important feature: the position of the PE peak is related to the net charge of polyions and is thus sensitive to the counterion condensation process. Above $f_{threshold}$, the chemical charge fraction f has to be replaced by the effective one f_{eff}. In addition to the structural characterization, a careful investigation of the PE peak position also permits the investigation of the counterion condensation phenomenon.

4.4 General structural organization

The PE peak observed in the semidilute regime reflects the polyion-polyion correlations in the solution. If it is clearly related to the electrostatic repulsions, its physical origin in term of structural organization is however difficult to determine.

This maximum could come from an ordered phase in which polyions (or part of them) form a lattice of aligned rigid rods [10]. In this model, q^* is related to the d-spacing ($q^*=2\pi/d$) and its variation with concentration results from a simple dilution law. However, in flexible PEs solutions, only one peak is visible and no higher-order reflexion is observed. PEs solutions are therefore certainly more disordered systems.

The semidilute solutions can also be described by an "isotropic model" [10] in which polyions form a disordered transient network of mesh size ξ (correlation length). The presence of a maximum can be interpreted in terms of a correlation hole [23, 43, 44, 46]. Due to the electrostatic repulsions, chains or chain segments cannot approach each other closer than a certain distance (of the order of Debye screening length κ_D^{-1}): each polyion is surrounded by a correlation tube in which other chains are expelled. From a scattering point of view (if we neglect the contribution of the counterions for simplicity), the intra- and intermolecular structure functions ($S_1(q)$, $S_2(q)$) are monotonic decreasing ($S_1(q)$) and increasing ($S_2(q)$) functions. Their combination $S(q)=S_1(q)+S_2(q)$ (and therefore the coherent scattering cross section measured in small angle scattering experiments) exhibits a maximum at a position $q^{*}=2\pi/\xi$ [10, 23, 43, 44, 46, 47]. Other interpretations attribute the PE peak to a liquid-like order between chains segments [44]. Contrary to the correlation hole description, this organization induces a characteristic maximum in the intermolecular structure function $S_2(q)$: chain segments have the tendency to stay at a finite distance ξ from each other. The distinction between these two descriptions is difficult to achieve on total structure functions ($S(q)$) and requires the separation between intra ($S_1(q)$) and intermolecular ($S_2(q)$) structure functions.

The problem is now to understand the relationship between the position q^* (or the correlation length ξ) and the physical parameters of the systems. A simple model is therefore needed for the chain conformation and their dispersion state in the solution.

We present in Fig. 9a, a global overview of q^* variations as a function of c_p for a large set of hydrophilic flexible PEs found in the literature, in the semidilute regime (considering different polyions, counterions, solvents and linear charge densities, above and below the counterion condensation threshold) [13]. They were measured by SANS, SAXS, and light scattering. If the general variations are still described by a "classical" $c_p^{1/2}$ behaviour, there is a wide dispersion of the experimental values. The concentration is indeed not the only parameter that governs the organization or the dispersion state of macroions.

In the scaling approach initially developed by de Gennes and Pfeuty for weakly charged PEs [10, 24] and recently revisited by Dobrynin et col. [25], polyions are viewed as an assembly of electrostatic blobs of typical size ξ_e. Inside each blob, the chain is not perturbed by the electrostatic interactions and adopts a conformation similar to that of a neutral chain. For larger length scale the electrostatic repulsions dominate and the electrostatic blobs strongly repulse each other and align. In the dilute regime, the average conformation of the macroions can be seen as a linear array of electrostatic blobs and is thus highly stretched. In the semidilute regime, macroions overlap and form a transient network.

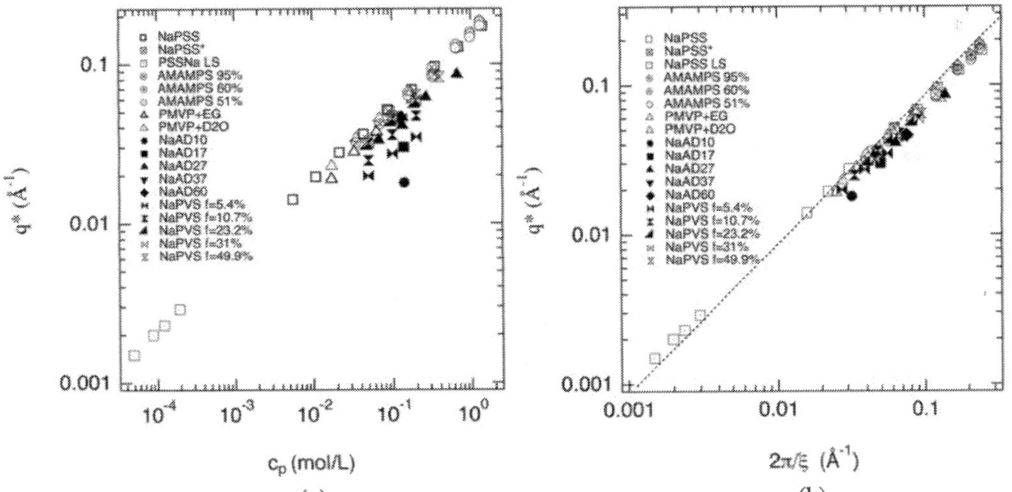

(a) (b)

Fig. 9. (a) q^* versus c_p for different macroions, monovalent counterions, solvents and chemical charge fractions above and below the counterion condensation threshold. (b) q^* versus $2\pi/\xi$. From Ref. [13].

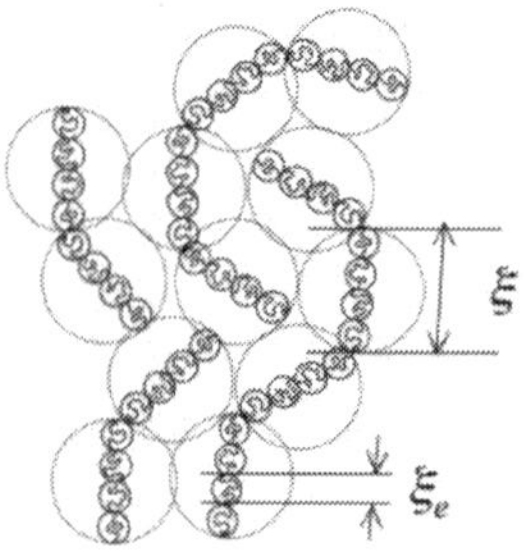

Fig. 10. Schematic representation of a semidilute PE solution according to the scaling approach within the isotropic model.

The isotropic model [10] postulates that a single characteristic length governs the system: the mesh size of the transient network ξ that also represents the screening length of the electrostatic interactions. For length scales smaller than ξ, the electrostatic interactions are dominant and the average conformation of the macroions remains highly stretched as in the dilute regime. For distances larger than ξ, the electrostatic interactions are screened: the average conformation of the macroions becomes a random walk of correlation blobs of size ξ. A schematic representation of semidilute solution of PE is given in Fig. 10.

By imposing a close packing condition for correlation blobs, scaling arguments allow the determination of the correlation length ξ. In good solvent conditions, it can be written:

$$\xi = f^{-2/7}(l_B/b)^{-1/7}(bc_p)^{-1/2} \tag{12}$$

From an experimental point of view, the scattering peak position q^* provides a direct measurement of ξ (through $q^* = 2\pi/\xi$) and gives insights into the structure of the semidilute solutions:

$$q^* = \frac{2\pi}{\xi} = 2\pi f^{2/7}(l_B/b)^{1/7}(bc_p)^{1/2} \tag{13}$$

The f-dependence of q^*, or ξ, is correlated to that of the electrostatic blob size. Eq. (13) perfectly accounts for the experimental observations: $q^* \propto c_p^{1/2}$ and $q^* \propto f^{\approx 0.3}$ (2/7=0.285) in the weak-coupling regime.

The scaling model was initially developed for weakly charged PEs, where the net charge fraction is the chemical charge fraction of the macroions. For highly charged PEs, the electrostatic blobs are very small and only concern a very limited number of monomers. However, the model is still assumed to apply provided that the macroions are treated as weakly charged PEs with an effective charge fraction f_{eff}. In the strong-coupling limit Eq. (13) then becomes

$$q^* = \frac{2\pi}{\xi} = 2\pi f_{eff}^{2/7}(l_B/b)^{1/7}(bc_p)^{1/2} \tag{14}$$

This theoretical expression can be directly compared with the measurements presented in Fig. 9a. If we plot q^* as a function of $2\pi/\xi$ (as defined in Eq. (13) or (14)), instead of the concentration c_p we observe a master curve (Fig. 9b). The isotropic model associated with the scaling approach therefore satisfactorily describes the structure of flexible hydrophilic PE solutions.

5 Intra- and intermolecular structure functions

The examples presented in the previous section involve the partial structure functions of the different components. These functions don't allow to draw direct and accurate conclusions about the average conformation and the intermolecular correlations of macroions. The determination of the intra- and intermolecular structure functions can be achieved through neutron scattering taking advantage of isotopic H/D substitution as described in Sec. 3.2. In the following, we will assume that the contribution of counterions to the scattered intensity is negligible. We present experimental mesasurements and a way to interpret the data. A more detailed presentation of the usual form factors of polymers is given in [26, 27].

5.1 Form factor

The intramolecular structure function, or form factor, characterizes the average shape of macroions and give insight into their degree of polymerization N, their radius of gyration R_g and their average conformation. Two q-ranges are usually considered according to qR_g values.

5.1.1 Guinier range

In the Guinier range ($qR_g<1$), the spatial resolution allows to probe distances on the macromolecular scale (larger than R_g). Measurements are used to determine the degree of polymerization N and the radius of gyration R_g of the chains[10].

The form factor of a particle is a function, which usually decreases monotonically and tends to zero for larger scattering vectors. In the Guinier range, it can be approximated by a series:

$$g_1(q) = N\left[1 - \frac{q^2 {R_g}^2}{3} + \circ(q^2)\right] \tag{15}$$

For linear macromolecules, we usually prefer Zimm [28] or Berry [29] representations:

$$Zimm: \frac{1}{g_1(q)} = \frac{1}{N}\left[1 + \frac{q^2 {R_g}^2}{3} + \circ(q^2)\right] \quad , \quad Berry: \frac{1}{\sqrt{g_1(q)}} = \frac{1}{\sqrt{N}}\left[1 + \frac{q^2 {R_g}^2}{6} + \circ(q^2)\right] \tag{16}$$

Plotting $1/g_1(q)$ or $1/\sqrt{g_1(q)}$ as a function of q^2 allows the determination of the radius of gyration (slope), whereas the extrapolation of $g_1(q)$ to q=0 gives the degree of polymerization N (provided that the measurements are in absolute scale)[11]. Since we know the molar mass of the monomer unit m, we can deduce the molecular mass of the polyion M=Nm.

Taking into account polydispersity, we measure the weight average (w) of the form factor $\langle g_1(q)\rangle_w$ [27]. Eq. (16) can then be written (Zimm representation)

---

[10] The formalism developed in this section only applies to homopolymers.

[11] In some cases, the determination R_g and N can be difficult. For relatively long chains, the Guinier range may be shifted to very low q, even out of the accessible values. Furthermore, the low q range may present strong upturns that prevent any accurate R_g or N measurements.

$$\frac{1}{\langle g_1(q)\rangle_w} = \frac{1}{N_w}\left[1+\frac{q^2\langle R_g{}^2\rangle_z}{3}+\circ(q^2)\right] \tag{17}$$

where N_w is the weight average degree of polymerization, and $\langle R_g{}^2\rangle_z$ is the "z average" mean square radius of gyration.

5.1.2 Asymptotic range

In the asymptotic q-range (qR_g>1), we probe the internal structure of the polyion. The effect of polydispersity is negligible. The data analysis is usually based on a comparison with theoretical models. Measurements performed on poly(styrene sulfonate) (NaPSS) [30] showed that the radius of gyration of the macroions was in between those of a Gaussian chain and a rigid rod, reflecting a global stretching of the macroions due to the electrostatic repulsions. The model usually adopted to describe the average conformation is that of the the wormlike chain (Porod-Kratky model).

The electrostatic repulsions between monomers along the chemical sequence of the macroions lead to an elongated average conformation. The chain conformation (monodisperse infinitely thin thread) is Gaussian at low resolution (large distances, small q values) and gradually evolves to that of rigid rod for typical distances smaller than the persistence length l_p (high q values), the usual length characterizing the local stiffness of the thinthread. The form factor can be expressed as a function of the effective contour length L (or as a function of the apparent monomer size b, using $L = Nb$) and the persistence length l_p. There is no analytical expression valid over the whole q range for a chain with a finite length. The global form factor $g_1(q)$ can however be obtained by juxtaposing three different functions $f_1(q)$, $f_2(q)$ and $f_3(q)$ according to the considered ql_p domain [27, 31-34].

For $ql_p \le 2$ and when the condition $L \ge 20l_p$ is satisfied (region 1), we can consider the form factor calculated by Sharp and Bloomfield [32]:

$$g_1(q) = f_1(q) = N\left(\frac{2}{x^2}\left[x-1+\exp(-x)\right]+\frac{2lp}{Nb}\left[\frac{4}{15}+\frac{7}{15x}-\left(\frac{11}{15}+\frac{7}{15x}\right)\exp(-x)\right]\right) \tag{18}$$

$$x = Nq^2bl_P/3$$

The form factor of an infinite wormlike chain has been calculated by des Cloizeaux [33]. For $2 \le ql_p \le 4$ (region 2), we can use the analytic expression[12]:

$$g_1(q) = f_2(q) = \frac{1}{q^2bl_p}\left(6+0.5470(ql_p)^2-0.01569(ql_p)^3-0.002816(ql_p)^4\right) \tag{19}$$

For $ql_p \ge 4$ (region 3), we can consider the asymptotic form derived by des Cloiseaux [33]:

$$g_1(q) = f_3(q) = \frac{\pi}{qb}+\frac{2}{3}\frac{1}{q^2bl_P} = \frac{\pi M_L}{mq}+\frac{2}{3}\frac{M_L}{q^2ml_P} \tag{20}$$

[12] Other numerical coefficients associated to slightly different ql_p domains may be found in the literature. The final results are however very similar.

with m, the molar mass of monomers and M_L the mass per unit length ($M_L = Nm/L = m/b$).

The form factor $g_1(q)$ of the wormlike chain model is shown in Fig. 11 using log-log and Kratky ($q^2 g_1(q)$ versus q) representations.

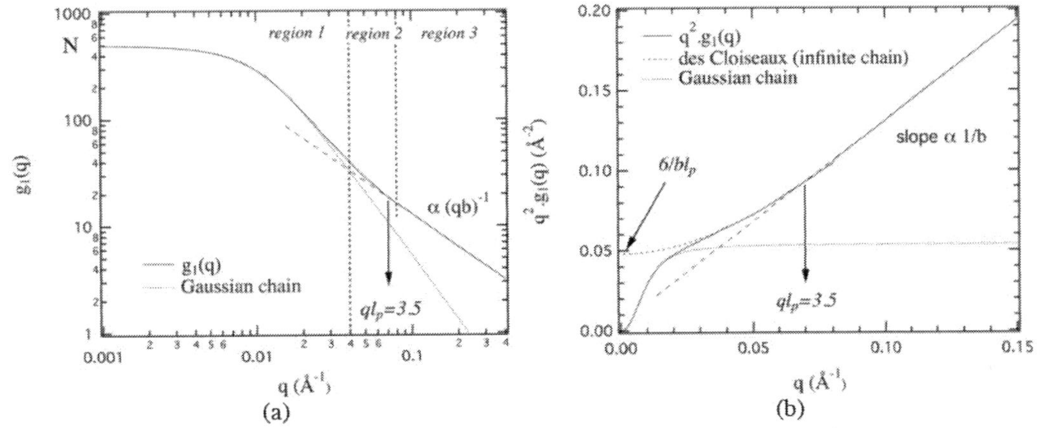

Fig. 11. Form factor of a wormlike chain according to Eq. (18), (19) and (20) in logarithmic (a) and Kratky (b) representations. N=500, b=2.5 Å and l_p=50 Å (R_g=136 Å). The three distinct regions associated with $f_1(q)$, $f_2(q)$ and $f_3(q)$ are separated by vertical dashed lines. The long dashed line corresponds to the asymptotic form of des Cloiseaux. The form factor of a Gaussian chain with the same radius of giration is also indicated.

In the high q range (right part of region 3), $g_1(q)$ shows a q^{-1} decay that is characteristic of rigid rods. In this domain, we probe the molecular structure on a typical length smaller than l_p and the chain is expected to be rigid and close to small cylinders. In the Kratky representation, this corresponds to a straight line with a slope related to the inverse of the monomer size ($q^2 g_1(q) = \pi/qb$), or to the mass per unit length M_L.

For $ql_p \leq 3.5$, $q^2 g_1(q)$ deviates from the asymptotic behaviour and initiates a plateau, characteristic of the Gaussian statistics, whose height is $6/bl_p$. For rather short chains, this plateau is never reached and $q^2 g_1(q)$ decreases again. The end of the q^{-1} scattering behaviour is located at $ql_p = 3.5$. Its location can be used to directly evaluate the persistence length.

For the lowest q values, the chain behaves like a Gaussian chain. We enter the Guinier range and $g_1(q)$ tends to a plateau. This regime enable the determination of the radius of gyration (Eq. (16)) while the extrapolation of $g_1(q)$ to q=0 gives the degree of polymerization N (Eq. (15)).

The Gaussian behaviour in region 1 extends beyond the Guinier range (in Fig. 11, the limit of the Guinier range is $q_{\text{Guinier limit}} = 7.10^{-3}$ Å^{-1}). The radius of gyration can then be determined by analyzing $g_1(q)$ on an extended q range ($qR_g \leq 3$) with the help of the Debye function (form factor of a Gaussian chain) [35]:

$$g_1(q) = N \frac{2}{x^2}\left[x - 1 + \exp(-x)\right] \qquad x = q^2 R_g^{\,2} \tag{21}$$

We can also note that the persistence length is sometimes determined from the crossover point $\tilde{q}$ between the Debye function and the asymptotic form of des Cloiseaux using the relation $\tilde{q} l_p = 1.91$ [36-37].

From an experimental point of view, the experimental determination of $g_1(q)$ is sometime difficult [34, 38-40]. Experiments under ZAC conditions do not scatter very much. Problem of statistic may appear more especially at large q values where $g_1(q)$ approaches to zero. Background subtraction must be done with care, since it can alter the the high q behaviour and introduce significant errors on the b parameter.

Beyond a certain q value, the chain can no longer be considered as infinitely thin. Describing macroions as cylinders with an axial radius of gyration R_c, the form factor, for qR_c<1, can be replaced by:

$$g_1(q)\exp(-\frac{q^2 R_c^{\ 2}}{2}) \tag{22}$$

The form factor of PSS macroions, in a salt-free aqueous solution at c_p=0.15 mol/L, is presented in Fig. 12 (D22 beamline, ILL, Grenoble, France). It has been measured in ZAC conditions, using a fraction x=0.5 of hydrogenated chains and a solvent containing 29% of H_2O and 71% of D_2O. Under these conditions, the contribution of the Na^+ counterions of the PEs is negligible [41].

The extrapolation of $g_1(q)$ to q=0 give a weight averaged degree of polymerization N_w=800, which is close to the value determined by size exclusion chromatography (N_w=830) on the parent H or D polystyrene chains. The influence of the cross section term related to the finite size of the monomers can no longer be neglected above 0.1 Å^{-1}. This really limits the observation of the characteristic q^{-1} scattering behaviour of g_1(q). The apparent monomer size b is slighlty lower than the expected geometrical value (≈2.5 Å) [18, 40]. If physical explanations can be put forward (such as a helical twist in the backbone structure, or the presence of the electrostatic blobs) [18], we must however emphasise the importance of the data treatment and the absolute scale determination on the final b value.

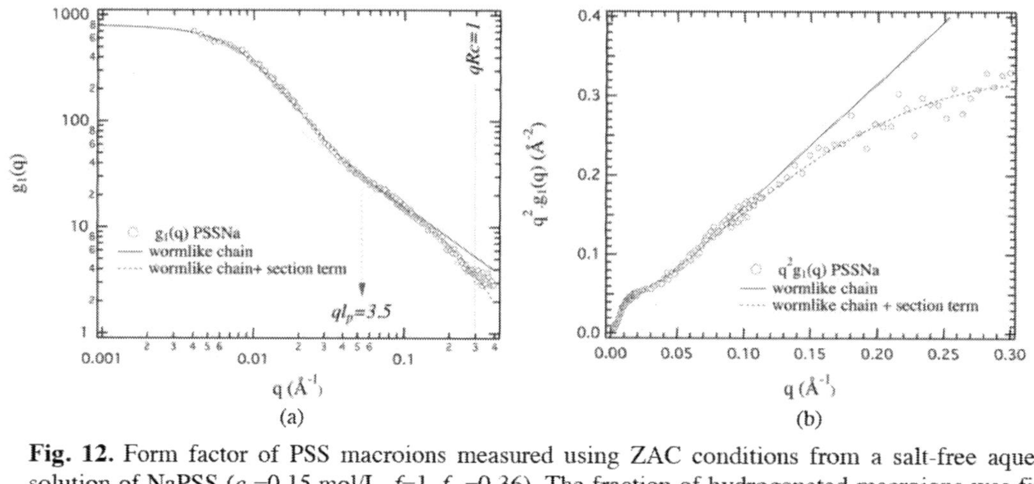

Fig. 12. Form factor of PSS macroions measured using ZAC conditions from a salt-free aqueous solution of NaPSS (c_p=0.15 mol/L, f=1, f_{eff}=0.36). The fraction of hydrogenated macroions was fixed to x=0.5. Solvent was composed of 29% of H_2O and 71% of D_2O. The dashed line and the full line correspond to the wormlike chain model with and without an additional cross section term (N=800, b=2.02 Å, l_p=65 Å, R_c=3 Å, R_g=176 Å). (a) logarithmic representation; (b) Kratky representation.

5.2 Intermolecular structure function

The knowledge of the correlation function $g(q,c_p)$ (or the total structure function $S_T(q,c_p)$) and the form factor $g_1(q,c_p)$ (or the intramolecular structure function $S_1(q,c_p)$) allow to determine the intermolecular correlation function $g_2(q,c_p)$ (or the intermolecular structure function $S_2(q,c_p)$).

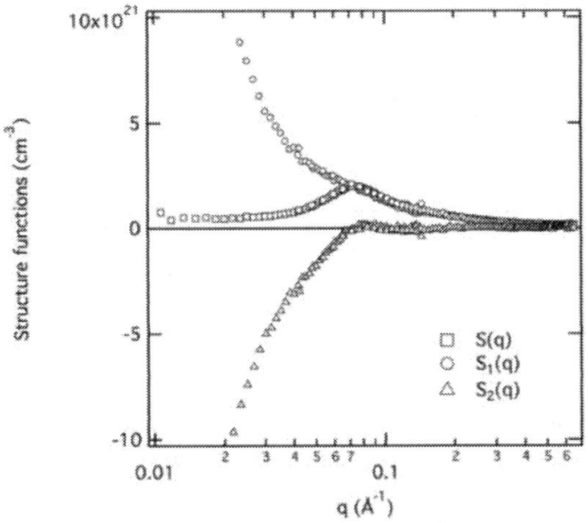

Fig. 13. Total $S(q)$, intramolecular $S_1(q)$ and intermolecular $S_2(q)$ structure functions measured from salt-free aqueous solutions of NaPSS at c_p=0.15 mol/L. $S_1(q)$ has been measured in ZAC conditions. $S(q)$ has been measured with fully hydrogenated PSS^- macroions in D_2O.

An example of total, intra- and related intermolecular structure functions ($S(q)$, $S_1(q)$ and $S_2(q)$) is presented in Fig. 13 for NaPSS (f=1, c_p=0.15 mol/L, D22 beamline, ILL, Grenoble, France). The intramolecular structure functions $S_1(q)$ has been measured in the same experimental conditions than described in the previous section. The total scattering functions $S(q)$ was measured from hydrogenated PSS^- macroions in D_2O. The intermolecular structure functions are then given by $S_2(q)= S(q)-S_1(q)$.

The analysis of the intermolecular structure functions is not simple to achieve mainly because the statistical character of the macromolecules. However, different properties can be clearly identified.

The intermolecular structure function $S_2(q)$ is negative for the lowest q range reflecting the repulsive character of electrostatic interactions and tends to zero for larger scattering vectors.

$S_2(q)$ presents a small positive maximum at a position very close to that found in $S(q)$. It reflects the high repulsive interaction between polyions segments and could indicate the premiss of a local periodic organisation or liquidlike correlations between chain segments. However, its contribution to the PE peak observed in the total scattering function $S(q)$ is quite low. Beyond a certain q value (around q^*), the total structure function $S(q)$ is very close to the intramolecular structure function $S_1(q)$ ($S_2(q) \approx 0$). The PE peak observed in the total structure function $S(q)$ therefore essentially results from the sum of a decreasing structure function ($S_1(q)$) and an increasing one ($S_2(q)$). All these observations are rather consistent with an isotropic disordered network and the "correlation hole" model described in Sec. 4.4.

These observation bring some comments on the partial structure functions presented in Sec. 4.2. The measure of S_{mm} should reproduce the form factor for large enough q values and especially, the theoretical q^{-1} behaviour. Actually, due to the additional cross section term, this is not the case. When the measurements are performed without special contrast condition (NaPSS, X-rays), all the partial structure functions contribute to the scattered intensity, the q^{-1} decay is never observed.

6 Conclusion

In this contribution, we presented an introduction to the application of small angle X-ray and neutron scattering to the study of PE solutions. We tried to highlight the main features of the PE scattering curves and showed the structural information that could be drawn in the case of salt-free aqueous solutions of flexible and hydrophilic PEs in the semidilute regime.

The experiments performed without any special contrast condition allow to determine the coherent differential scattering cross section per unit volume $\Sigma^{coh}(q)$ which contains all the information about the structure of PEs. It is a linear combination of partial structure functions S_{mm}, S_{cc} and S_{mc} weighted by the contrast lengths of both components: monomers and counterions.

The coherent scattering cross section present a broad maximum (PE peak). Its position q^* depends on the PE concentration c_p and the linear charge density and is directly related to the electrostatic repulsive interactions. In the semidilute regime, flexible PEs form an isotropic disordered network. The macroions repel each other and cannot get closer than a minimum distance ξ ("electrostatic correlation hole"). It is precisely this characteristic length that is probed through q^* measurements. These measurements therefore allow a fine characterization of the dispersion state of the chains. The $c_p^{1/2}$ variation of q^* over a large concentration range is characteristic of hydrophilic PEs where the interactions are dominated by the electrostatic repulsions.

The study of the PE peak also presents other interests. Its sensitivity to the net charge provides an attractive way to investigate the counterion condensation process when gradually increasing the chemical charge fraction, but also in more complex solutions with mixtures of counterions of distinct valence [13, 42].

SANS, through isotopic substitution and contrast matching provides significant advantages over other techniques (as SAXS) and allows to reveal or to mask the contribution of each component (monomers or counterions). In particular, the measurement of the partial structure function S_{cc} associated with counterions is rich in information. In the strong-coupling limit, the presence of a maximum is a direct consequence of the accumulation of counterions in the close vicinity of the macroions.

SANS can also be used to separate out, without any theoretical model, the form factor and the intermolecular structure function of macroions. The average conformation of highly charged macroions can be described by the wormlike chain, or Porod-Kratky model. They behave like Gaussian chains at low spatial resolution (large distances, small q values) and gradually evolves to rigid rods for typical distances smaller than the persistence length l_p (small distances, high q values). Measurements in the Guinier range allow the determination of their radius of gyration and their degree of polymerization. In the asymptotic q-range, a comparison with the wormlike chain model can be used to extract the monomers size, or the mass per unit length, and the persistence length. However, it is usually necessary to account for an additional term related to the finite chain cross section to carefully reproduce the data.

The decomposition of the scattered intensity in intra- and intermolecular contributions help us to understand the origin of the PE peak. It mainly results from the combination of an increasing function (intermolecular term) and a decreasing term (form factor).

I would like to thank the organisers of the French-Swedish "Winter School in Neutron Scattering: Applications for Soft Matter" Max Wolff, Fabrice Cousin, Martin Månsson and Isabelle Grillo for their kind invitation to participate to this special issue. I also warmly thank my colleagues Michel Rawiso, François Boué, Philippe Lorchat, Isabelle Grillo, Bruno Démé and Cyrille Rochas for discussions and their help during SANS and SAXS experiments. The Institut Laue Langevin (ILL, Grenoble, France) and the European Synchrotron Radiation Facility (ESRF, Grenoble, France) are acknowledged for beam time allocation.

References

1. A.V. Dobrynin, M. Rubinstein, Prog. Polym. Sci. 30 (2005) 1049–1118
2. J. Combet, P. Lorchat, M. Rawiso, Eur. Phys. J. Special Topics **213**, 243–265 (2012)
3. G.S. Manning, J. Chem. Phys. **51**, 954 (1969).
5. F. Cousin, EPJ Web of Conferences **104**, 01004 (2015)
4. F. Oosawa, Biopolymers **6**, 134 (1968)
6. L. Van Hove, Phys. Rev. **95**, 249 (1954).
7. J.P. Cotton, M. Moan, J. Phys. Lett. **37**, L-75 (1976)
8. K. Kaji, H. Urakawa, T. Kanaya, R. Kitamaru, J. Phys. France **49**, 993 (1988)
9. M. Nierlich, C.E. Williams, F. Boué, J.P. Cotton, M. Daoud, B. Farnoux, G. Jannink, C. Picot, M. Moan, C. Wolf, M. Rinaudo, P.-G. de Gennes, J. Phys. France, **40**, 701 (1979)
10. P.-G. de Gennes, P. Pincus, R.M. Velasco, F. Brochard, J. Phys. France **37**, 1461 (1976)
11. K. Nishida, K. Kaji, T. Kanaya, J. Chem. Phys. **114**, 8671 (2001)
12. P. Lorchat, I. Konko, J. Combet, J. Jestin, A. Johner, A. Laschewski, S. Obukhov, M. Rawiso, EPL, 106 (2014) 28003
13. J. Combet, F. Isel, M. Rawiso, F. Boué, Macromolecules, **38**, 7456 (2005)
14. W. Essafi, F. Lafuma, C.E. Williams, ACS Symposium Series 548, *Macro-ion Characterization. From Dilute Solutions to Complex Fluids*, chp 21, edited by K.S. (1994)
15. V. Dobrynin, M. Rubinstein, Macromolecules, **32**, 915 (1999)
16. J.R.C. van der Maarel, L.C.A. Groot, J.G. Hollander, W. Jess, M.E. Kuil, J.C. Leyte, L.H. Leyte-Zuiderweg, M. Mandel, J.P. Cotton, G. Jannink, A. Lapp, B. Farago, Macromolecules, **26**, 7295 (1993)
17. L.B. Bhuiyan, C.W. Outhwaite, J.R.C. van der Maarel, Physica A **231**, 295 (1996)
18. K. Kassapidou, W. Jesse, M.E. Kuil, A. Lapp, S. Egelhaaf, J.R.C. van der Maarel, Macromolecules **30**, 2671 (1997)
19. K. El Brahmi, Ph.D Thesis, University of Strasbourg, 1991
20. K. Nishida, K. Kaji, T. Kanaya Macromolecules, **28**, 2472 (1995)
21. W. Essafi, F. Lafuma, C.E. Williams, *Eur. Phys. J. B*, **9**, 261 (1999)
22. W. Essafi, F. Lafuma, D. Baigl, C.E. Williams, Europhys. Lett. **71**, 938, (2005)
23. R. Koyama, Macromolecules **17**, 1594 (1984)
24. P.J. Pfeuty, J. Phys. Colloques **39** C2-149 (1978)
25. A.V. Dobrynin, R.H. Colby, M. Rubinstein, Macromolecules **28**, 1859 (1995)
26. J.S. Higgins, H.C. Benoît, *Polymers and Neutron Scattering* (Clarendon Press, Oxford 1994)
27. M. Rawiso, J. Phys. IV France **9**, Pr-147 (1999)
28. B.H. Zimm, J. Chem. Phys. **16**, 1099 (1948)
29. G.C. Berry, J. Chem. Phys. **44**, 4550 (1966)
30. M. Nierlich, F. Boué, A. Lapp, R. Oberthur, J. Phys. France **46**, 649 (1985)

31. J.P. Cotton, J. Phys. IV France **9**, Pr-21 (1999)
32. P. Sharp, V.A. Bloomfield, Biopolymers **6**, 1201 (1968)
33. J. Des Cloiseaux, Macromolecules **6**, 403 (1973)
34. A. Brûlet, F. Boué, J.P. Cotton, J. Phys. II France **6**, 885 (1996)
35. P. Debye, J. Chem. Phys. **14**, 636 (1946)
36. O. Glatter and O. Kratky, *Small Angle X-ray Scattering* (Academic Press, London, 1982)
37. K. Nishida, H. Urakawa, K. Kaji, B. Gabrys, J.S. Higgins Polymer 38, 6083 (1997)
38. M. Ragnetti, R.C. Oberthür, Colloid Polym. Sci. **264**, 32 (1986)
39. M.N. Spiteri, Ph.D Thesis, University of Paris XI, 1997
40. E. Dubois, F. Boué, Macromolecules **34**, 3684 (2001)
41. M. Heinrich, Ph.D Thesis, University of Strasbourg, 1998
42. J. Combet, M. Rawiso, C. Rochas, S. Hoffmann, F. Boué, Macromolecules **44**, 3039 (2011)
43. J. Hayter, G. Jannink, F. Brochard-Wyart, P.-G. de Gennes, Journal de Physique Lettres **41** (18), 451 (1980)
44. M. Sedlak, *Physical chemisty of polyelectrolytes*, M. Sedlák, in Physical Chemistry of Polyelectrolytes (Dekker, New York, 2001), Chap. 1, pp.1-58
45. V. M. Prabhu, E. J. Amis, D. P. Bossev, N. Rosos, J. Chem. Phys., **121**, 4424 (2004)
46. M. Dymitrowska, L. Belloni J. Chem. Phys. **11**, 4659 (1998)
47. S. Förster, M. Schmidt, M. Antonietti Polymer **31**, 781 (1990)

Block copolymer electrolytes for fuel cells and secondary batteries, the small angle neutron scattering inputs

*Laurent Rubatat**

CNRS/ UNIV PAU & PAYS ADOUR/ E2S UPPA, Institut des Sciences Analytiques et de Physico-Chimie pour l'Environnement et les Matériaux, IPREM UMR5254, 64000, Pau, France

Abstract. This paper aims at giving an overview on the importance of scattering, and more specifically neutron scattering, for probing the nano-morphology of polymer electrolytes made of block copolymers. Two types of self-assembled polymer electrolyte materials will be discussed: (i) the ionomer membranes used in fuel cell and (ii) the solid polyelectrolytes used in secondary batteries. Both are used to physically separate the electrodes in the respective electrochemical devices and are expected to have a high ion transport capacity so as good chemical and mechanical stabilities. Unfortunately, in most cases improving one property leads to the degradation of the others. Nonetheless, through block copolymers self-assembly it is possible to tackle this issue; indeed, antagonist properties can be decoupled and associated within controlled nano-morphologies. This aspect will be discussed and supported by examples based on published studies; in parallel useful scattering analytical tools and models will be presented along the paper and detailed in annex.

1 Introduction

Nowadays ion containing polymers materials are of major interest in devices such as fuel cells and secondary batteries. Ionomer membranes such as Nafion have proved their high potential to transport efficiently protons in low temperature fuel cells; and polymer electrolytes, such as polyethylene oxide (PEO), have demonstrated their capacity to well dissociate lithium salts and transport ions between batteries electrodes. The relatively low cost and ease to process, make polymer materials of great interest for such applications. Nevertheless, improvements are still necessary; for instance, the stability of ionomer membranes needs to be improved when cycled – swelling/drying cycles – in order to extend the fuel cell lifetime; also, the mechanical modulus of polymer electrolyte should be increased toward solid polyelectrolytes for a safer use of the batteries. Unfortunately, in most cases improving one property leads to the degradation of the others. One rout to decouple antagonist properties is through block copolymers self-assembly, where specific polymers with their one desirable property (*e.g.* ion conductivity, mechanical strength, etc.)

* Corresponding author: laurent.rubatat@univ-pau.fr

are associated along a single polymer chain and later phase segregated and self-assembled into nano-domains to form the electrolyte material.
Block copolymer (BCP) self-assembly mechanisms are well observed experimentally[1] and well described theoretically [1-4] (*cf.* figure 1). The nano-morphology is controlled essentially through the Flory-Huggins parameter, χ, the total number of repeating units along the macromolecule, *N,* and the volume ratio of one of the blocks, *f.* Also, the macromolecular architecture (*e.g.* linear, star-like, comb-like, blocks number, blocks dispersity, etc.) has an impact on the self-assembly.[5, 6] The variation of those parameters leads to a variety of structures such as lamellar, hexagonal packing of cylinders, spheres, and bicontinuous morphologies, for the most common as depicted on figure 1.[1] In the case of a charge free block copolymers, a single Flory-Huggins parameter, χ, describes the properties of the system in both the ordered and disordered phases, corresponding respectively to the strong ($\chi.N >> 10$) and the weak ($\chi.N < 10$) segregation limit, which are unified through the self-consistent mean field theory.[4] Considering a diblock copolymer AB constituted of a block A and a block B, χ_{AB} depends on temperature and on the chemical nature of both blocks and quantify their degree of incompatibility. Then χ_{AB} can be expressed as follow: [1]

$$\chi_{AB} = \left(\frac{z}{k_B T}\right)\left[\epsilon_{AB} - \frac{1}{2}(\epsilon_{AA} + \epsilon_{BB})\right] \quad (1)$$

where z is the number of nearest neighbors per repeat units in the polymer, k_B is the Boltzman constant, T the temperature, and ε_{AB} is the interaction energy per repeat unit of A-B, ε_{AA} for A-A and ε_{BB} for B-B. A negative χ_{AB} indicates a free-energy drive towards mixing of A and B, on the other hand a positive one indicates a clear repulsion between A and B. Experimentally, rough estimations of χ can be obtained through contact angle measurements [7] or through solubility parameters.[8] Scattering techniques, such as small angle X-ray scattering (SAXS) and small angle neutron scattering (SANS) can also be exploited to extract the parameter; whether by the domain spacing in the ordered phase,[7] or by the behaviour of the structure factor in the disordered phase.[2, 9] This particular point will be discussed in the solid electrolytes section and developed in annex. Neutron reflectometry can be used to characterize the interface between both layers of homopolymers A and B which also depends on the χ_{AB} parameter.[7]

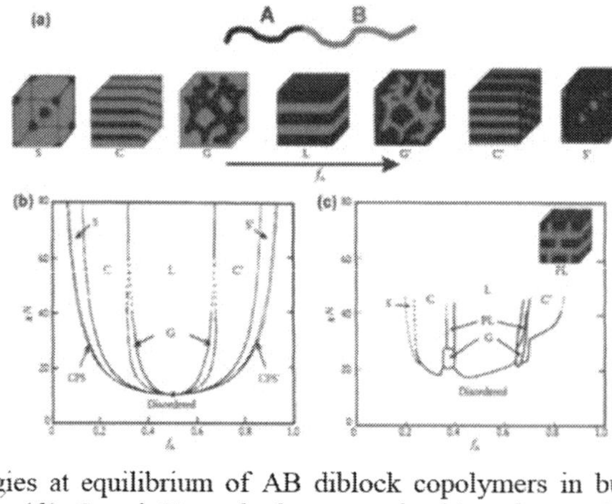

Fig. 1. (a) Morphologies at equilibrium of AB diblock copolymers in bulk as function of the A blocks volume fraction (f_A). S and S' are body-centered-cubic spheres, C and C' are hexagonally packed cylinders, G and G' are bicontinuous gyroids, and L is the lamellar morphology. (b) Self-consistent mean-field theory phase diagram of an AB diblocks; with CPS and CPS' closely packed spheres. (c)

Experimental phase diagram of polyisoprene-*block*-polystyrene diblock copolymers, in which f_A represents the volume fraction of polyisoprene; with perforated lamellae, PL. Reproduced with permission from reference [10] and [1], for reference [1] with the permission of the American Institute of Physics.

Self-assembled BCPs can then combine and concentrate properties in controlled nano-morphologies, which present a double interest. (i) The first one is to optimize the global efficiency of the electrolyte materials, and (ii) the second one is to use BCPs as model systems to gain knowledge and ultimately develop more efficient electrolyte materials. Thus, in order to control and improve the polyelectrolyte performances it is essential to characterize the morphologies at the nanometer length scale and pointing out the structure / properties relationship.
Microscopy, with transmission electron microscopy (TEM) and atomic force microscopy (AFM) are methods of choice for the structural characterisation. Microscopy approach has the advantage of a direct visualization of the morphology and the disadvantage of a limited representativeness of the collected pictures. To complement microscopy, scattering techniques such as SAXS and SANS, are frequently performed. Indeed, both SAXS and SANS permit to collect structural averaged data in a large volume of sample; the difficulty is the need of mathematical models to transfer the data from the reciprocal space to the real space.[11] Also, SAXS and SANS are well complementary to each other; for instance, the high brilliance of the X-ray sources allows a fast data acquisition (below the second) together with a high resolution which is essential to investigate fast structural kinetic; on the other hand, neutron contrast variation approach allows to selectively enhance or match the contrast in multiphasic systems, which is essential for a detailed description of complex morphologies. Indeed, the neutron coherent scattering length of an atom gives its scattering efficiency and depends on its nucleus composition. Nevertheless, two isotopes of the same element present different scattering lengths; consequently, isotope exchange can be performed in a material to tune selectively the coherent scattering length densities (SLD) of one phase, while preserving the structural, physical and chemical properties. This substitution is regularly used between hydrogen and deuterium, leading for example in the case of water to SLDs of -0.56 and 6.39 $\times$ 10^{10} cm^{-2} for H_2O and D_2O respectively. The wide range of SLDs allows then mixtures of H_2O-D_2O to match the majority of polymers SLD's. In this paper, the scattering curves are plotted as a function of the scattering vector q defined as followed: $q = 4\pi/\lambda \sin(\theta/2)$ with θ the scattering angle. Performing small angle scattering on phase segregated BCPs bulk samples usually permit to discriminate the different morphologies thanks to the relative structural peak q-positions and permit also to determine the lattice parameters.[12] More details on this particular point are given in annex A.
The present paper points out the essential input of scattering, and especially SANS, in the morphology characterization of fuel cell ionomer membranes and battery polymer electrolytes made of BCPs. Nevertheless, this paper does not aim to be an exhaustive review on fuel cell or battery materials, neither on BCPs but will give an overview based on selected published papers. Finally, few useful analytical tools and models are presented all along the paragraphs and detailed in annex.

2 BCP ionomer membranes for fuel cell

Ionomer membranes are currently employed as a separator in prototype and commercialized proton exchange membrane fuel cells (PEMFCs) running at low temperature, targeting a vast range of applications: from small portable to large stationary power sources. The basic principle of hydrogen fuel cell is schematized on figure 2. (i) Hydrogen is decomposed in protons and electrons on the anode side, then (ii) the protons are transported through the hydrated membrane to the cathode, (iii) to finally produce water after recombination with the oxygen from air and electrons transported through the external electrical circuit. The ionomer membrane plays a central role and as a consequence it is a key element toward PEMFC performances improvement. The membrane must be chemically and electrochemically stable, have good mechanical strength, have high proton conductivity and must be synthesized at low cost. In the last few decades, a significant number of ionomers have been synthesized to meet these requirements, such as sulfonated aromatic hydrocarbon polymers, *e.g.* polyimides [13], poly(ether ether ketone) [14], poly(phenylene) [15], poly(phenylene sulfone) [16], or polybenzimidazole [17]. Nonetheless, the membranes of reference, in terms of durability and performance, remains perfluoro-sulfonated membranes, of which Nafion commercialized in the 60's by DuPont de Nemours is the first example. More precisely, Nafion polymer is made of poly-tetrafluoroethylene backbone with perfluorinated side chains each ended by a sulfonic acid function. However, these fluorinated ionomer membranes are still expensive to produce, due to a complex synthesis. In addition, they are not suitable for high-temperature PEMFC applications (typically above 130 °C) because of the drop of proton conductivity above 100 °C, correlated with the membrane drying. Also, Nafion membranes are not good candidates for direct methanol fuel cells (DMFC) due to their low methanol crossover resistance. Therefore, the synthesis of new alternative ionomer materials so as their structural and diffusion characterization are still necessary. For that purpose, neutron and X-ray scattering techniques are particularly well-suited tools. [18, 19]

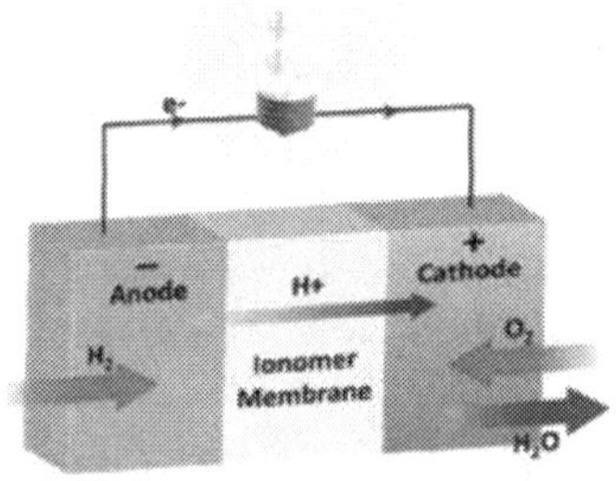

Fig. 2. Schematic representation of proton exchange membrane fuel cell principle.

Nafion remains also a major reference among the ionomer membranes in term of morphology. Indeed, it was intensively investigated by SAXS and SANS, to point out the role of morphology on the remarkable macroscopic properties. Nonetheless, the studies were leading to contradictory structural conclusions, mainly as a matter of fact of the delicate interpretation of scattering data. The Nafion typical scattering curve (*cf.* figure 3)[20] presents peaks at large angles corresponding to the amorphous and crystalline peaks, signature of the macromolecular chains arrangement. The well-known ionomer peak positioned at a scattering vector q about 0.15 $Å^{-1}$ and systematically observed on ionomer material is correlated to the segregation between the hydrated ion rich and the hydrophobic polymer phases. In the intermediate q-range a large bump named the matrix-knee, commonly correlated to the crystallinity is sited on a general q^{-1} slope. Finally, in the ultra-small angles region, *i.e.* below 0.005 $Å^{-1}$, an upturn associated to large length scale morphologies was observed and analyzed with Debye-Bueche [21] and Voronoï [22] models.[20, 23]

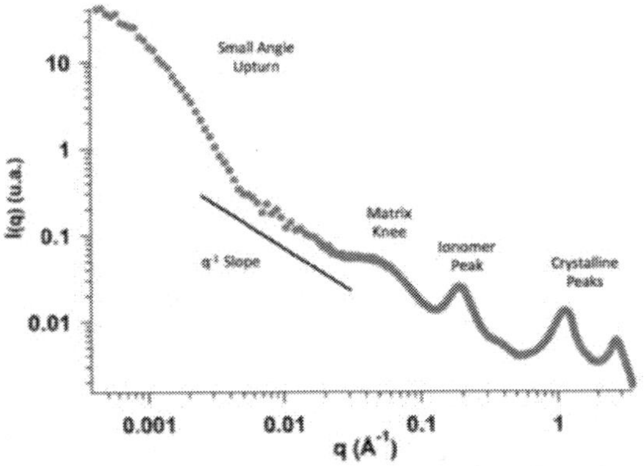

Fig. 3. Wide, Small and Ultra Small Angles X-ray Scattering data (respectively WAXS, SAXS and USAXS) measured on a hydrated Nafion membrane. Figure reproduced with permission from reference [20].

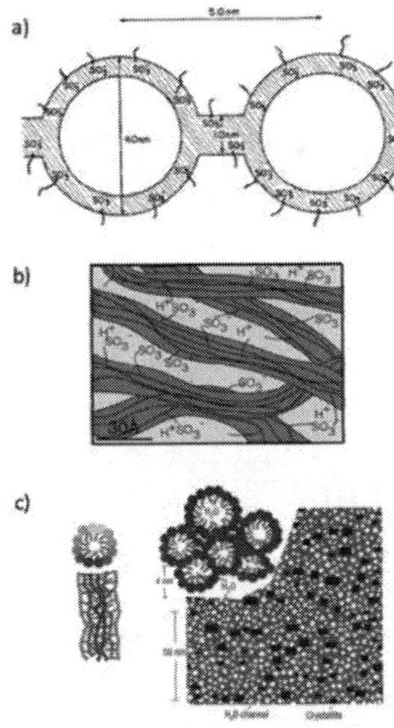

Fig. 4. (a) The Gierke's cluster-network model.[24-26] (b) The direct cylinder model.[23, 27] (c) The Schmidt-Rohr *et al.* inverted cylinder model.[28] Figures reproduced and modified with permission from references [18, 26, 28].

The pioneer Nafion structural model was the Gierke's cluster-network model, [24, 25] with spherical clusters of sulfonic acid groups (*i.e.* inverted micelles), connected by narrow channels within an amorphous fluorocarbon matrix (*cf.* figure 4a). In 2002, a direct fibrillar model was proposed, considering the aggregation of the fluorocarbon chains into elongated objects surrounded by the ionic groups and water. This model is based on SAXS and SANS data collected at much smaller q compare to Gierke's model, *i.e.* information on the larger dimension (*cf.* figure 4b) and also over a wide range of water content.[23, 27] The systematic presence of a q^{-1} slope at small q was a key clue, suggesting elongated scattering objects. In addition, this model is able to describe the orientation of the structure under an uniaxial stretching (*cf.* figure 5).[20, 29] In 2008, an inverted cylinder model was proposed based on new analysis of existing SAXS curves using numerical Fourier transform techniques.[28] This model describes Nafion as parallel cylinders decorated inside with ionic groups and water within a fluorocarbon matrix (*cf.* figure 4c). More recently, Nafion was described by locally flat and narrow water domains (about 1 nm) in a fluorocarbon matrix.[30] To conclude, all those models agree at least on one point, there is a co-continuous phase separation between the hydrophobic fluorinated polymer and the hydrated ionic group phase at the nanometer length scale, even if they disagree on the exact shape of those phases.

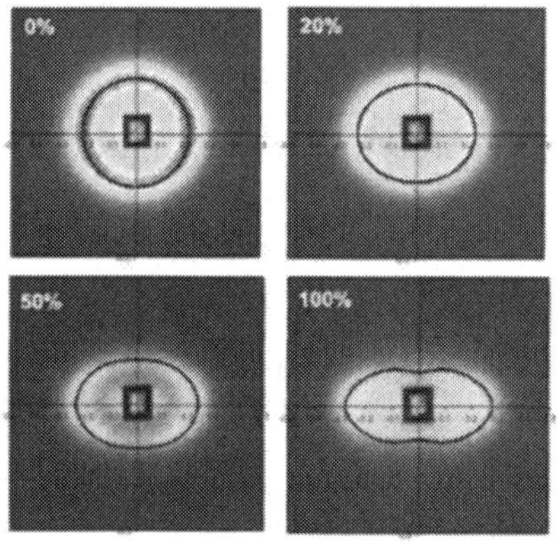

Fig. 5. SANS experimental results collected on a drawn Nafion film, with deformation 0, 20, 50 and 100%. The black line corresponds to the ionomer peak position calculated through the cylinder deformation model[29] with the deformation ratio as a single adjustable parameter. Figure extracted with permission from reference [20].

SANS with contrast variation is a powerful method to investigate the nano-morphology of Nafion and other ionomer membranes.[18] Indeed, the structures are extremely dependent on water content and in most cases they are observed hydrated such as fuel cell running conditions. Then fully or partially deuterated water (H_2O-D_2O mixtures) can be used for contrast matching or enhancing. To illustrate this point, on figure 6 are presented SANS curves collected on the same Nafion membrane highly hydrated with four different H_2O-D_2O mixtures. The impact on the scattering curves is dramatic and can be predicted with a core-shell cylinder form factor assuming that the scattered intensity can be expressed as the product of a structure factor $S(q)$ and a form factor $P(q)$:[31]

$$I(q) = K.S(q).P(q) \tag{2}$$

where K is a scaling factor related to the number of scattering objects. The form factor describes the specific shape of the scattering objects, and the structure factor their spatial ordering. More conditions on the validity of this expression are discussed in the next paragraph. Here, it is assumed also that the structure factor can be neglected due to the large hydration of the membrane (*i.e.* polymer volume fraction is 16%). The solid black lines on figure 6 correspond to the core-shell cylinder form factor using identical structural parameters for the four curves while adjusting the SLDs of the water phase and the hydrated ionic corona.

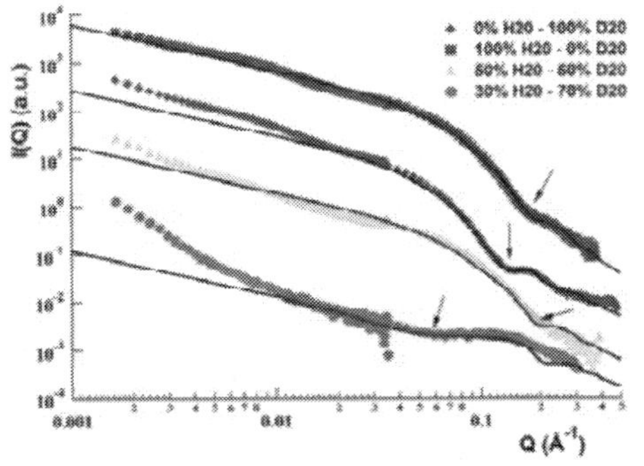

Fig. 6. Contrast variation SANS curves measured on a single swollen Nafion membrane exchanged with $N(CH_3)_4$ counter ion at four different H_2O-D_2O ratios. The membrane polymer volume fraction is 16%. The solid black curves correspond to a core-shell cylinder form factor fits taking into account the contrast variation from the solvent. The four arrows indicate the position of the first minimum for each curve; this minimum is related to the counterions shell interferences. Figure reproduced with permission from reference [27]. Copyright (2002) American Chemical Society.

BCP ionomer membranes

The design and synthesis of model polymers, such as BCP, is a key element of any research program aimed at improving ionomer membranes. It is expected that properties, such as proton conductivity, is influenced by many factors including the chemical structure (*i.e.* composition and nature of the polymer chain), ion exchange capacity (IEC), water content, and also by the multi-length-scale morphology. The study of model materials can then provides valuable knowledge on proton transport mechanisms and ultimately will lead to the design of future generation of membranes. Several non-fluorinated copolymers containing a partially sulfonated polystyrene block had been synthesized, such as sulfonated poly(styrene-*b*-[ethylene-*co*-butylene]-*b*-styrene),[32-37] sulfonated poly(styrene-*b*-isobutylene-*b*-styrene),[38, 39] sulfonated hydrogenated poly(butadiene-*b*-styrene),[40, 41] sulfonated poly(styrene-*b*- [ethylene-*co*-propylene]),[42, 43] sulfonated poly(styrene- *b*-ethylene/propylene-*b*-styrene),[42] block copolymers sulfonated poly-(styrene-*b*-methylbutylene).[44] In most cases, the structural investigations on ionomer BCPs reveal hierarchical structures. Indeed, the regular BCPs self-assembly occurs at the length scale of few tens of nanometers and the charges clustering arises within the ionomer phase at the length scale of few nanometers; that was demonstrated by Weiss and coworkers by small angle scattering.[34]

In order to take advantage of the chemical stability of fluorinated polymers, BCPs such as sulfonated poly-([vinylidene difluoride-*co*-hexafluoropropylene]-*b*-styrene) block copolymers,[45-47] were also explored.

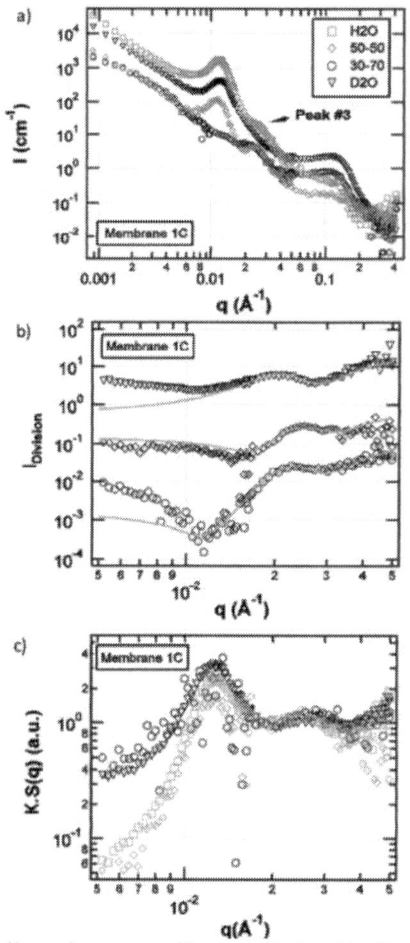

Fig. 7. (a) SANS spectra collected on one film swelled with four different H_2O-D_2O mixtures: 100-0 (squares), 50-50 (diamonds), 30-70 (circles) and 0-100 (triangles). (b) SANS spectra divided by the H_2O spectra: D_2O/H_2O (triangles), 30-70/H_2O (circles) and 50-50/H_2O (diamonds). The solid lines represent the theoretical results for the normalized spectra P_{div}. For the sake of clarity, the curves are offset. The fits are performed in absolute intensity. (c) Calculated structure factor, *K.S(q)*, for the film swollen in H_2O (square), 50-50 (diamonds), 30-70 (circles) and D_2O (triangles). Figures reproduced with permission from reference [45]. Copyright (2006) American Chemical Society.

Holdcroft and coworkers have synthetized original block copolymers ionomer with fluorinated poly (vinylidene difluoride-*co*-hexafluoropropylene) and sulfonated polystyrene ionomer blocks.[46-48] The blocks incompatibility and low dispersion lead to well-defined structures as revealed by TEM imaging and SANS over a wide range of length scales while varying blocks dimension and ion exchange capacity (IEC).[45] Taking advantage of the SANS contrast variation, the films were soaked in four different H_2O-D_2O mixtures: 100-0, 50-50, 30-70 and 0-100. These mixtures present SLDs of -0.56, 2.915, 4.305, and 6.39 $\times 10^{10}$ cm^{-2}, respectively. On the other hand, the polystyrene has a SLD of 1.41 $\times 10^{10}$ cm^{-2} and the fluoropolymer 3.22 $\times 10^{10}$ cm^{-2};[45] in consequence no contrast matching was expected. To analyze the large length scale SANS data (*cf.* figure 7a), it was assumed that the scattered intensity can be expressed as the product of a structure factor $S(q)$ and a form factor $P(q)$ (*cf.* equation 2). It is important to note that Equation 2 is valid in two cases: (i) the scattering objects are centro-symmetric (typically spheres) or (ii) the scattering objects are anisotropic (*e.g.* cylinders, disks) but locally oriented.[49] Also the expression is valid in the q-range where $q.L > 1$, with L the characteristic dimension of the structure. In addition, it is assumed that changing the solvent H_2O-D_2O ratio has no effect on the morphology,[50] *i.e.* the structure factor is expected to be invariant at all contrast. Finally, it is assumed that the geometry of the scattering objects remains the same but the SANS form factor is expected to vary due to the variation of the SLDs. Under all these assumptions, if two spectra collected on the same membrane but soaked in two different solvents are divided by each other, then the $S(q)$ and K components will be eliminated, resulting to a ratio of two form factors (*cf.* equation 3). Then this ratio can be compared to curves calculated for different object shapes.[23, 51] Experimentally, the spectra collected on a membrane soaked in 50-50, 30-70, and D_2O mixtures (solvent1) were divided by that soaked in H_2O (solvent2) to obtain the $P_{div}(q)$.

$$I_{div}(q) = \frac{I_{solvent1}(q)}{I_{solvent2}(q)} = \frac{\cancel{K.S(q)}.P_{solvent1}(q)}{\cancel{K.S(q)}.P_{solvent2}(q)} = P_{div}(q) \tag{3}$$

Finally, once the theoretical $P(q)$ are determined, then the products $K.S(q)$ are calculated by dividing the measured spectra by the corresponding $P(q)$ as expressed below for solvent 1:

$$K.S(q) = {I_{solvent1}(q)}/{P_{solvent1}(q)} \tag{4}$$

where K and $S(q)$ are both contrast independent, then all the $K.S(q)$, calculated from the different solvents, should all collapse on each other. Figure 7b presents the results for divided experimental spectra. The fact that the divisions are not flat means that the scattering objects cannot be modeled as a basic two-phases system, but at least three phases. This first observation is a motivation to consider core-shell form factors. Then the shape of the scattering object has to be determined; spherical and cylindrical core-shell form factors gave rise to unrealistic parameters or low fit quality, only the core-shell discoidal form factor led to a reasonable analysis (*cf.* figure 7b). Indeed, each scattering object is considered as a disk-like core with an external layer (shell) on both sides. The core is defined by its radius r, thickness t, thickness polydispersity σ. The shell is defined by its thickness, s. The SLDs were calculated considering a core made of hydrated sulfonated PS, a pure PS shell, and a surrounding matrix of fluoropolymer. In that case the unknown fitting parameters are ϕ_p, the polymer volume fraction of the disk core, t, the core thickness with σ its polydispersity, and s, the shell thickness. The same set of structural parameters was used to fit the three normalized spectra collected on the same samples, only the SLD of the swollen core was varying and calculated with the proper solvent's SLD. Curves showing the fits are presented in figure 7b. The fit approach is good enough to confirm the core-shell disk-like form factor, considering that most of the parameters are strongly constrained and considering also the fit agreement in relative intensity for the three curves simultaneously. Finally, the $K.S(q)$ were calculated, the results are presented in figure 7c, with a first peak at q^* and second order at $2q^*$ consistent with a lamellar ordering of the disk-like objects. All four $K.S(q)$ collapse on each other which supports the analysis assumptions. To conclude the SANS contrast variation analysis was a key element in this study, pointing out the precise hierarchical structure brought by the block copolymer self-assembly and the sulfonic acid groups clustering in the sPS phase. As shown on figure 8, the fine SANS analysis points out the presence of a thin dead zone, in term of conductivity, of pure PS in-between the sPS and the fluorinated lamellas.

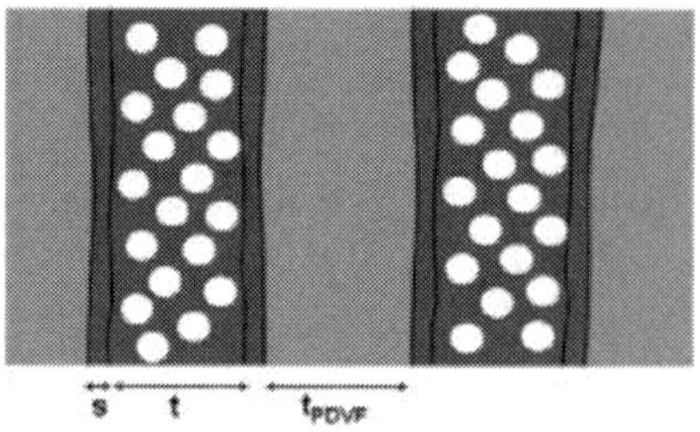

Fig. 8. Sketch of the membrane structure. The lighter gray corresponds to the fluorous lamellae with a thickness t_{PVDF}, and the darker gray corresponds to the PS lamellae. Within the PS lamellae there are two different regions: the central region contains hydrated sulfonated polystyrene, with a thickness *t*; at the interfaces are located thin layers of pure polystyrene, with a thickness *s*. The white solid circles within the sulfonated polystyrene domains represent hydrated ionic aggregates. Figure reproduced with permission from reference [45]. Copyright (2006) American Chemical Society.

A similar SANS approach was performed on grafted block copolymers perfluorinated ionomer membranes made of partially sulfonated poly-([vinylidene difuoride-*co*-chlorotriuoroethylene]-*g*-styrene) on which the grafting densities and the polystyrene side chains length and sulfonation rate was modulated.[52] X-ray scattering at large and small angles permitted to collect data on crystallinity and nano-morphology on the non-hydrated films. To complete the study a SANS experiment with contrast variation was performed on hydrated samples using four different H_2O-D_2O mixtures, *i.e.* 100-0, 70-30, 40-60 and 0-100.[53] The respective SLDs are -0.56, 1.52, 2.91, and 6.37×10^{10} cm^{-2}. The fluorinated polymer SLD was estimated at 2.87×10^{10} cm^{-2} and the polystyrene one 1.41×10^{10} cm^{-2}; consequently, contrast matching can be expected. This SANS experiment allowed quantifying (i) the swelling of the polymer films, (ii) the size and order of the ion-rich regions, and (iii) the size and order of the fluorinated domains. Figure 9a presents the scattering curves and reveal two broad peaks located within two *q*-ranges 0.04 to 0.06 Å^{-1} and 0.1 to 0.3 Å^{-1}. Those peaks are analyzed as signatures of a hierarchical morphology consisting of a disordered and partially phase-separated system, with two characteristic features and dimensions (*cf.* figure 9b). Thanks to contrast variation and matching both peaks were attributed to a structure consisting of fluorinated domains in a partially sulfonated PS matrix at large length scale and ion-rich domains within the PS matrix at smaller length scale. Practically, the SANS data were interpreted by fitting the Kinning & Thomas model [54] (*cf.* annex B) to the fluorinated peak and the Teubner & Strey model [55] (*cf.* annex C) to the peak associated to the ion-rich domains. This combination of models provided the best fits to the data with a reasonable number of parameters. The analysis of the structural parameters permitted to point out the impact of the morphology on water uptake and proton conductivity. It was found that the structural swelling observed at the nanometer length scale was consistent with the membrane water uptake, indicating consistency between microscopic and macroscopic length scale swellings in these materials. Also, it was demonstrated that phase separation between fluorinated polymer and PS phase is limited because of the competition with the sulfonated acid groups aggregation mechanism. The similarities in water uptake and conductivity values for membranes of radically different morphology were also pointed out, particularly when volume-based quantities from references [45] and [53] are compared. To summarize, the present SANS study points out that the control of water uptake through the macromolecular structure is a key feature in designing efficient materials.

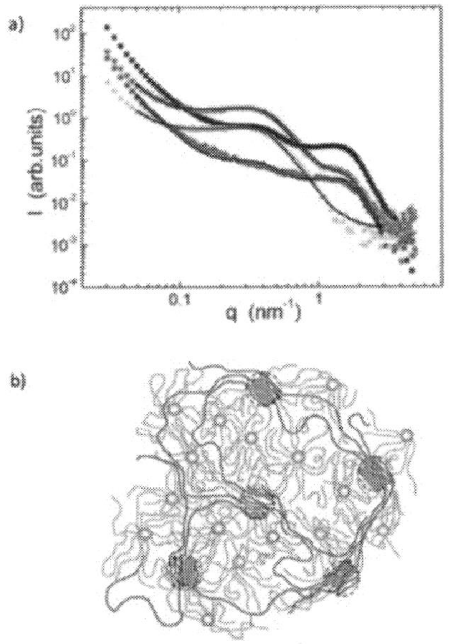

Fig. 9. (a) SANS spectra measured on an ionomer film with an IEC of 2 mmol/g swollen in four H_2O-D_2O mixtures: D_2O (squares), H_2O (downward triangles), 40-60 (circles), and 70-30 (upward triangles). The water content was 36%. The solid black lines correspond to the fits using the combined models: Kinning & Thomas and Teubner & Strey model at small and larger q respectively. Data sets have been shifted vertically for sake of clarity. (b) Schematic representation of the structural model used to depict the morphology and fit the SANS data. In black the fluorinated backbones, with PVDF rich spherical domains, dispersed in a continuous PS matrix (green). At smaller length scales, the ion-rich domains in red. Figures reproduced with permission from reference [53]. Copyright (2013) American Chemical Society.

Finally, to illustrate the structure / properties relationship using model ionomer diblock copolymers, we are now focussing on a study performed on partially sulfonated poly-(styrene-*b*-methyl methacrylate) (PS-*b*-PMMA) that point out the role of controlled morphologies on the proton conductivity and the water uptake.[56] The advantage of this system is that series of polymers presenting similar sulfonated PS blocks were used while varying the PMMA block dimension. The structure and properties of the sulfonated PS (sPS) phase are supposed to be identical. By combining TEM and SANS the different morphologies were described. Depending on the molecular weight of the copolymers and the degree of sulfonation of the PS block, disordered (DIS), lamellar (LAM), hexagonal cylindrical packing (HEX) and hexagonally perforated lamellae (HPL) phases were observed. The swelling of the films exposed to water was significantly different according to the morphology. For instance, the samples exhibiting a lamellar structure undergo a consequent increase in the inter-lamellar dimension, whereas no swelling was observed for the samples presenting the HEX or HPL morphologies. It is worth to note that for those morphologies the matrix is made of PS. As expected, the water uptake of the films is accompanied by a significant increase in proton conductivity. Nevertheless, this increase was also dependent on the nano-metric morphology. In order to highlight the role of structure, the authors have compared the conductivities normalized by the volume fractions of the conductive phase (*i.e.* the domains formed by sPS, PS, and water). Figure 10 presents the normalized conductivity which increases according to the following sequence of morphologies: DIS, HEX, HPL and LAM, in agreement with hopping, 1D and 2D charges transport mechanisms; *i.e.* hopping (DIS), 1D (HEX) and 2D (HPL and LAM).

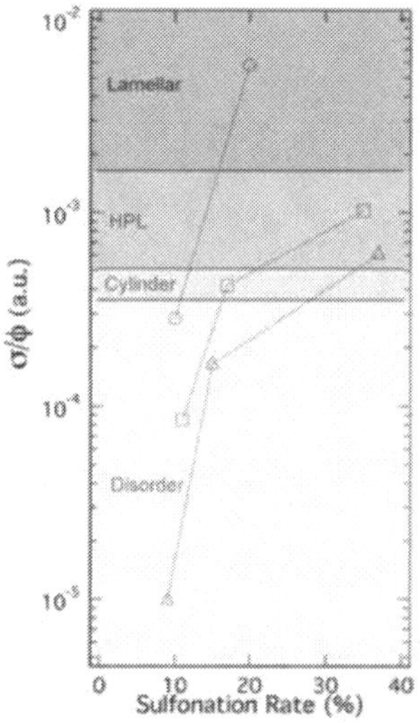

Fig. 10. Normalized proton conductivity (σ / Φ) as a function of the sulfonation rate for the three series of diblock copolymers varying the ratio of PS (circles 53 %, squares 22% and triangles 15%). Here Φ refers to the volume fraction of the conducting phase, *i.e.* PS + sPS + water domains. Figure reproduced with permission from reference [56]. Copyright (2008) American Chemical Society.

3 BCP solid polyelectrolytes for battery

Polymer electrolytes can be used as separator between the electrodes of secondary batteries. The two principal expected properties are: the ability to dissociate the salt and also to efficiently transport the ions during charge and discharge. The basic secondary battery principle is schematized in figure 11. During the charge, the lithium ions migrate through the electrolyte toward the anode and get intercalated or inserted in the electrode material (typically carbon graphite or lithium metal). During the discharge, the lithium ions diffusion toward the cathode, where they get inserted in the electrode material (typically cobalt oxide), is accompanied by an electron current in the external circuit.

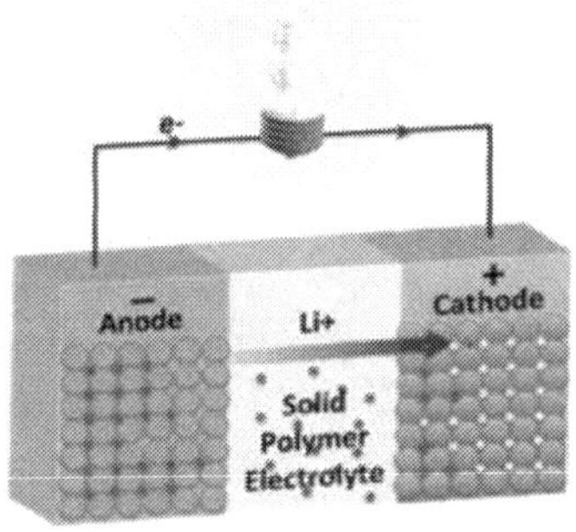

Fig. 11. Schematic representation of lithium solid polymer electrolyte battery principle during discharge.

Most of the batteries currently commercialized are prepared with electrolyte containing solvents, such as ethylene or propylene carbonate. Those solvents facilitate the ion transport; but unfortunately, they may cause critical safety issues: chemical leakage and risk of flammability. Nowadays, this is major concern in the development of secure

batteries for nomad and transportation applications. A suitable alternative to solvent containing electrolyte, is solid polymer electrolyte (SPE), despite poorer ion conductivity; indeed, important efforts are done to reach the objective of 10^{-3} to 10^{-2} $S.cm^{-1}$ at working temperature. In addition, it is expected that the high mechanical modulus of the solid polyelectrolytes (> 6 GPa) prevent the formation of dendrites on lithium anode during the charge.[57] Another issue that polymer electrolytes can overcome is the anions concentration gradient between the electrodes occurring during discharge and limiting the maximum current accessible; only a fraction of the current is carried by the lithium ions. This current fraction can be measured and expressed in term of transport number, t_+, which can range from 0 to 1. In PEO blended with lithium salt, t_+ is ranging generally between 0.1 and 0.2,[58] meaning an extremely low fraction of the current is carried by the lithium ions. Recently, single ion polymers were synthesized, with anions grafted on the polymer chain, leading to a significant increase of the transport number toward the unity.[59, 60]
The classical SPE polymers are poly(ethylene oxyde) (PEO), poly(vinylidene fluoride) (PVDF) or polycarbonates based polymer presenting similar chemical function than the carbonate solvents, such as polytrimethylene carbonate.[61] On figure 12 extracted from reference [62] are presented the ionic conductivity versus temperature of different polymer based electrolytes. The most efficient (highest conductivity at lowest temperature) are the organic electrolyte (containing solvent), then the gel electrolyte (solid containing solvent), then the SPE containing free lithium salts. The poorer performances are for single ion conductor[59, 60] and solid lithium phosphorous oxynitride electrolyte LiPON.[63]

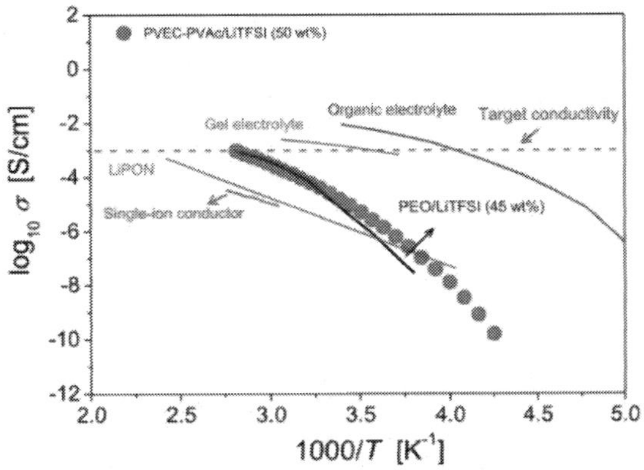

Fig. 12. Ionic conductivity versus temperature curves for selected representative lithium based electrolytes: PVEC-PVAc/LiTFSI (50 wt%) block copolymer,[62] a gel electrolyte made of 1 M $LiPF_6$/EC-PC (50:50 vol%) in 10 wt% of polyvinylidene difluoride-hexafluoropropylene,[64] an organic electrolyte made of 1 M $LiPF_6$/EC-PC (50:50 vol%),[65] a single-ion electrolyte made of P(STFSILi)-*b*-PEO-b-P(STFSILi),[60] and a lithium phosphorous oxynitride electrolyte (LiPON). Figure extracted and modified with permission from reference [62].

So far, PEO is the most extensively investigated polymer because of its high dielectric constant and its strong lithium ion (Li^+) solvating ability, brought by the Li^+ coordination on the ether oxygen. Actually, each Li^+ is potentially coordinating several ether oxygen atoms (on the same or different polymer chains) thus creating transient cross-linking points decreasing the segmental relaxation of the polymer chains. Also, the charges transport involves the segmental motion of the polymer chain and the re-coordination of the ions with neighboring ether oxygen (EO). Meaning that the ratio of EO units per lithium ion is a major parameter (EO:Li), so as the polymer glass transition temperature, *Tg*, which should be low to facilitate the charges transport (*e.g.* PEO Tg = - 60 °C). PEO is a semi-crystalline polymer with a melting temperature at the vicinity of 60 °C. The crystallites preserve the mechanical strength of the low *Tg* polyelectrolyte and simultaneity tend to lower the ionic transport, as the conducting phase is quasi exclusively in the amorphous PEO / lithium phase.[66] Nevertheless, it was shown that PEO blended with lithium salt can present crystalline PEO / lithium complex phases,[67] including the $PEO_6:LiAsF_6$ phase that potentially present an enhanced conductivity.[68] In this particular complex phase, Bruce and coworkers claim that the charges diffuse in nano-channels of crystalline PEO, on the other hand the counter ion are located outside the channel and are partially immobile; thus, ameliorating the electrolyte transport number, t_+. A detailed structural and dynamical study, using SANS and quasi elastic neutron scattering (QENS) was performed by Maranas and coworker on $PEO/LiClO_4$ as function of temperature and ratio EO:Li to investigate further the structure / dynamics relationship.[69] The analysis of the SANS data, taking advantage of the SLD contrast resulting from crystallization, indicates that the $(PEO)_3:LiClO_4$ phase presents a cylindrical morphology, with a radius of 125 Å and a length of 700 Å. Implying that multiple $(PEO)_3:LiClO_4$ helices align to form those large cylindrical structures. The authors have also localized the $LiClO_4$ nearby the amorphous domains, inducing a SLD contrast between the amorphous $PEO/LiClO_4$ bilayer and the crystalline lamellae detectable by SANS. The dimensions of the crystalline lamellae and the amorphous bilayer are both maximized at a particular salt concentration (*i.e.* 14:1) so as ion conductivity, despite the fact that this sample is partially crystalline. Nonetheless, the most prevailing view is that PEO crystallization is detrimental to the ion diffusion, because of structural tortuosity and chain dynamics limitation (tethered chains confinement). [66] Inelastic neutron scattering (*e.g.* QENS) is a technique of choice for dynamical investigations; nevertheless, as the incoherent scattering cross-section of lithium is negligible compared to hydrogen, inelastic neutron scattering is quasi exclusively sensitive to the dynamics of the hydrogenous polyelectrolyte chains and not to the mobile Li^+.[70] This technique is then complementary to dielectric or impedance spectroscopy in which the signal is dominated by the dynamics of the Li^+. The correlation between the segmental relaxation in polymer electrolytes and the dynamics of the ionic charges, at the nanosecond and picosecond time scale, is extremely informative: the general trend is that the Li^+ ions aggregate with multiple polymer chains and slow down the molecular and ionic mobility, *i.e.* creation of transient cross-links as mention previously.

BCP solid polyelectrolytes

In the perspective of developing solid polyelectrolytes, BCPs present the major advantage of fully decoupling ion conduction and mechanical properties. In addition, as already mentioned, BCPs electrolytes can be exploited as a model system to investigate the structure / properties relationship. Recently, Bouchet and coworkers have explored a wide range of block copolymers architectures made of polystyrene (PS) and PEO, *i.e.* diblock, triblock (ABA & BAB) and comb-like copolymers.[71] They conclude that the triblock PS-*b*-PEO-*b*-PS structure presents the best compromise between ion conductivity and mechanical properties. Also, the salt repartition is a crucial point to understand and improve the diffusion mechanisms. Wang and coworker have demonstrated theoretically that the ion distribution profile, so as the electrostatic potential, depend on the blocks dielectric contrast and also on the capacity of lithium ions to redistribute within the ion-dissolving phase.[72] In parallel, Bouchet and coworkers have demonstrated the presence of a dead zone – in term of conductivity – at the interphases PS/PEO in PS-*b*-PEO diblock copolymers.[73] This dead zone is constituted by 4 to 5 units of EO and can be explained by the constraint imposed by the PS block limiting the mobility of the PEO and more particularly on the first EO units. So far, the BCP solid polyelectrolyte self-assembly was intensively studied by SAXS and microscopy techniques, such as TEM. It is worth to note that in many of those studies the main fundamental objective is the determination of the Flory–Huggins parameter, χ, in order to point out the role of the lithium salt on the self-assembly mechanism.

Nevertheless, one of the pioneer works on BCP solid polyelectrolyte has demonstrated the great potential of SANS in the domain.[74] The system investigated by Ruzette *et al.* was a poly (methyl methacrylate -*b*- oligo oxyethylene methacrylate) BCP blended with $LiCF_3SO_3$ salt. The advantage of oligo oxyethylene methacrylate compared to regular PEO is the absence of crystallinity while preserving the same chemical functions. The neat copolymer presents a disordered phase, *i.e* segmentally mixed at temperatures ranging between 100 and 200 °C. The order to disorder transition temperature, T_{ODT}, is below the *Tg* of the whole copolymer. The authors have demonstrated that increasing the salt content is raising the T_{ODT} of the block copolymer bringing a microphase separation. Also, there is preferential complexation of $LiCF_3SO_3$ with the ether linkages of POEM compared to the carbonyl groups of methacrylate, inducing a weakening of the segmental interactions.

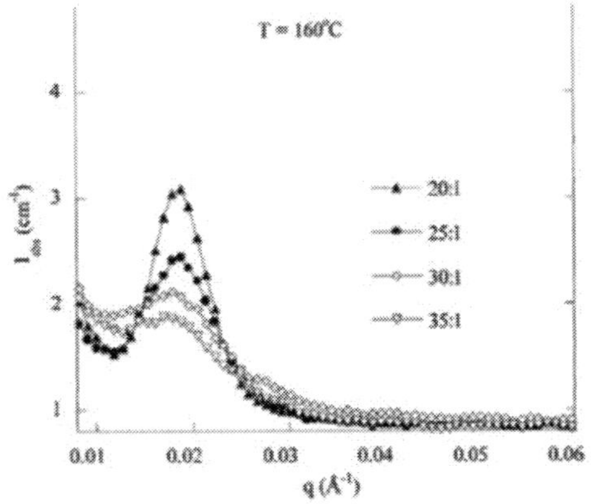

Fig. 13. SANS scattering intensity curves collected on salt-doped PMMA-*b*-POEM at 160 °C and four different salt concentrations (EO:Li): (20:1), (25:1), (30:1) and (35:1). Reproduced with permission from reference [74].

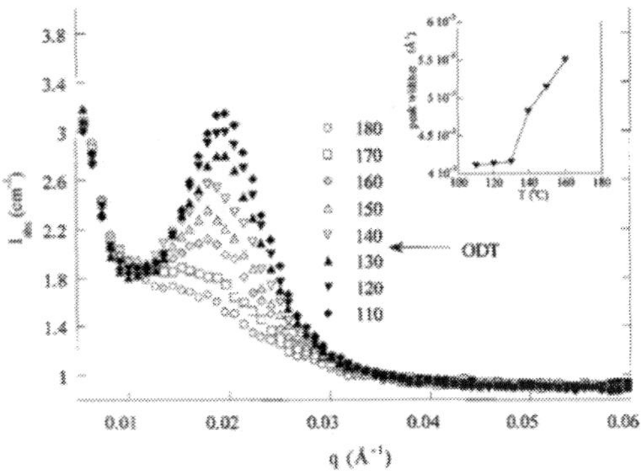

Fig. 14. SANS scattering intensity curves collected on PMMA-*b*-POEM blended with $LiCF_3SO_3$ (30:1), as a function of temperature. The inset presents the variation in peak width with temperature. Reproduced with permission from reference [74].

Figure 13 presents the SANS spectra collected on the BCP with different amount of salt at 160 °C. At higher salt content 20:1 and 25:1 (solid markers) sharp intense scattering maximums are observed, which are the first order Bragg reflections. Those are characteristic of the ordered micro-phase separated state. On the other hand at lower salt content 30:1 and 35:1 (empty markers) broad low intensity scattering maximums are observed, which indicate blocks segmentally mixed characteristic of a correlation hole, first described by Liebler.[2] More classically, temperature was change in the range 110 to 180 °C for a given salt concentrations (*cf.* figure 14), in the inset the peak width is used to point out the T_{ODT}. The results are presented in table 1, and show that the T_{ODT} increases significantly with the salt content. Both extreme systems: (i) below 40:1 are always disordered and (ii) 20:1 are systematically ordered. This result indicates that increasing the salt content raise the Flory–Huggins parameter, χ.

Table 1. Phase behavior and order to disorder transition temperatures, T_{ODT}, of salt-doped PMMA-*b*-POEM as a function of salt concentration. Reproduced with permission from reference [74].

Salt concentration [EO]:Li⁺	Type of phase behavior	Transition temperature T_{ODT} (°C)
40:1 and lower	always disordered	<80
35:1	ODT	110
30:1	ODT	130
25:1	ODT	150
20:1	always ordered	>200

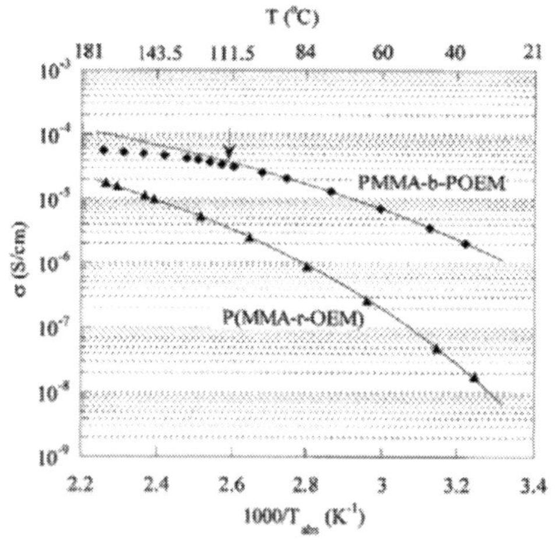

Fig. 15. Ionic conductivity versus temperature measured on salt-doped PMMA-*b*-POEM and P(MMA-*r*-OEM) with a salt concentration of (30:1). The solid lines correspond to the fits with the Vogel-Tammann-Fulcher (VTF) relation which best fits the non-Arrhenius behaviour of the data. Reproduced with permission from reference [74].

Conclusively, on figure 15, Ruzette *et al.* compare the BCP conductivity to a random copolymer with an identical composition, *i.e.* an identical amount of conducting polymer. The ordered BCP presents a significantly higher conductivity especially at lower temperature; also, the conductivities are increasing and converging with temperature. This increasing rate slows-down dramatically for the BCP above the ODT, *i.e.* 110 °C (arrow on figure 15). As expected, at the highest temperature, 170 °C, when the block copolymer is no longer microphase separated, its conductivity approaches that of the random copolymer. More recently, Bergfelt and coworkers have used SANS on partially deuterated triblock copolymer of poly[(oligo ethylene glycol) methyl ether methacrylate] (PMMA-*b*-

POEGMA-*b*-PMMA) blended with LiTFSI (9:1)(EO:Li).[75] In this system the PMMA block was selectively deuterated. The neutron experiment was performed at temperatures used during battery tests, *i.e.* 25, 60 & 95 °C. Considering the elevated salt ratio, those temperatures were expected to be below the system T_{ODT}, according to Ruzette *et al.* observations on similar polymers.[74] Nevertheless, SANS spectra revealed block copolymers in a mixed state, with broad and low intensity peaks, which were interpreted as the result of the correlation hole effect.[2] In order to extract the Flory-Huggins interaction parameter, the random phase approximation (RPA, *cf.* annex D) model was used to fit the data. This approach will be detailed in the next section.

As previously mentioned, one of the main fundamental interrogations when considering BCP is the determination of the Flory–Huggins parameter, χ, value. Adding salt changes considerably the interaction between the blocks, as demonstrated by Ruzette *et al.*; then equation 1 can be adapted, assuming that the ions are dissolved in only one of the phases. The Flory-Huggins parameter χ_{eff} for block copolymers blended with salt becomes:

$$\chi_{eff} = \frac{A(r)}{T} + B(r) \quad (5)$$

with r the molar ratio between the ions and the monomers of ion-containing blocks. Experimentally, χ_{eff} can been determined from scattering data, considering (i) the structure factor changes in the disordered phase, (ii) the change in the domain spacing of the ordered phases, or also (iii) the shift in the phase boundaries.
Russell and coworkers [76] have fitted the changes in structure factor of the disordered phase using the RPA theory (*cf.* annex D). Their experimental data was collected by SANS on PMMA-*b*-PS diblock copolymers blended with LiCl salt (*cf.* figures 16 & 17). Considering the following expression of the Flory-Huggins parameter with an enthalpic, χ_H, and entropic components, χ_S:[76]

$$\chi = \frac{\chi_H}{T} + \chi_S \quad (6)$$

They have shown that the entropic contribution χ_S for the salt complexed copolymer increases and on the other hand the enthalpic contribution, χ_H decreases, resulting in a nearly temperature-independent χ.

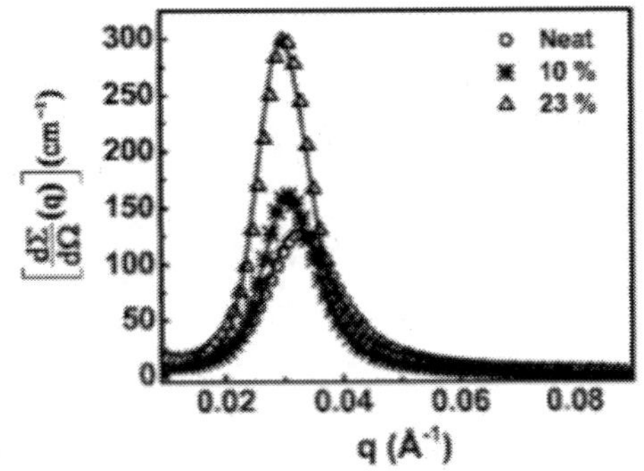

Fig. 16. SANS spectra (symbols) measured at 240 °C for the neat dPS-*b*-PMMA copolymer and 2 different salt concentrations, 10%, and 23% of PMMA carbonyl groups coordinated to lithium ions (CO:Li). The solid lines correspond to the calculated profiles with χ values of 0.03616, 0.03647, and

0.03702 for the neat copolymer and the copolymers with 10% and 23% (CO:Li), respectively. Reproduced with permission from reference [76]. Copyright (2008) American Chemical Society.

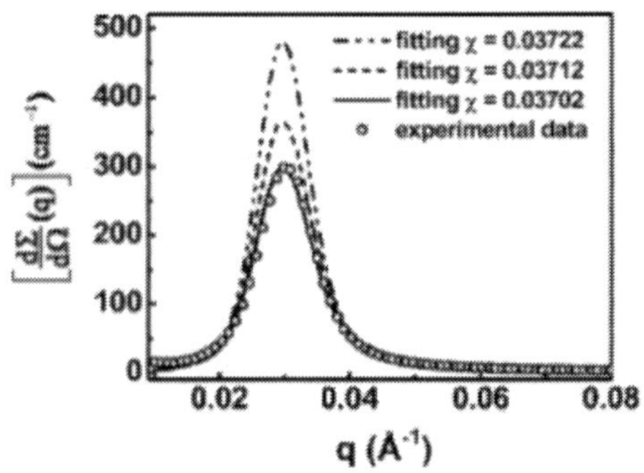

Fig. 17. SANS spectra (open circles) measured at 240 °C for the dPS-*b*-PMMA copolymer with 23% (CO:Li). The lines correspond to the calculated profiles with 3 different χ values, 0.03702, 0.03712, and 0.03722. Reproduced with permission from reference [76]. Copyright (2008) American Chemical Society.

Epps and coworker have exploited the change in the domain spacing, d^*, in the ordered phases versus salt concentration to estimate χ_{eff}.[77] Their experimental data were collected by SAXS on PEO-*b*-PS diblock copolymers blended with $LiCF_3SO_3$, $LiClO_4$ and $LiAsF_6$ salts. Considering the strong segregation condition, *i.e.* $\chi N > 100$, the theory predicts that:

$$d^* \sim \chi_{eff}{}^{1/6} \tag{7}$$

Note that χ for the neat copolymer was obtained from the literature. In this approximation, the volume occupied by the salt is accounted but changes in statistical segment length upon salt addition are not. Also, the neat sample is in the intermediate regime ($\chi N \sim 18$ at room temperature) where d^* is expected to vary from $\chi^{1/5}$ to $\chi^{1/6}$, inducing a slight over-estimation of χ_{eff}. Nonetheless, a linear dependence of χ_{eff} versus salt concentration is found, which is in agreement with the observations of Balsara and coworkers while exploiting the shift in the phase boundaries.[78] The experimental data of Balsara and coworkers, were obtained by SAXS on PEO-*b*-PS diblock copolymers blended with LiTFSI salt. In that case, the Leibler's theory is used to predict the χ at the ODT using the following expression:

$$\left(\chi_{eff}N\right)_{ODT} = S \tag{8}$$

where $N=\varphi_{PS}N_{PS}+\varphi_{PEO}N_{PEO}$, with φ_x and N_x respectively the volume fraction and the number of monomers of the specified phase x. S is a constant that depends on φ_{EO} (*e.g.* $S = 10.5$ for $\varphi_{EO} = 0.5$). It is important to note that this approximation ignores the temperature dependence of χ. In a second paper the same authors improved their approximation taking into account for composition fluctuations.[79]

In summary, the studies dedicated to the salt concentration effect on χ_{eff} reveal systematically a linear relationship which can be expressed as:[80]

$$\chi_{eff} = \chi + m.r \tag{9}$$

where χ is the intrinsic Flory–Huggins parameter for the salt-free BCP, r is the molar ratio between ions and the solvating monomers. The slope m depends on the anion and block chemical natures. For a PS-*b*-PEO diblock copolymer blended with $LiAsF_6$, $LiClO_4$ and $LiCF_3SO_3$ m are 5.70, 5.53 and 3.90 respectively. [77] For PS-*b*-PMMA diblock copolymers blended with LiCl, m is about 10^{-4}, [76] and for PS-*b*-PEO diblock copolymers blended with LiTFSI m is 1.56. [78, 79]. Moreover, m can be estimated considering the dielectric constants of the two phases, the size of the ions (m decreases with increasing anion radius) and the statistical segment lengths of the polymer chains.[81]

4 Conclusion

This paper gives an overview on the essential inputs of scattering techniques, with an emphasis on small angle neutron scattering, on the structural investigations of block copolymers based electrolytes. Two kinds of system were discussed: (i) block copolymers ionomer membranes used in fuel cell and (ii) block copolymers solid polyelectrolytes used in secondary batteries. The literature on neutron scattering is relatively rich concerning the ionomer membranes, which can be explained by the possibility to perform easily contrast variation with various H_2O-D_2O mixtures. On the other hand, SANS was rarely exploited to probe the morphology of solid polyelectrolytes used in batteries, SAXS is generally preferred. That opens up a vast field of SANS investigations using, for instance, precise polymer deuteration to probe the morphologies, the salt location, the dead zones, etc. in classical polymer / salt blend electrolytes; but also on more recent and promising systems, such as single ion polymer electrolytes.

Annex

A. Small angle scattering spectra collected on the BCP samples reveal structural peaks, whose q-positions are the signatures of the morphologies. Here, we give the expected q-positions for three current morphologies: (i) the 1D lamellar structure, (ii) the 2D hexagonal packing of cylinders, and (iii) 3D cubic arrangement of spheres. The relative peak position is defined as q_{peak}/q^* with q_{peak} the q-position of a specific higher order Bragg diffraction peak and q^* the q-position of the first order Bragg diffraction peak.
For the lamellar structure the peaks position are given by:[12]

$$q_h = {}^{2\pi h}\!/_{a} \tag{10}$$

with h the first Miller indice and a the lattice dimension. Then the relative peak positions are expected to follow the sequence: 1, 2, 3, 4, 5, etc. and the lamella-to-lamella distance is expressed as:

$$a = {}^{2\pi}\!/_{q^*} \tag{11}$$

For the hexagonal packing of cylinders the peak positions are given by the expression:[12]

$$q_{hk} = {}^{4\pi\sqrt{h^2+hk+k^2}}\!/_{\sqrt{3}a} \tag{12}$$

with h and k the first and second Miller indices and a the lattice dimension. The relative peak positions are expected to follow the sequence: 1, √3, √4, √7, √9, etc. and the cylinder-to-cylinder distance is expressed as:

$$a = {}^{4\pi}/_{\sqrt{3}q^*} \tag{13}$$

For the cubic arrangement of spheres, the peak positions are given by the expression:[12]

$$q_{hkl} = {}^{2\pi\sqrt{h^2+k^2+l^2}}/_{a} \tag{14}$$

with h, k and l, the Miller indices and a, the lattice dimension. The relative peak positions are expected to follow the sequence: 1, √2, √3, √4, √5, etc. and the sphere-to-sphere distance is given by:

$$a = {}^{2\pi}/_{q^*} \tag{15}$$

B. The Kinning and Thomas model[54] is expressed as a product of the form factor for polydisperse spheres, P, and the Percus-Yevick structure function, S_{PY}, describing their short-range order in a matrix:[82]

$$I_{KT}(q) = \Delta^2 . N . P(q; \bar{R}, \sigma) . S_{PY}(q; R_{HS}, \phi_{HS}) \tag{16}$$

where Δ^2 is the square of the contrast (*i.e.* SLDs difference between the spheres and the matrix) and N the number of the scattering particles. It is important to note that in the present scattering intensity formalism, the contrast parameter, Δ^2 is not contained in the form factor unlike presented in expression (2). This form factor simplification is allowed since homogeneous spheres are considered. $\bar{R}$ is the mean sphere radius and σ is the root-mean-square deviation from the mean size. R_{HS} and ϕ_{HS} are respectively the hard sphere radius and the hard spheres volume fraction. $2R_{HS}$ corresponds to the closest approach between two spheres.

P is expressed as

$$P(q; \bar{R}, \sigma) = \frac{\int_0^\infty p(R; \bar{R}, \sigma) f^2(q, R) dR}{\int_0^\infty p(R; \bar{R}, \sigma) dR} \tag{17}$$

with f (q,R) the sphere form factor of radius R, given by

$$f(q; R) = \left(\frac{4\pi}{3}\right) R^3 \frac{3}{(qR)^3} [\sin(qR) - qR\cos(qR)] \tag{18}$$

and $p(R)$ is a probability function considering a Schultz distribution,

$$p(R; \bar{R}, Z) = \left(\frac{Z+1}{\bar{R}}\right)^{Z+1} \frac{R^Z \exp\left[-\frac{Z+1}{\bar{R}} R\right]}{\Gamma(Z+1)} \tag{19}$$

in which $\Gamma(x)$ is the Gamma function and $\mathrm{Z} + 1 = \left(\frac{\bar{R}}{\sigma}\right)^2$.
Then,

$$P(q; \bar{R}, Z) = 8\pi^2 \bar{R}^6 (Z+1)^{-6} Q^{Z+7} G_1(Q) \quad (20)$$

with $Q = \frac{(Z+1)}{q\bar{R}}$ and G_1 is defined as

$$G_1(Q) = Q^{-(Z+1)} - (4+Q^2)^{-\frac{Z+1}{2}} cos\left[(Z+1)tan^{-1}\frac{2}{Q}\right]$$
$$-2(Z+1)(4+Q^2)^{-(Z+2)/2} sin\left[(Z+2)tan^{-1}\frac{2}{Q}\right]$$
$$+(Z+2)(Z+1)\left(Q^{-(Z+3)} + (4+Q^2)^{-(Z+3)/2} cos\left[(Z+3)tan^{-1}\frac{2}{Q}\right]\right) \quad (21)$$

The Percus-Yevick structure function, S_{PY}, is given by

$$S_{PY}(q; R_{HS}, \phi_{HS}) = \frac{1}{1+24\phi_{HS}G_2(A)/A} \quad (22)$$

in which $A = 2qR_{RH}$ and G_2 is given by the following expression,

$$G_2(A) = \frac{\alpha}{A^2}(sinA - AcosA) + \frac{\beta}{A^3}(2AsinA + (2-A^2)cosA - 2)$$
$$+\frac{\gamma}{A^5}(-A^4 cosA + 4[(3A^2-6)cosA + (A^3-6A)sinA + 6]) \quad (23)$$

where α, β and γ are

$$\alpha = \frac{(1+2\phi_{HS})^2}{(1-\phi_{HS})^4} \quad (24)$$

$$\beta = \frac{-6\phi_{HS}(1+\phi_{HS}/2)^2}{(1-\phi_{HS})^4} \quad (25)$$

$$\gamma = \frac{\frac{1}{2}\phi_{HS}(1+2\phi_{HS})^2}{(1-\phi_{HS})^4} \quad (26)$$

C. The Teubner-Strey model[55] is used to fit broad peaks which are the signature of irregularly shaped domains with short-range order. This model has two parameters defining two characteristic lengths: d a characteristic separation length of the domains and ξ a correlation length that describes the extent of correlation of the domain arrangement. Then the Teubner-Strey scattered intensity is given by

$$I_{TS}(q; d, \xi) = \frac{8d^4\pi}{\xi[16\pi^4 + 8d^2\pi^2(\xi^{-2}-q^2) + d^4(\xi^{-2}+q^2)]} \quad (27)$$

D. The Flory-Huggins parameter, χ, can be extracted from the scattering data through the random phase approximation theory (RPA), *i.e.* the correlation hole in the disordered regime.[2] To do so, the correlation hole peak is fitted with a structure factor, $S(q)$, which accounts for the concentration fluctuations:[9]

$$\frac{1}{S(q)} = \frac{S_{AA}+S_{BB}+2S_{AB}}{S_{AA}S_{BB}-S_{AA}^2} - \chi \quad (28)$$

where S_{ij} correspond to the Fourier transform of the density-density correlation function for the component i and j. The S_{ij} can be expressed in term of Debye functions, when considering large block length, then:

$$S_{AA} = Ng(f, x) \tag{29}$$
$$S_{BB} = Ng(1 - f, x) \tag{30}$$
$$S_{AB} = \frac{N}{2}(g(1, x) - f(f, x) - g(1 - f, x)) \tag{31}$$

with $g(f, x) = \frac{2}{x^2}(fx + e^{-fx} - 1)$, x is given by $x = q^2 R_g^2$, where R_g denotes the whole copolymer radius of gyration (*i.e.* $R_g = Nb^2/6$, if a unique statistical segment length b is considered). The number fraction of A segments f, is expressed $f = N_A/(N_A+N_B)$. The general trend is that the χ value dictates the peak intensity maximum and also the peak width. On the other hand, the statistical segment length b determines the q-position of the peak.

Acknowledgements

I am grateful to M. Wolf, F. Cousin, M. Månssson and I. Grillo for their kind invitation to the French & Swedish Winter School in Neutron Scattering (Uppsala, 2016). Also, I would like to thanks my supervisors and colleagues O. Diat, G. Gebel, B. Frisken, S. Holdcroft, R. Mezzenga, J. Ruokolainen, D. Mecerreyes, T. Bowden and D. Brandell for our fruitful collaborations. Finally, I thank LLB, ILL, SINQ, NIST-CNR and ISIS for the neutrons beamtime allocations.

References

1. F. S. Bates, G. H. Fredrickson, Physics Today **52**, 32 (1999)
2. L. Leibler, Macromolecules **13**, 1602 (1980)
3. F. S. Bates, G. H. Fredrickson, Annual Review of Physical Chemistry **41**, 525 (1990)
4. M. W. Matsen, F. S. Bates, Macromolecules **29**, 1091 (1996)
5. M. W. Matsen, Physical Review Letters **99**, 148304 (2007)
6. M. W. Matsen, The European Physical Journal E **36**, 44 (2013)
7. C. J. Clarke, A. Eisenberg, J. La Scala, M. H. Rafailovich, J. Sokolov, Z. Li, S. Qu, D. Nguyen, S. A. Schwarz, Y. Strzhemechny, B. B. Sauer, Macromolecules **30**, 4184 (1997)
8. A. V. G. Ruzette, P. Banerjee, A. M. Mayes, M. Pollard, T. P. Russell, R. Jerome, T. Slawecki, R. Hjelm, P. Thiyagarajan, Macromolecules **31**, 8509 (1998)
9. B. Stuehn, A. R. Rennie, Macromolecules **22**, 2460 (1989)
10. Y. Mai, A. Eisenberg, Chem Soc Rev **41**, 5969 (2012)
11. R. Borsali, R. Pecora, *Soft-Matter Charactezation* (Springer, 2008)
12. S. Förster, A. Timmann, M. Konrad, C. Schellbach, A. Meyer, S. S. Funari, P. Mulvancy, R. Knott, The Journal of Physical Chemistry B **109**, 1347 (2005)
13. Y. Yin, Q. Du, Y. Qin, Y. Zhou, K.-i. Okamoto, Journal of Membrane Science **367**, 211 (2011)
14. D. X. Luu, E.-B. Cho, O. H. Han, D. Kim, The Journal of Physical Chemistry B **113**, 10072 (2009)
15. X. Zhang, L. Sheng, T. Hayakawa, M. Ueda, T. Higashihara, Journal of Materials Chemistry A **1**, 11389 (2013)
16. E. A. Weiber, S. Takamuku, P. Jannasch, Macromolecules **46**, 3476 (2013)
17. F. Ng, D. J. Jones, J. Rozière, B. Bauer, M. Schuster, M. Jeske, Journal of Membrane Science **362**, 184 (2010)
18. G. Gebel, O. Diat, Fuel Cells **5**, 261 (2005)

19. S. Lyonnard, École thématique de la Société Française de la Neutronique **11**, 177 (2010)
20. L. Rubatat, PhD thesis, *New structural model for Nafion membranes, the reference polymer for low temperature fuel cells* (UJF, Grenoble, 2003).
21. P. Debye, J. H.R. Anderson, H. Brumberger, J. Appl. Phys. **28**, 679 (1957)
22. H. Brumberger, J. Goodisman, J. Appl. Crist. **16**, 83 (1983)
23. L. Rubatat, G. Gebel, O. Diat, Macromolecules **37**, 7772 (2004)
24. T. D. Gierke, G. E. Munn, F. C. Wilson, J. of Poly. Sc. **19**, 1687 (1981)
25. T. D. Gierke, W. Y. Hsu, Am. Chem. Soc., 283 (1982)
26. W. Y. Hsu, T. D. Gierke, Journal of Membrane Science **13**, 307 (1983)
27. L. Rubatat, A. L. Rollet, G. Gebel, O. Diat, Macromolecules **35**, 4050 (2002)
28. K. Schmidt-Rohr, Q. Chen, Nature Materials **7**, 75 (2008)
29. L. Rubatat, O. Diat, Macromolecules **40**, 9455 (2007)
30. K.-D. Kreuer, G. Portale, Advanced Functional Materials **23**, 5390 (2013)
31. J. Baruchel, J. L. Hodeau, M. S. Lehmann, J. R. Regnard, C. Schlenker, *Neutron and synchrotron radiation for condensed matter studies* (Springler-Verlag, 1993)
32. R. A. Weiss, A. Sen, C. L. Willis, L. A. Pottick, Polymer **32**, 1867 (1991)
33. R. A. Weiss, A. Sen, L. A. Pottick, C. L. Willis, Polymer **32**, 2785 (1991)
34. X. Lu, W. P. Steckle, R. A. Weiss, Macromolecules **26**, 5876 (1993)
35. X. Y. Lu, W. P. Steckle, B. Hsiao, R. A. Weiss, Macromolecules **28**, 2831 (1995)
36. K. A. Mauritz, R. F. Storey, D. A. Reuschle, N. B. Tan, Polymer **43**, 5949 (2002)
37. K. A. Mauritz, R. F. Storey, D. A. Mountz, D. A. Reuschle, Polymer **43**, 4315 (2002)
38. Y. A. Elabd, C. W. Walker, F. L. Beyer, J. Membr. Sci. **231**, 181 (2004)
39. Y. A. Elabd, E. Napadensky, J. M. Sloan, D. M. Crawford, C. W. Walker, J. Membr. Sci. **217**, 227 (2003)
40. A. Mokrini, C. D. Rio, J. L. Acosta, Solid State Ionics **166**, 375 (2004)
41. A. Mokrini, J. L. Acosta, Polymer **42**, 9 (2001)
42. S. Mani, R. A. Weiss, C. E. Williams, S. F. Hahn, Macromolecules **32**, 3663 (1999)
43. D. Gromadzki, P. Cernoch, M. Janata, V. Kudela, F. Nallet, O. Diat, P. Stepanek, European Polymer Journal **42**, 2486 (2006)
44. M. J. Park, K. H. Downing, A. Jackson, E. D. Gomez, A. M. Minor, D. Cookson, A. Z. Weber, N. P. Balsara, Nano Letters **7**, 3547 (2007)
45. L. Rubatat, Z. Q. Shi, O. Diat, S. Holdcroft, B. J. Frisken, Macromolecules **39**, 720 (2006)
46. Z. Shi, S. Holdcroft, Macromolecules **38**, 4193 (2005)
47. Z. Q. Shi, Phd, *Synthesis and Characterization of Proton Conducting, Fluorine-Containing Block Copolymers* (Simon Fraser University, Burnaby, 2004).
48. Z. Shi, S. Holdcroft, Macromolecules **37**, 2084 (2004)
49. J. B. Hayter, J. Penfold, J. Phys. Chem. **88**, 4589 (1984)
50. G. Gebel, *Elaboration et etude structurale de membranes ionomeres perfluorees a partir de solutions* (Universite Joseph Fourier, Grenoble, 1989).
51. A. L. Rollet, O. Diat, G. Gebel, The J. of Phys. Chem. **106**, 3033 (2002)
52. A. C. C. Yang, R. Narimani, Z. Zhang, B. J. Frisken, S. Holdcroft, Chemistry of Materials **25**, 1935 (2013)
53. R. Narimani, A. C. C. Yang, E. M. W. Tsang, L. Rubatat, S. Holdcroft, B. J. Frisken, Macromolecules **46**, 9676 (2013)
54. D. J. Kinning, E. L. Thomas, Macromolecules **17**, 1712 (1984)
55. M. Teubner, R. Strey, The Journal of Chemical Physics **87**, 3195 (1987)

56. L. Rubatat, C. Li, H. Dietsch, A. Nykänen, J. Ruokolainen, R. Mezzenga, Macromolecules **41**, 8130 (2008)
57. C. Monroe, J. Newman, Journal of The Electrochemical Society **152**, A396 (2005)
58. R. Bouchet, S. Maria, R. Meziane, A. Aboulaich, L. Lienafa, J.-P. Bonnet, T. N. T. Phan, D. Bertin, D. Gigmes, D. Devaux, R. Denoyel, M. Armand, Nature Materials **12**, 452 (2013)
59. R. Meziane, J. P. Bonnet, M. Courty, K. Djellab, M. Armand, Electrochimica Acta **57**, 14 (2011)
60. R. Bouchet, S. Maria, R. Meziane, A. Aboulaich, L. Lienafa, J.-P. Bonnet, T. N. T. Phan, D. Bertin, D. Gigmes, D. Devaux, R. Denoyel, M. Armand, Nature Materials **12**, 452 (2013)
61. J. Mindemark, L. Imholt, D. Brandell, Electrochimica Acta **175**, 247 (2015)
62. Y. Wang, F. Fan, A. L. Agapov, T. Saito, J. Yang, X. Yu, K. Hong, J. Mays, A. P. Sokolov, Polymer **55**, 4067 (2014)
63. X. Yu, J. B. Bates, G. E. Jellison, F. X. Hart, Journal of The Electrochemical Society **144**, 524 (1997)
64. J. Y. Song, Y. Y. Wang, C. C. Wan, Journal of The Electrochemical Society **147**, 3219 (2000)
65. P. E. Stallworth, J. J. Fontanella, M. C. Wintersgill, C. D. Scheidler, J. J. Immel, S. G. Greenbaum, A. S. Gozdz, Journal of Power Sources **81–82**, 739 (1999)
66. S. Cheng, D. M. Smith, C. Y. Li, Macromolecules **47**, 3978 (2014)
67. C. D. Robitaille, D. Fauteux, Journal of The Electrochemical Society **133**, 315 (1986)
68. Z. Gadjourova, Y. G. Andreev, D. P. Tunstall, P. G. Bruce, Nature **412**, 520 (2001)
69. S. K. Fullerton-Shirey, J. K. Maranas, Macromolecules **42**, 2142 (2009)
70. H.-G. Peng, M. Tyagi, K. A. Page, C. L. Soles, in *Polymers for Energy Storage and Delivery: Polyelectrolytes for Batteries and Fuel Cells.* (American Chemical Society, 2012), vol. 1096, chap. 5, pp. 67.
71. D. Devaux, D. Glé, T. N. T. Phan, D. Gigmes, E. Giroud, M. Deschamps, R. Denoyel, R. Bouchet, Chemistry of Materials **27**, 4682 (2015)
72. I. Nakamura, Z.-G. Wang, Soft Matter **8**, 9356 (2012)
73. R. Bouchet, T. N. T. Phan, E. Beaudoin, D. Devaux, P. Davidson, D. Bertin, R. Denoyel, Macromolecules **47**, 2659 (2014)
74. A.-V. r. G. Ruzette, P. P. Soo, D. R. Sadoway, A. M. Mayes, Journal of The Electrochemical Society **148**, A537 (2001)
75. A. Bergfelt, L. Rubatat, R. Mogensen, D. Brandell, T. Bowden, Polymer **131**, 234 (2017)
76. J.-Y. Wang, W. Chen, T. P. Russell, Macromolecules **41**, 4904 (2008)
77. W.-S. Young, T. H. Epps, Macromolecules **42**, 2672 (2009)
78. N. S. Wanakule, A. Panday, S. A. Mullin, E. Gann, A. Hexemer, N. P. Balsara, Macromolecules **42**, 5642 (2009)
79. N. S. Wanakule, J. M. Virgili, A. A. Teran, Z.-G. Wang, N. P. Balsara, Macromolecules **43**, 8282 (2010)
80. F. Xue, S. Jiang, RSC Advances **3**, 23895 (2013)
81. I. Nakamura, N. P. Balsara, Z. G. Wang, Phys Rev Lett **107**, 198301 (2011)
82. J. K. Percus, G. J. Yevick, Physical Review **110**, 1 (1958)

 https://doi.org/10.1051/epjconf/201818804001

Neutron reflectivity for soft matter

Fabrice Cousin and Alexis Chennevière

Laboratoire Léon Brillouin, CEA-CNRS, CEA-Saclay, 91191 Gif sur Yvette, France

Abstract. Specular neutron reflectivity is a technique enabling the measurement of coherent neutron scattering length density profile perpendicular to the plane of a surface or interface, and thereby the profile of chemical composition. The characteristic sizes that are probed range from around 5Å up 5000 Å. It is a scattering technique that averages information over the entire surface and it is therefore not possible to obtain information on correlations in the plane of the interface. The specific properties of neutrons (possibility of tuning the contrast by isotopic substitution, negligible absorption, low energy of the incident neutrons) makes it particularly interesting in the fields of soft matter and biophysics. This course is composed of three parts describing respectively its principle, the experimental aspects (diffractometers, samples), and some scientific examples of neutron reflectometry focusing on the use of contrast variation to probe polymeric systems.

Résumé. La réflectivité spéculaire des neutrons est une technique permettant de mesurer le profil de densité de longueur de diffusion cohérente de neutrons, et ce faisant le profil de composition chimique, perpendiculairement à une surface ou une interface plane. Les tailles caractéristiques sondées sont de l'ordre de 5Å à 5000Å. C'est une technique de diffusion qui moyenne l'information sur l'ensemble de la surface et qui ne permet donc pas d'obtenir d'informations sur les corrélations dans le plan de l'interface. Les propriétés spécifiques des neutrons (possibilité de moduler le contraste par substitution isotopique, absorption négligeable, faible énergie des neutrons incidents) la rendent particulièrement intéressante dans les domaines de l'étude de la matière molle et de la biophysique Ce cours est composé de trois parties décrivant respectivement son principe, les aspects expérimentaux (appareils de mesure, échantillons) et quelques exemples d'expériences axés sur l'utilisation de la variation de contraste pour l'étude de systèmes polymériques.

1 Introduction

Neutron reflectivity (NR) is a technique enabling to measure the thickness and the chemical composition of one or several thin layers at a surface or at an interface. Typically, length scales in the range from 5 - 5000 Å can be probed.. The principle is to measure the coefficient of reflection R of a neutron beam sent in grazing incidence on the surface that is to be studied. In this course, we will focus only on the *specular*[*] reflectivity that provides only the profile of chemical composition *perpendicular to the surface*. This specular reflectivity corresponds to the case where the incident and reflected neutron

[*] From the Latin adjective *specularis* that stands for "pertaining to mirrors, mirror-like".

beams are symmetric with respect to the surface normal. In such geometry, NR does not provide any information on the possible in-plane structure at the surface. It is thus usually very valuable to combine NR with a technique enabling to determine such in-plane structure information, either a scattering method (Grazing Incidence Small Angle Neutron Scattering, surface diffraction, off-specular measurements) or a microcopy technique (Atomic Force Microscopy, Transmission Electron Microscopy, Brewster microscopy…).
The lower limit of thickness that can be measured is directly linked to the minimal reflectivity that can be experimentally obtained. Given the actual brightness of the neutrons sources, such minimal reflectivity is of the order of 10^{-6} to 10^{-7} on samples of a few cm^2. This is of the same order of magnitude of what can be measured by X-Ray reflectivity, whose principle is similar to NR, on laboratory instruments. Such minimal X-Ray reflectivity can be extended down to lower reflectivities on synchrotron sources due to the high brilliance of such sources. X-rays measurements have thus to be preferred in many experimental cases and are in any case very complementary to NR measurements as they provide the electron density profile.

Neutrons have, however, some unique specific properties with respect to photons that enable some measurements impossible with X-rays, such as the characterization of buried interfaces, the study of materials containing hydrogen or for magnetic materials. Such specificities have made NR experiments very popular since the early 90's:
-First, **the neutron-matter interaction directly occurs between the neutron and the nuclei of atoms**, and not with the electronic cloud, as for X-rays. It differs from one isotope of a given atom to another and varies randomly from one atom to another along the periodic table of elements. In particular, hydrogen ^1H has a very different value than the ones of all other atoms constitutive of organic molecules (C,O,N..) and, last but not least, than those of deuterium ^2H. This enables to achieve experiments on organic materials because the neutron refraction index is essentially dependent on its content on hydrogen atoms. In particular, the replacement of some hydrogen atoms of a molecule by deuterium atoms will strongly modify its refraction index without altering significantly its physical and chemical properties. This opens the way to **contrast variation experiments** where some molecules are labeled by deuterium to create a neutron contrast in the system. It is also possible to continuously tune the neutron refractive index of a solvent in a complex system by mixing hydrogenated and deuterated solvents in order to *match* the one of a component, making it invisible from the neutron point of view. Such contrast variation experiments, which have been very successfully applied in soft matter in the field of surfactants [1] and polymers [2] or in biophysics [3][4].
- Second, neutrons are **very weakly absorbed** by matter, except few atoms of periodic table that have a huge neutron absorption cross-section (boron, gadolinium, lithium, ...). The imaginary part of the neutron refractive indexes is usually negligible, and not considered in neutron reflectivity calculations, contrarily to calculations of X-Ray reflectivity. This weak interaction opens the way to the study of **buried interfaces** or to achieve ***in situ* experiments in various sample environments**.
-Third, the energy of neutrons that are used for experiments whose wavelengths lay typically in the range 1 Å – 20 Å have an energy of a few meV, contrarily to X-rays that have an energy of a few keV for the same wavelengths. **The NR experiments are thus absolutely not destructive**.
Please note that neutrons have also another very important **specificity**, namely **their ability to interact with the atoms that bear a magnetic moment because they** bear a spin ±½. This enables the measurement of the magnetization profile in thin magnetic films. This aspect is out of scope of this introduction course focused on soft matter. A very good description of the magnetic neutron reflectivity principles can be found in the following reference by Fermon *et al*[5], in particular the treatment of magnetic neutron reflectivity that necessitates a formalism more complicated than those that will be described in the following.

This course is divided in three parts. We will present in the first part the principle of NR by an approach where the spin of neutrons is not taken into account. Besides, the other aspects of *grazing incidence scattering* (off-specular scattering, Grazing Incidence Small Angle Neutron Scattering) are not presented here but are extensively presented in the course devoted to it by Max Wolff in this thematic volume [6]. We will describe in the second part how it is possible to realize a NR experiment from practical point of view (description of instruments and of possible geometries of measurements, sample environments, accessibility to neutrons beams…). The third part is devoted to the presentation of some examples dedicated to the use of neutron NR for polymeric systems, with a specific emphasis on the different strategies of contrast variation that can be envisaged for such studies.

2 Principle of neutron reflectivity

The formalism described in this part is a wave formalism very close to the one usually used for reflection of light. It is limited to specular reflectivity and does neither consider the absorption of neutrons nor incoherent scattering.

2.1 Neutron-matter interaction and calculation of refractive index for neutrons

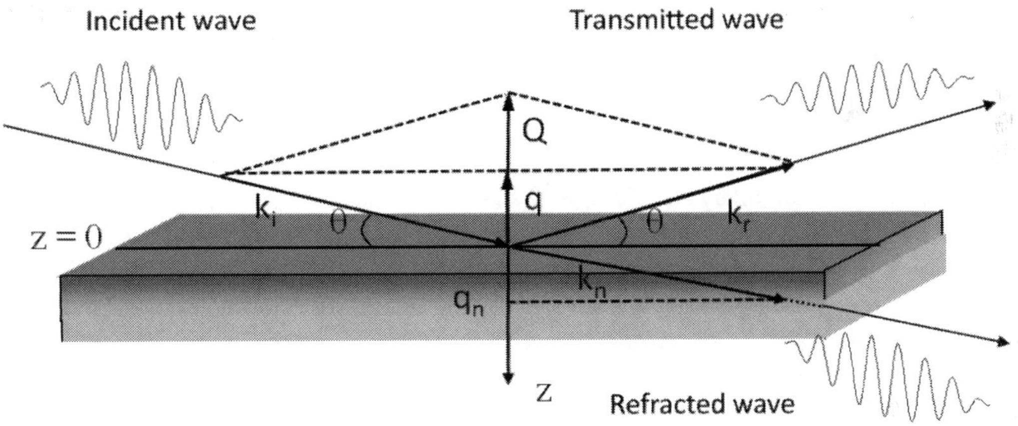

Figure 1. Reflection on an infinite planar surface.

Let us consider a neutrons beam reflecting on an infinite flat surface with an incident angle θ (see Figure 1). This surface is defined by the interface between vacuum (*n*=1) and a media with a refraction index *n*. In vacuum, the modulus of the wave vector of an incident neutron of wavelength λ is defined by:

$$k = \frac{2\pi}{\lambda} \tag{1}$$

The energy E of the neutron is:

$$E = \frac{\hbar^2}{2m} k^2 \tag{2}$$

where $\hbar$ is the reduced Planck constant and m the mass of the neutron.

In the media of refractive index *n*, the neutron interacts with the nuclei of atoms.
This interaction has a very short range and neutron-matter interaction is described by the Fermi pseudo-potential considering a point like potential V(r):

$$V(r) = b\left(\frac{2\pi\hbar^2}{m}\right)\delta(r) \tag{3}$$

where δ(r) is the Dirac function and *b* the coherent scattering length that describes the amplitude of interaction between neutron and the nuclei of the atom. This coherent scattering length has a real part and an imaginary part that accounts for absorption of neutrons by the nuclei. In the following *b* is always reduced to its real part because absorption is neglected. In the vacuum or in the media of index *n* the wave function ψ of the neutron is described by the Schrödinger equation:

$$\frac{\hbar^2}{2m}\Delta\psi(r) + [E - V(r)]\psi(r) = 0 \tag{4}$$

By considering that the structure of the media is invariant within the *xy* plan parallel to the surface, it is possible to separate the space variables *x,y* and *z* in equation (4) since V(r) is only dependent on *z*:

$$\frac{\hbar^2}{2m}\frac{d^2\psi_z}{dz^2} + [E_z - V_z)]\psi_z = 0 \tag{5}$$

The mean potential V_z within the layer is obtained by integration of the Fermi pseudo-potential:

$$V_z = \frac{1}{V}\int_V V(r)dr = \frac{2\pi\hbar^2}{m}Nb \tag{6}$$

where N is the number of atoms per volume unit.

In the framework of this course, one considers that neutrons do not exchange energy when then get reflected from the media of index *n* (**elastic scattering**). The conservation of energy at z = 0 enables to write the wave vector of neutrons in media k_n as a function of k and V_z:

$$\frac{\hbar^2k^2}{2m} = \frac{\hbar^2k_n^2}{2m} + V_z \tag{7}$$

One finally obtains from equation (6) and (7):

$$k_n^2 = k^2 - 4\pi Nb \tag{8}$$

This last equation allows to express the index of refraction *n* of the homogeneous media, defined as the ratio of the wave vectors in the material and in vacuum (6) :

$$n^2 = \frac{k_n^2}{k^2} = 1 - \frac{\lambda^2}{\pi}Nb \tag{9}$$

The product *Nb* is the **neutron coherent Scattering Length Density (SLD)**. Table 1 shows the coherent scattering length of some atoms and the SLD of some materials [7]. Please note that ***b* can be either positive or negative**, contrarily to the case of X-rays for which it is always positive and proportional to the number of electrons. Some full neutron tables of elements can be found on internet [8]. The SLD of the substrate is a crucial information in a NR experiment. It has to be ideally calculated prior to the experiment. Its calculation is very simple for crystals or solvents[†] but may raise difficulties for glasses

[†] An example of calculation of scattering length density: the case of water. This density is obtained through *Nb* = $(\Sigma_i b_i)/V$ where the b_i correspond to the atoms constitutive of the molecule (table 1) and V the volume of the molecule. Since the mass molar of H_2O is 18g/mol and the density of H_2O is 1g/cm^3, the volume of a molecule of water is 18/6.02 10^{23} = 2.99 10^{-23} cm^3 per molecule. The volume of a molecule of D_2O is the same. On gets :
Nb(H_20) = (-0.374*2+0.58).10^{-12} / 2.99 10^{-23} = -0.56 10^{10} cm^{-2} (– 0.56 10^{-6} Å^{-2})
Nb(D_20) = (0.667*2+0.58).10^{-12} / 2.99 10^{-23} = 6.38 10^{10} cm^{-2} (6.38 10^{-6} Å^{-2})

or polymers when the exact value of the specific volume of the molecule is not known with a high precision.

Table 1. Coherent scattering length of some atoms and scattering length density of some molecules. PS refers to polystyrene.

Coherent scattering length b (10^{-12} cm)					
^{1}H	D (^{2}H)	C	O	N	Si
-0.374	0.667	0.665	0.580	0.936	0.415
Scattering length density (Å^{-2})					
H_2O	D_2O	PS_H	PS_D	SiO_2	Si(crystal)
-0.56 10^{-6}	6.38 10^{-6}	1.41 10^{-6}	6.5 10^{-6}	3.41 10^{-6}	2.07 10^{-6}

The characteristic wavelengths of the incident neutrons being of the order of a few Å and the *Nb* of the order of 10^{-6} Å^{-2}, the values of the neutron refractive index n are always very close to 1 (~ 1± 10^{-3}). Contrarily to the case of visible light, **the effects of reflection/refraction for neutrons on surfaces occur only at very grazing angles, of the order of ~ 1°**. Please note that the refractive index can be smaller or larger than 1 for neutrons.

At the interface between the vacuum and the media of index n, one writes the refraction law of Descartes:

$$\cos\theta = n\cos\theta_n \tag{10}$$

Total reflection is achieved when $\theta \leq \theta_c$, where the critical angle θ_c is defined so that $\theta_n = 0$. If n is larger than 1, the conditions of total reflection are never fulfilled, which happens for example when the substrate is H_2O.

This leads to:

$$\cos\theta_c = n \tag{11}$$

The above equation is usually expressed as the sinus of the total reflection angle:

$$1 - (\sin\theta_c)^2 = 1 - \frac{\lambda^2}{\pi}Nb \tag{12}$$

Or equivalently:

$$\sin\theta_c = \sqrt{\frac{Nb}{\pi}}\lambda \tag{13}$$

When total reflection occurs, **the measurement of the critical angle θ_c for a given wavelength is a very precise measurement of the substrate's *Nb* and therefore of its chemical composition**.

2.2 Reflection on a succession of layers

The reflectivity is a function that is only dependent on the scattering vector $\vec{Q}$ defined by:

$$\boxed{\begin{aligned}\vec{Q} &= \vec{k_r} - \vec{k_\iota} \\ Q = \left\|\vec{k_r} - \vec{k_\iota}\right\| &= \frac{4\pi\sin\theta}{\lambda}\end{aligned}} \tag{14}$$

It is equal to twice the projection of the incident wave vector k_i on the z axis perpendicular to the surface. In the following we will use in the calculation the variable q, projection of the wave vector on z such that q =Q/2.

If the system under study is made of several layers, each of them having a refractive index n_p, the plane wave functions describing the neutron within the layer p (ψ_p) and the layer p+1(ψ_{p+1}) can be written like:

$$\psi_p(z) = A_p \exp(iq_p z) + B_p \exp(-iq_p z) \quad (15)$$

$$\psi_{p+1}(z) = A_{p+1} \exp(iq_{p+1} z) + B_{p+1} \exp(-iq_{p+1} z) \quad (15')$$

where i^2=-1, and A_p and B_p are respectively the amplitudes of the wave propagating towards the inner and the outer of the material. From equation (8) we can also write:

$$q_p^2 = q^2 - 4\pi N b_p \quad (16)$$

$$q_{p+1}^2 = q^2 - 4\pi N b_{p+1} \quad (16')$$

The conditions of continuity of the wave and of its derivative at the interface p/p+1 are:

$$\psi_p(z_p) = \psi_{p+1}(z_p) = u(z_{p/p+1}) \quad (17)$$

$$\psi_p'(z_p) = \psi_{p+1}'(z_p) = u'(z_{p/p+1}) \quad (17')$$

The reflectivity at $z_{p/p+1}$ is defined by the ratio of the intensity of the beam reflected by the layer p+1 by the intensity of the incident beam on such a layer:

$$R = \frac{|B_p|^2}{|A_p|^2} = \left| \frac{1 - \frac{u'(z_{p/p+1})}{iq_p u(z_{p/p+1})}}{1 + \frac{u'(z_{p/p+1})}{iq_p u(z_{p/p+1})}} \right|^2 \quad (18)$$

where $u(z_{p/p+1})$ and $u'(z_{p/p+1})$ are functions that depend both on $z_{p/p+1}$ and q_{p+1}.

From this last equation it is possible to calculate the reflectivity at the last interface between the last layer and the bulk substrate material, and then to calculate recursively the reflectivity at each interface up to the outer surface in order to obtain the whole reflectivity of the system. This method is known as the optical matrix method that uses the Abeles formalism [9].

2.3 The ideal interface and the Fresnel reflectivity curve

Let us consider here a neutron beam propagating in vacuum (or air) and reflecting from an ideal planar substrate without any layer at the surface. If the interface is perfect without roughness, such a system is a diopter. The reflectivity of such an interface is called **the Fresnel reflectivity** (R_F). Its calculation can be achieved with last equation (18) by replacing the media p and p+1 by vacuum (n=1) and the substrate of refractive index n. In this case, within the bulk material substrate, $B_{p+1}=B_s=0$ in equation (15) because there is no intensity returning from z=∞. The Fresnel reflectivity writes:

$$R_F = \left|\frac{B}{A}\right|^2 = \left|\frac{q - q_s}{q + q_s}\right|^2 \quad (19)$$

or:

$$R_F = \left| \frac{1-\left(1-\left(\frac{q_c}{q}\right)^2\right)^{1/2}}{1+\left(1-\left(\frac{q_c}{q}\right)^2\right)^{1/2}} \right|^2 \quad (20)$$

where q_c and q_s are obtained from:

$$q_s^2 = 4\pi N b_s \quad (21)$$

$$q_s^2 = q^2 - 4\pi N b_p = q^2 - q_c^2 \quad (21')$$

q_c is the critical wavevector that separates total reflection (below q_c) from partial reflection where a part of the beam is refracted (above q_c). The measurement of q_c enables the measurement of Nb_s (this is another way of writing equation (13) in section 2.1).

Please note that, **far from the critical angle ($q>>q_c$), $R_F \propto(1/q)^4$**. This means that the decay of the reflectivity is always very steep whatever the layer at the surface (see the representation of figure 2 in linear scale). The NR curves R=f(q) are thus traditionally represented in linear-logarithmic scales or logarithmic-logarithmic scales, as in the inset of Figure 2. Another convenient way of representation is the *Fresnel representation* $R(q)q^4 = f(q)$ for which the q^4 term compensates the intrinsic q^{-4} decay of the NR curve to highlight the features coming from the layers at the interface (see the experimental data in the part 4 devoted to the examples). A last possible useful representation is to show $R(q)/R_F(q) = f(q)$.

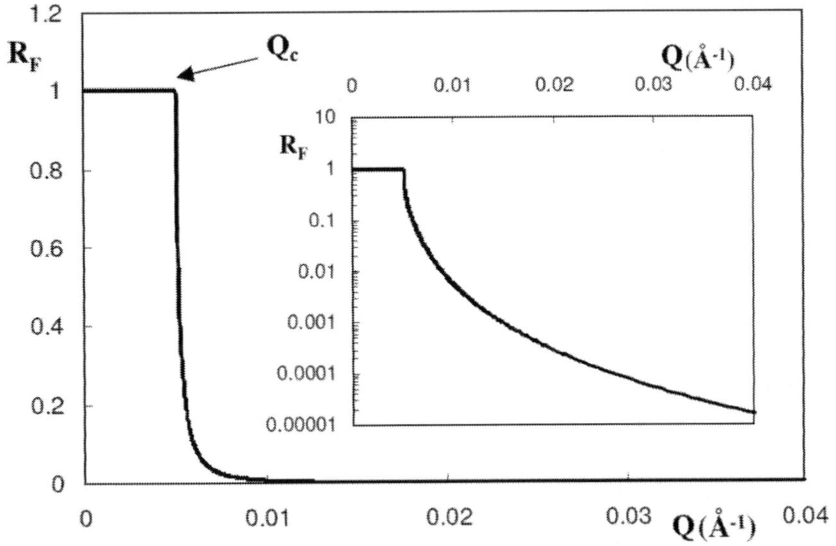

Figure 2 : Linear-linear representation and linear- logarithmic representation (inset) of the Fresnel reflectivity as function of Q for $Nb_s = 2.07\ 10^{-6}$Å^{-2} (interface air/silicon).

2.4 The case of an homogeneous layer on a substrate

Let us now consider that there is a single homogeneous layer of thickness *d* at the surface of the perfectly flat substrate whose thickness is considered as infinite (Figure 3). The three media are respectively the air (n=1), the layer of refractive index n_l and scattering length density Nb_l and the substrate of index n_s and scattering length density Nb_s. The air/layer interface is set at z=0 and the layer/substrate interface is set at z=d.

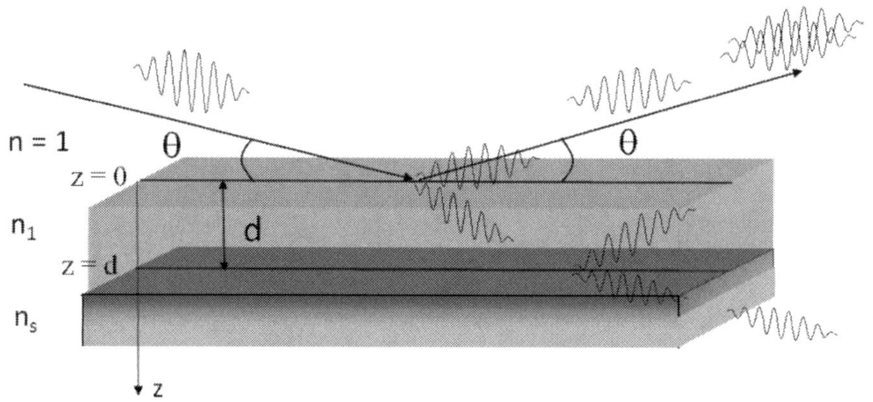

Figure 3. Homogeneous layer on a substrate.

On such a system a part of the incident wave will be reflected on the first interface at z = 0 and a part will be transmitted for $q>q_c$. Also a part of this last transmitted wave will be reflected on the second interface at z = d. The two reflected waves will interfere constructively or destructively depending on their phase which is directly linked to their difference of optical path and thus to the thickness of the layer. **It therefore appears qualitatively that the reflectivity curve will exhibit interference fringes that enable the measurement of the layer**.

After a few calculations the reflectivity writes:

$$R = \frac{\cos(2q_1 d)\left[1+\left(\frac{q_s}{q}\right)^2-\left(\frac{q_1}{q}\right)^2-\left(\frac{q_s}{q_1}\right)^2\right]+1-4\frac{q_s}{q}+\left(\frac{q_s}{q}\right)^2+\left(\frac{q_1}{q}\right)^2+\left(\frac{q_s}{q_1}\right)^2}{\cos(2q_1 d)\left[1+\left(\frac{q_s}{q}\right)^2-\left(\frac{q_1}{q}\right)^2-\left(\frac{q_s}{q_1}\right)^2\right]+1+4\frac{q_s}{q}+\left(\frac{q_s}{q}\right)^2+\left(\frac{q_1}{q}\right)^2+\left(\frac{q_s}{q_1}\right)^2} \tag{22}$$

It appears on such an equation that the curve of R as function of q will exhibit oscillations whose frequency is given by $2q_1d=m \times 2\pi$ (m is an integer). This is the Bragg relationship corrected for the critical angle:

$$2d\sqrt{\sin\theta_m{}^2 - \sin\theta_c{}^2} = m\lambda \tag{23}$$

These oscillations are called the Kiessig fringes. The inter-fringes distance is linked to the thickness of the layer and their amplitude come from the difference of scattering length densities existing between the monolayer and both air and substrate. Figure 4 presents the characteristic NR curve of a system made of a single homogeneous layer on a substrate.

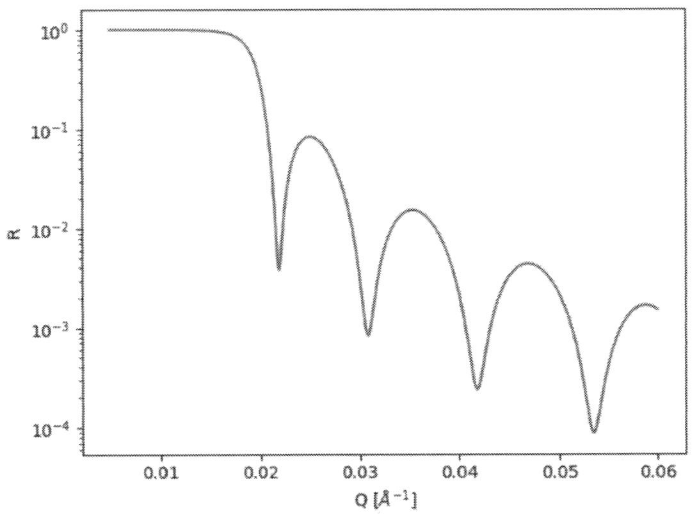

Figure 4. Reflectivity R calculated for a layer of deuterated PS (Nb=6.5 10^{-6}Å^{-2}) of thickness 500Å on silicon (Nb_s=2.07 10^{-6}Å^{-2}). The calculation has been done by taking into account an experimental resolution, assuming a typical resolution of neutron reflectometer (see part 3.1).

2.5 Roughness and interdiffusion

Up to now, we have considered that the interface between the layer p and the layer p+1 was perfect without any heterogeneity. There was thus a discontinuity in Nb(z) when passing from Nb_p to Nb_{p+1} (step function). In practice, the interfaces are never so abrupt because of two different physical phenomena: *(i)* the **roughness** and/or *(ii)* **the interdiffusion** (see Figure 5). The roughness accounts from the fact that the passage from one layer to the next is not always at the same distance from the surface z from one area of the surface to another. The interdiffusion accounts from the fact that two materials making two successive layers often mutually diffuse slightly one to each other, which creates a smooth transition.

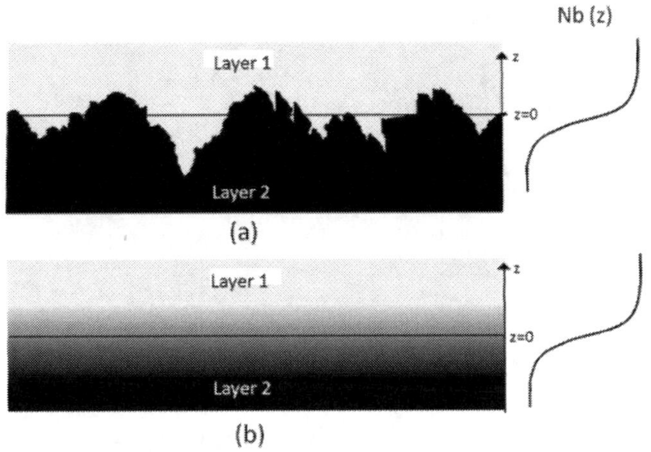

Figure 5. (a) Schematic representation of roughness; (b) schematic representation of interdiffusion. The profiles represented on the right show the effective Nb(z) profile averaged over the whole surface.

The lone measurement of the specular reflectivity does not allow to distinguish between roughness and interdiffusion. If the size of the defects is smaller than the spatial coherence length of the neutrons (of the order of a few micrometers), off-specular measurements allow to discriminate between the two cases. Roughness is indeed equivalent to have reflections by numerous surfaces with different angular

orientations which make some neutrons being reflected out of the specular plane. This does not happen in case of interdiffusion.
Assuming that both kinds of interfaces have a Gaussian profile, they will be simulated by the same error function:

$$erf\left(\frac{z-z_{p/\,p+1}}{\sigma_{p/\,p+1}}\right) = \frac{2}{\sqrt{\pi}}\int_0^{(z-z_{p/p+1})/\sigma_{p/p+1}} e^{-t^2}\,dt \tag{24}$$

where p/p+1 is the interface between two successive layers. The curve that shows such an error function has an inflection point in $z_{p/p+1}$. $\sigma_{p/p+1}$ is the inverse of the tangent's slope of the curve at $z_{p/p+1}$. The thickness of the interface is given by $2\sigma_{p/p+1}$.
It is possible to show that the introduction of such roughness is equivalent to multiply the reflectivity R obtained in case of a perfect flat interface between p and p+1 (equation 19) by a Debye-Waller factor that writes:

$$DW = \exp(-4q_p q_{p+1}\sigma^2_{p/p+1}) \tag{25}$$

where q_p and q_{p+1} are given by equation (16). This approach is only valid when the roughness is smaller than the thicknesses of the adjacent layers and the lateral fluctuations of the roughness are smaller than the coherence length of the beam.

Another method to take into account roughness is to discretize the density profile in a large number of layers having a low variation of the *Nb* from one layer to another. However, the calculation time may become very long.
Figure 6 shows the effect of roughness/interdiffusion on the NR curve calculated in case of the single layer reflectivity of deuterated PS of thickness 500Å on silicon substrate already presented in Figure 4.

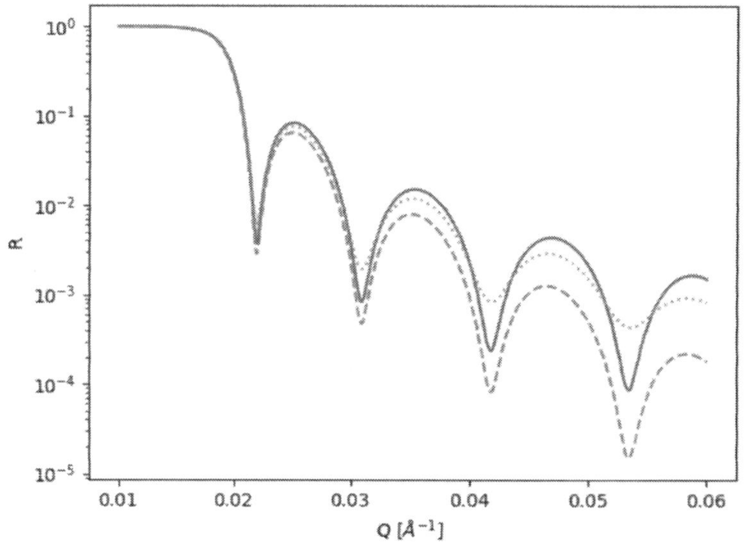

Figure 6. Influence of the roughness on the NR curve calculated on a 500Å deuterated PS deposited on a silicon substrate. Thick line: $\sigma_s = \sigma_{air} = 0$Å; dashed line: $\sigma_s > 0$; Dotted line: $\sigma_{air} > 0$. The calculation has been done by taking into account an experimental resolution, assuming a typical resolution of neutron reflectometer (see part 3.1).

2.6 Data modeling

2.6.1 Calculation of a reflectivity curve

As mentioned in paragraphs 2.2 to 2.4, **it is possible to numerically calculate recursively the reflectivity of a system with n layers if the thickness and the scattering length density of each layer is known, taking into account the roughness at each interface between two adjacent layers** by the optical matrix method. The data modeling is then achieved by fitting numerically the calculated NR curve to the experimental curve. The parameters of the fits are the respective thicknesses and scattering length densities of the layers of the multilayers system probed, as well as the roughness of the interfaces. The experimental resolution has also to be taken into account in the calculation (see paragraph 3.2). When the scattering length density profile perpendicular to the surface Nb(z) is continuous (as in the case of the adsorption profile of a polymer in solvent at an interface for example), such profile can simply be discretized. The scattering length density of a layer that contains several types of objects is the average of the scattering length density of the objects weighted by their volume fraction within the layer. It is assumed in this case that the average on the coherence area of the beam is representative of the average of the SLD on the whole surface. For instance, in the case of a two-phase layer made of polymers in a solvent, the *N*b of the layer is $Nb(\text{layer}) = \Phi_{poly} Nb_{poly} + (1 - \Phi_{poly}) Nb_{solvent}$.

2.6.2 Data fitting

Although the calculation of a NR curve from a known profile is easy, the fitting of NR is in fact tricky. This stems from the fact that the NR curves cannot be inverted in direct space because only the intensity is measured and not the phase. **Therefore different profiles in direct space may give the same NR curve**. Besides, owing to the rather poor fluxes provided by the neutron sources, the error bars are often rather large in an experiment, which makes the distinction between two close profiles sometimes impossible. A blind treatment of the raw data by a fitting program that will minimize a χ^2 between calculated and experimental NR curve starting from a random profile is thus very risky, as it can converges towards a non-physical profile. It is thus important to start with a reasonable physical profile. In particular, the SLD of the different components of the systems have to be evaluated prior to fitting when it is possible in order to constrain the parameters of the fit, or at least to mark them off. Then, it is mandatory to check that the obtained profiles are consistent from a physical point of view. For instance, when studying a responsive thin polymeric layer as function of a stimulus, mass conservation has to be verified (see for instance the example presented in part 4.3). Ideally, results have to be confronted with complementary results obtained with other techniques (isotherms, QCM, etc..). This sometimes enables to discriminate between several possible profiles.

There exists also a very popular strategy to obtain unambiguous results based on a specificity of NR from the contrast point of view. It consists in achieving the experiment by using several different neutron contrasts for a given physical profile. For instance, when probing a thin solvated organic layer at the air/liquid or solid/liquid interfaces, the SLD of the solvent can be easily changed by total or partial deuteration. At the present times, most studies performed at the air/water interface involve at least 3 different contrasts conditions for the aqueous subphases (H_2O, D_2O and a mixture of both). Obtaining of a unique profile that fits all experiments simultaneously in the different contrasts conditions enables then to assess the real profile with confidence [1]. This strategy have some minor drawbacks, namely the necessity to make reproducible samples in various contrast conditions as well as the possible change of physical properties of the system upon deuteration, that generally only occurs when H-bonding plays an important role.

There exists also another experimental strategy to refine modelling based on the insertion of a magnetic layer below the layer to be studied and to perform polarized NR. The advantage here is to get two contrasts conditions on a unique sample but the price to pay is the addition of a layer that make the system more complex [10], that can however be fully precisely characterized by X-Ray reflectivity prior to neutrons, and the use of polarized neutrons that decrease the flux. Although very elegant, this strategy has only be very sparsely used in literature.

3 Experimental aspects of neutron reflectivity measurements

3.1 Instruments

3.1.1 The Different kinds of instruments

There exists two ways to measure the reflectivity of a sample as a function of the scattering vector Q = 4πsinθ/λ (see Figure 7).

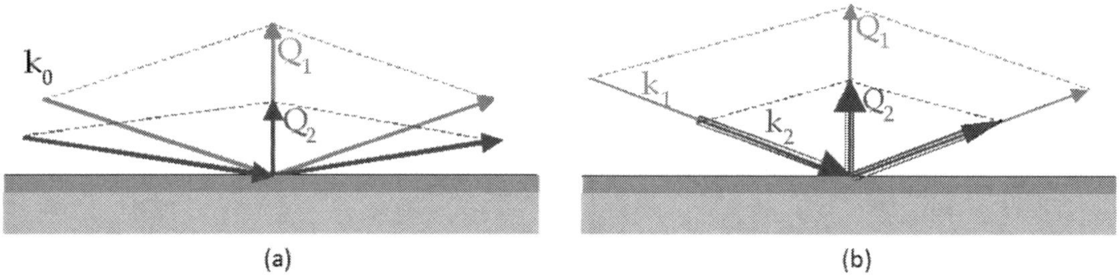

Figure 7. Principle of neutron reflectivity: (a) θ, 2θ: measurement at variable θ and fixed λ; (b) time of flight: measurement at fixed θ and variable λ.

The most simple **(θ, 2θ)** is similar to a 2-axis diffractometer and uses a fixed wavelength λ, that typically lies between 1.5Å and 6Å. It is perfectly suited to reactors that provide a continuous neutron flux. The neutron beam, polychromatic at the reactor source, is made monochromatic upstream of the sample by a monocrystal, collimated, and sent on the sample in grazing incidence. One measures then the reflected beam, step by step, by rotating the sample to an angle θ and the detector of 2θ, similarly as in a diffraction 2-axis experiment. The illumination of the sample varies from one angle to another and has to be taken into account in the data reduction. Such a geometry is also used for measurements with X-rays.

The relatively low neutron velocities allow to use another method of measurement called **time of flight** to measure NR curves, enabling to work at fixed θ and to measure as a function of λ. This method is especially suited to pulsed sources (spallation sources) and is so convenient for NR that it is very often used with reactor sources, too. The neutron moves at a finite speed, linked to its wavelength *via* the De Broglie relation $v = h/m\lambda$ where h is Planck constant and m the neutron mass. For the cold neutrons used in the experiments (with a typical wavelength that ranges between 2 and 20 Å), the speed of neutrons lies between a few 100 m/s and a few 1000 m/s. This permits to determine the wavelength of a given neutron within a polychromatic neutron beam by separating the neutrons of different wavelengths of a neutron pulse through their speed, the fastest neutrons moving faster than the slower ones. The pulse, which is spontaneously created on a spallation source, is created by a chopper on a reactor source. The neutrons beam is sent in grazing incidence on the sample and the chopper defines the origin of times $t_0 = 0$ (the "starting line"). One measures then the time t taken by the neutron to reach the detector located at the distance L from the chopper to determine the wavelength (via $\lambda = h/mL*t$)[‡]. The rotation of the chopper is chosen so that the slowest neutrons of a pulse arrive sooner on the detector than the fastest neutrons of the next pulse in order to avoid the overlap of the pulses. In some setups, an other chopper is placed upstream of the beam to avoid this "frame overlap", which makes it possible to

[‡] One recalls that the value of the Planck constant h is $6.62\ 10^{-34}$ J.s and that the value of the neutron mass is $1.67\ 10^{-27}$ kg. The expression can then be rewritten like λ(Å) = t(μs)/(252.7*L(m)). The travel time is then around 1ms/Å. With a flight path length of a few meters, it is possible to discriminate neutrons with a very good resolution with conventional electronics. For instance, on EROS spectrometer at LLB, it was possible to distinguish between neutrons whose wavelength varied from 0.2 Å by counting neutrons by slices of 200 μs with electronics with the flight path of 4.4m.

increase the speed of rotation of the first chopper and in turn to increase the flux on the sample for the longest wavelength. Measuring the coefficient of reflection is simple: The white incident polychromatic beam has to be first determined by measuring the incident beam directly on the detector without any reflection. The intensity measured for the reflected beam has then to be divided by this incident white beam for every value of λ.
The time-of-flight method is extremely convenient for some geometries of measurement, in particular those at the air/liquid, because it does not force to rotate the sample (and to make it flow…). Moreover, it enables to measure all the points of a reflectivity curve at the same time, or at least one decade in q, which removes any incertitude on the possible aging of the sample during measurement, a feature that often occurs in soft matter. Along the same lines, this method allows **kinetics measurements** that have become very popular these last years with the development of high-fluxes reflectometers such as Figaro at ILL that enable to measure a given NR spectrum in a few minutes [11].
In an NR experiment, the incident angle on the surface to study is always very small, ranging typically from 0.5 to 5°. Given the typical wavelengths of the incident neutrons (~ 3Å-25Å), the range of scattering vectors lies between 0.003 $Å^{-1}$ and 0.3 $Å^{-1}$. The typical thicknesses that can be measured in standard conditions vary then from around 10 Å to 2000 Å.

3.1.2 An example of Time-of-Flight Spectrometer: EROS at LLB

A full description of this horizontal spectrometer can be found in reference [12]. It is shown on Figure 8. Briefly, the neutrons produced by the Orphée reactor (CEA Saclay) are thermalized by a liquid hydrogen cold source (T=20K). A bended neutron guide conveys them up to the spectrometer located at around 30m from the source. A "white" beam containing a broad spectra of wavelengths is then available at the chopper position (in practice from 3 to 25Å). The chopper with a raw $\delta\lambda/\lambda$ constant resolution is made of two-rotating wheels covered by gadolinium, a neutron absorber, pierced by two large slits. The distance and phase between the two wheels enable to change the $\delta\lambda/\lambda$ from 1% to 13% depending on the resolution request. The beam passes then through a collimator under vacuum that is 1.8m long. The aperture of the horizontal entrance and exit slits can be opened from 0.2 mm up to several cm (in standard conditions, one uses apertures lying from 1mm to 3mm). Two neutron supermirrors can be inserted within the collimator to deviate the beam for air/liquid measurements. The beam is then reflected by the sample tilted of an angle θ and measured by a ^{3}He single detector.
The maximum angle that can be reached is 5°, which enables to cover scattering vectors ranging from 0.003 to 0.3$Å^{-1}$ (measurements below 0.3° are practically almost impossible because the reflected beam has to be separated from the incident non-deviated direct beam). A measurement at one angle allows to cover around one decade in q. Sometimes this is sufficient for the desired measurement, however, in most cases 2 or 3 different angles are used to cover a broader q-range. The actual minimal reflectivity in air/solid geometry that can be obtained is typically $R \sim 10^{-7}$. For measurements involving liquids, this limit is superior to 10^{-6} because of incoherent scattering (see reference [12] for more details). The vertical position has to be adjusted as a function of the chosen angle. If h_0 are h_D are the respective position of the direct beam and of the detector at the distance L from the chopper, the angle of a measurement θ is given by: $\tan 2\theta = \frac{h_D - h_0}{D_{S\_D}}$ where D_{S_D} is the distance from the sample to the detector (figure 8).
The principle of alignment is simple. The angle of work has to be chosen first such as a part of the neutron beam is in total reflection (if the neutron contrast conditions enable it) thanks to the formula (13). The height position is then fixed. The sample has then to be placed within the beam and rotated with the goniometer up to recording the maximum of neutrons on the detector. This is exactly similar to what children do when they rotate their watch to reflect the light (the neutrons) in the eye of someone (the detector)… For air/liquids measurements, the angle of work is fixed by the deviating supermirrors within the collimator (see Figure 8.b).

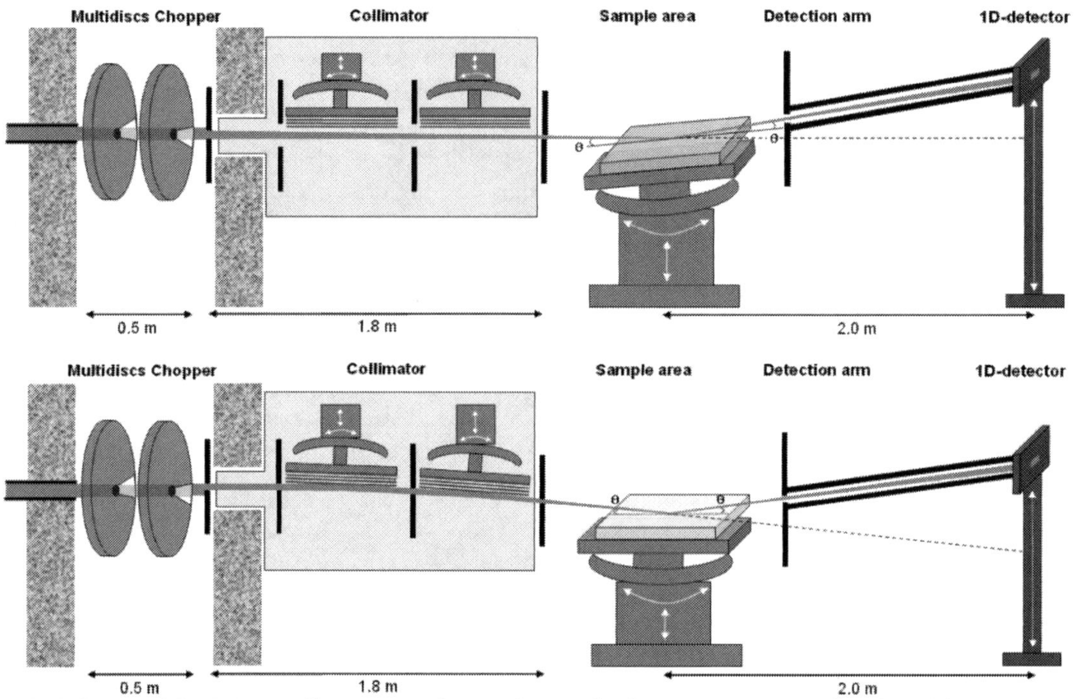

Figure 8. Scheme of the EROS reflectometer, from reference [12]. Top: Geometry for air/solid measurements; Bottom: Geometry for air/liquid measurements.

The typical duration of the acquisition time depends on the minimal reflectivity that is requested. If it is of the order of 10^{-5}, 30 min to 1h are necessary to obtain a correct statistics but this time has to be extended up to ~ 4 hours to reach a few 10^{-6}. These typical measurement times are given for sample of around 10 cm^2. Such large surfaces can be indeed used because the collimations used are large (see below).

3.1.3 Angular resolution

Owing to the collimation, the incident beam on the sample has a divergence $\delta\theta$ that induced an enlargement of the beam and that has to be taken into account in the calculation, the expressions presented in the precedent part for the reflectivity being calculated for an incident beam without divergence. This is achieved by the convolution of the calculated function by an apparatus function $f(\theta,\delta\theta_0)$ centered in 0. This function can be either a triangle function or a square function.

$$R_{\delta\theta/\theta}(q) = \int R\left(q\frac{\sin\theta_0}{\sin\theta}\right) f(\theta_0-\theta,\delta\theta)d\theta \qquad (26)$$

The divergence of the incident beam $\delta\theta$ and the angular resolution $\delta\theta/\theta$ are set by the apertures of the entrance and exit slits of the collimator and by the collimator length D_c. If the earth gravity is neglected, the trajectory of the beam is supposed straight and the divergence by: $\tan\delta\theta = (S_{ent}+S_{exit})/2D_c$. The calculation of the resolution for an angle of 1°, for slits $S_{ent} = 2$ mm and S_{exit} 1mm and for D_c=1800 mm give : $\delta\theta$= 0.047° and $\delta\theta/\theta$ = 4.7% (characteristic values used on EROS).

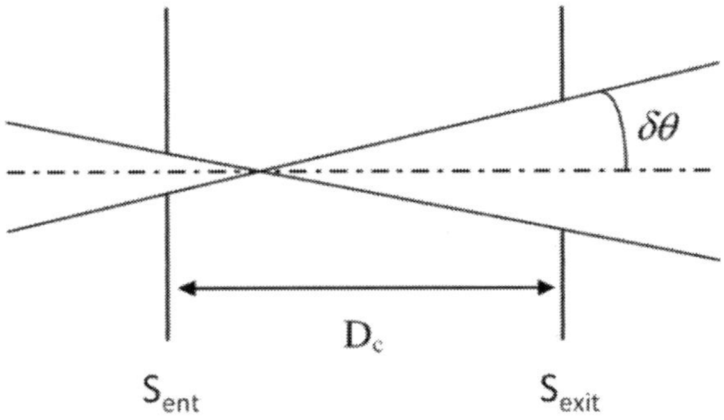

Figure 9. Divergence of incident beam when the effect of the earth gravity on the neutrons path is neglected.

3.2 Samples and possible geometries of measurements

The geometry of a neutron reflectivity experiment imposes three imperative conditions that have to be absolutely fulfilled to make it feasible. First, the media in which the neutron beam is travelling upstream the interface and after reflection has to be totally or partially transparent to neutrons, otherwise signal is lost. Second, **the flatness of the interface to be probed has to be as best as possible**, otherwise if it has a curvature this would have an effect equivalent to an increase of the divergence $\delta\theta$ of the beam. Although this divergence could be taken into account by an increase of the effective $\delta\theta$ within the resolution function during the data modeling, it has to be as limited as possible. This imposes some limitations on the different possible geometries of measurements and on the samples themselves. Third, the roughness of the surface has to be very limited, otherwise it would induce a very steep drop of reflectivity that leads to a loss of information on the scattering length density profile.

3.2.1 The different possible geometries of measurements

In soft matter, four types of interfaces are interesting to be studied: **air/solid** (spin-coated monolayers, polymeric nanostructures, etc..), **air/liquid** (surfactants monolayers, adsorption behavior of hydrophobic compounds, etc..), **solid/liquid** (polymeric brushes in solvent, interactions of biological species with membranes deposited on solid substrate, etc…) and **liquid/liquid** interfaces (model systems mimicking oil/water interfaces found in emulsions, etc..).
Air is almost transparent to neutrons. Indeed, the attenuation of a neutron beam for the wavelengths used is of the order of a few percent per meter. The length of the sample stage being of the order of a few tens of centimeters on a NR spectrometer, this attenuation is negligible when working in air/solid or air/liquid geometries. These two geometries are thus very easy to implement (figure 10).
Conversely, the transparency of media in which the neutron beam is travelling is a key point to consider for measurements in solid/liquid or liquid/liquid geometries (considered here as either water or an organic solvent as found in typical *Soft Matter* systems). In practice, two processes attenuate the beam: absorption and incoherent scattering. As mentioned earlier, only a very few atoms of the periodic table [7] have an important cross-section of absorption (^{3}He, ^{6}Li, B, Gd…), therefore absorption usually never hampers the design of an experiment. It has thus to be checked prior to the experiment that such atoms are not present in large amounts in the sample. On the contrary, incoherent scattering is a real problem to deal with when designing experiments. Incoherent scattering is a scattering that occurs with a constant probability over 4π independently from the position of scatterers (see introduction courses on neutron scattering for more information). For a given atom, the incoherent scattering cross-section comes from the different isotopes of the element and its different spin states. It is null or very weak for most elements of the periodic table, but unfortunately, rather important for ^{2}H deuterium and very high for hydrogen

^{1}H[§]. The incoherent scattering of a liquid is thus very important, which in turn decreases the transmission of the beam. For example, the transmission by a neutron beam at $\lambda = 6$ Å is only 0.5 through 1 mm of H_2O and 0.8 through 2 mm of D_2O. As the neutron beam has to travel on several centimeters in a NR experiment within the sample because angles of incidence are grazing (less than 5°), crossing a liquid is almost impossible. For measurement in solid/liquid geometries, the poor transmission of liquid is easy to circumvent. The neutron beam has simply to pass through the solid as shown in figure 10.c, provided that the solid chosen is transparent to neutrons, which is easy to achieve with silicon wafers (see next section) or quartz crystals. Such geometry of measurement is thus very wide spread in experiments related to Soft Matter.

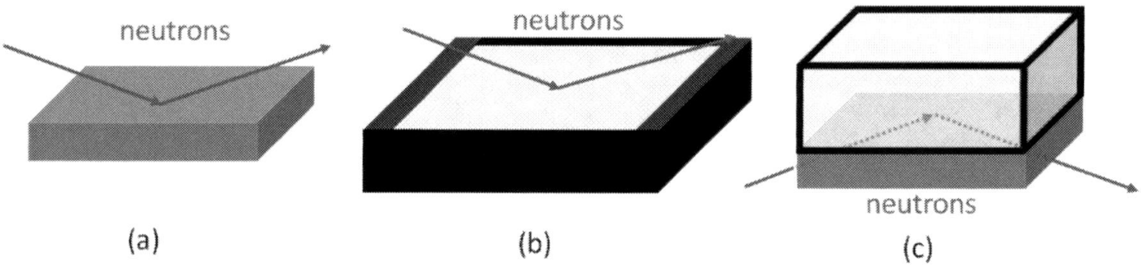

Figure 10. Standard geometries of measurements in Neutron Reflectivity: (a) air/solid interface; (b) air/liquid interface; (c) solid/liquid interface.

For **liquid/liquid geometries**, there are not good solutions and measurements are very sparse. Historically, the first strategy used to overcome the problem was to work in *pseudo* air/liquid geometry by using a ultrathin oil layer of well controlled thickness of ~ 3000 Å on water surface to probe a surfactant layer sandwiched between oil and water [13]. The limited loss of neutron through the oil made the experiment possible. Very nicely, the use of an hydrogenated surfactant and deuterated oil and water of same SLD enabled the determination of both the thickness and the roughness of a surfactant layer at the interface.

Later, the same kind of strategy was adapted at the solid/liquid interface. A thin uniform oil layer was trapped between a solid substrate and a bulk aqueous subphase. The thin oil layer was first spread onto the surface by a spin-coating deposit, then frozen below its melting temperature, sandwiched between the solid and the water, and maintained after thawing the oil phase [14].

Finally, true liquid/liquid measurements are now emerging owing due to gain performances of the present reflectometers compared to those operating thirty years ago. For instance, in reference [15], water-toluene interface was probed in a cell specifically designed for the purpose. The beam crossed a sapphire before travelling through several hundreds of microns of water before hitting the toluene interface. The SLD of water was adjusted to those of sapphire, so the first sapphire/water interface that was reached by neutrons was transparent to them. Such measurements remain however very tricky, in particular obtaining a flat interface over a large surface between two liquids is very difficult. In the above mentioned cited example, the cell was designed so that the lower part was hydrophilic and the upper part was hydrophobic, so as to force the interface between the two liquids.

[§] We regret here that ^{1}H has a very high incoherent scattering cross-section (80 barns) for what concerns neutron reflectivity. It has however to be noted that it is a true advantage in other neutron techniques such as quasi-elastic and inelastic scattering (see for example the course of Berrod *et al* in the current volume for more information [34]).

3.2.2 Samples

Besides the required qualities of transparency towards to neutrons depicted previously, it is mandatory that samples fulfill other conditions.
First, the surface of the sample has to be optimized. In practice, the raw measured reflectivity is directly proportional to the surface of the sample illuminated by neutrons, prior to the data reduction step where it is normalized to unity for total reflection. If possible, the surface of the sample has to be maximized in order to get a good signal-to-noise ratio with reasonable acquisition times. Ideally, the best is to have an overall surface of sample that is higher than the area illuminated by neutrons during experiment (some cm^2 depending on the collimation and samples). The surfaces can be decreased at cost of increasing counting rate but one must avoid to decrease surfaces area down to 1 cm^2.
Moreover, when a sample is studied at the air/liquid interface, it is necessary that the surface of the sample is sufficiently large so that the beam illuminates a perfectly flat area of the surface, thus located away enough from the edges (meniscus). This might be a problem if one studies the adsorption of an expensive molecule at such an interface because a minimal quantity is required.
Second, for solid samples, the surface curvature has also to be considered. It is indeed mandatory that such samples are rigid enough not to deform mechanically under internal stresses and have a very small *rms* roughness. Most common solid substrates used in neutron reflectometry are sapphire, quartz crystals and silicon wafers. These latter are especially suited to design model substrates for NR and are used in ~90% of experiments on solid substrates in soft matter, both at air/solid interface and solid/liquid interface because they have several properties that make them unique for NR: *(i)* they are nearly transparent to neutrons because their incoherent scattering cross-section is null and their absorption cross-section is very weak (7 cm of silicon are required to attenuate a neutron beam at λ = 6Å by a factor 2 !); *(ii)* their lack of incoherent scattering do not produce any additional background scattering; *(iii)* they have an excellent mosaicity with an *rms* roughness of a few Å; *(iv)* their chemical surface (Si-OH groups) lends itself to many chemical modifications which are easy to achieve at laboratory without any specific material in order to design hydrophilic surfaces, hydrophobic surfaces, charged surfaces, functionally-terminated surfaces...; *(v)* last but not least, they are rather cheap and easy to purchase at silicon manufacturers owing to their importance in electronic industry.
Third, one must also ensures that samples do not contain atoms that strongly absorb neutrons (Li, B, Gd…) or have an important incoherent scattering cross-section (V...) to avoid the decrease of flux in general and the radiological activation of samples in case of absorption. This is especially important for the solid incoming media that has to be crossed in solid/liquid geometries.

3.2.3 Sample environment

Thanks to the poor absorption of neutrons by matter and in particular by some metallic elements, it is easy to design various samples environments for *in situ* experiments: liquids cells, cells with solvent circulation baths, Langmuir trough, ovens, pressure cells, temperature controllers, etc.. It is advised to contact people from facilities prior to experiments to check that the equipment can be used with the kind of sample to study or to design a dedicated sample environment for a specific experiment.

3.3 Access to the existing reflectometers

As NR is a growing expanding technique, there are reflectometers in all of the large neutron facilities. A state of the art of new offers on reflectometry has been described in a recent special issue dedicated on neutron instrumentation that focused on the reflectometers recently released or upgraded, as well as the new techniques of instrumentations in NR [16]. In France, there exist two neutron facilities, the European source *Institut Laue Langevin* (ILL) located in Grenoble [17] and the French national source *Laboratoire Léon Brillouin* (LLB), located in Saclay 30 km from Paris [18]. There are actually 3 reflectometers at ILL (FIGARO dedicated to soft matter and biophysics, D17 for multipurpose measurements and SUPERADAM for magnetic studies) and 2 at LLB (HERMES, a horizontal versatile multipurpose instrument, and PRISM for magnetic studies). In the near future, the new *European*

Spallation Source will be operating. Two reflectometers are planned: KESTIA, a vertical polarized reflectometer that should be opened to users in mid-2023 and FREIA, a high flux horizontal reflectometer very suited to soft matter that is forecast for 2025 [19]. All of these instruments are available free of charge to the whole scientific community. The beamtime is allocated by a committee of experts on the basis of experimental proposals that can generally be submitted twice a year. Please visit the websites of the facilities for all information on formalities. Fast access or paying beamtime for industrial applications for which data have to remain confidential are also possible. The beamtime must then be negotiated directly with the facilities.

4 Examples of use contrast variation in neutron reflectivity for soft matter

Over the last 30 years, the unique properties of neutron reflectivity revealed to be an invaluable and an indispensable tool for the study of *Soft Matter*, in particular for the study of surfactants, polymers, adsorption of binary mixtures and membranes. This abundant literature has given rise to numerous reviews since the famous ones by Penfold and Thomas on surfactants [1] and those by Russel on polymers in the early 90's [2], for instance those on solid/liquid measurements [4] or on supported membranes [3][20]. We thus do not aim here to make a review on the topic, and we refer back to the existing ones. Among the very recent ones, those by Naranayan *et al* is very comprehensive [21] and those by Braun *et al* is very complete on possible experiments at the air/water interface [22]. In the following, we present some examples that depict the strategies based on contrast variation that can be used in neutron reflectivity, in order to guide the reader to design his/her own future experiments. We have selected some examples on polymers but they are generally transposable to other fields of *Soft Matter*.

4.1 Nanostructured polymer thin films

There is nowadays a strong interest in polymer science in developing nanostructured thin films from the layer-by-layer (LBL) technique. This technique consists in depositing alternatively and cyclically two (or more) thin layers of polymers or nanoparticles that interact one with each other to design nanomaterials with a high output value. The initial coating processes involved dip coating and has been extended to spray. For all these processes, the interaction between adjacent layers (electrostatic, hydrogen bonding.) play a significant role. Two important questions have to be answered when designing such multilayered system: Is the final system really nanostructured after synthesis, or is it partially or completely interdiffused? If the system is really nanostructured, does it keep such nanostructure when immersed in a solvent? NR combined with contrast variation is the best technique to answer such questions owing to the possibility of probing buried interface, especially when both components are polymers, as in most cases. Usually the elementary polymer layer is very thin, typically 10-20 Å, so it is not possible to distinguish the two kinds of polymers within the architecture, whether they are both hydrogenated or deuterated, because there exist always partial interdiffusion between adjacent layers. Only the overall thickness of the layer is then measured. On the reverse, if one of the two polymers is deuterated at regular intervals, it is possible to assess if the system is truly nanostructured. There are then 3 typical sizes in the system, which are easy to decouple if the experimental sample has been intelligently conceived: the thickness of the deuterated layer, which enables in particular to estimate the typical range of interdiffusion of a polymer layer with the adjacent one, the distance between two deuterated layers and the overall thickness of the multilayered system. This is illustrated in figure 11 that shows some data obtained on a poly(styrene sulfonate) (PSS)/poly(allyl amine hydrochloride) (PAH) multilayered system obtained by spray coating [23]: the top panel shows data when both PAH and PSS are hydrogenated. Only the Kissieg fringes associated with the overall thickness are visible. Conversely, when some of the PSS layers are deuterated, Bragg peak and its harmonics appear, revealing the true nanostructured nature of the system (lower panel). Please note that Decher has used such an approach to demonstrate that the LBL technique allows to make very well defined nanostructures in his very famous seminal paper on the topic[24], which is maybe the most cited paper ever published on soft matter that contains neutron scattering experiments**.

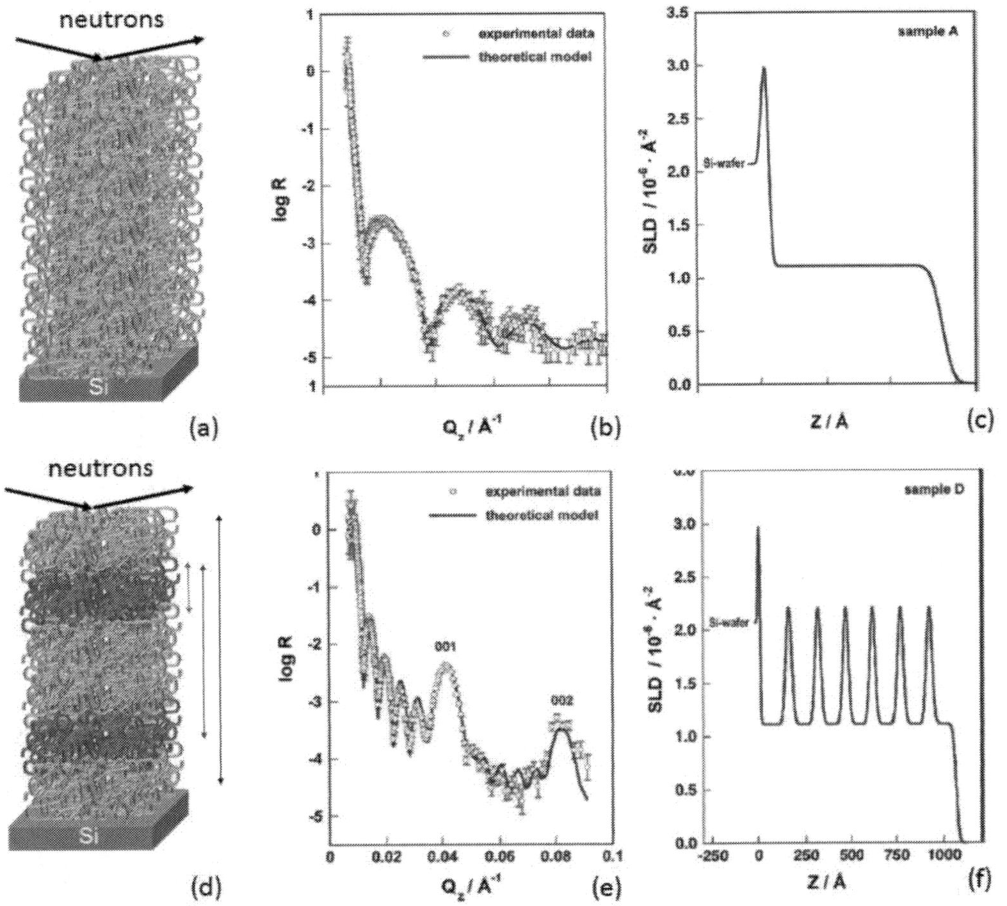

Figure 11. Top panel: structure of a nanostructured PSS/PAH polyelectrolyte multilayers when both polyelectrolyte are hydrogenated. (a) Sketch of the system from neutron contrast point of view; (b) Experimental neutron reflectivity obtained curve and fit with the SLD profile presented in (c) [23]. Lower panel: structure of a nanostructured PSS/PAH polyelectrolyte multilayers when the targeted architecture is: $[(PSS_{H7}/PAH)_5/(PSS_{D7}/PAH)_1]_6/(PSS_{H7}/PAH)_5$. (d) Sketch of the system from neutron contrast point of view; (e) Experimental neutron reflectivity obtained curve and fit with the SLD profile presented in (f)[23].

4.2 Dense brushes: growth of polymerization

Playing with contrast also allows to get an insight on a polymerization process. The idea is to change the contrast during the process so that the system keeps a "memory" of the process from the contrast

---

** At the date 8/2018, there are more than 7500 citations in Isi Web of Knowledge or more than 9500 in Google scholar.

point of view after its synthesis. This is illustrated here on the characterization of a PS brush grown on a silicon wafer through a *grafting from* approach were both the grafting density σ and the polymerization index N were precisely tuned [25]. The grafting density σ was controlled thanks to the deposit onto the surface of a monolayer of initiators which density was adjusted by Langmuir-Blodgett techniques. The polymerization index N was controlled by the NMP controlled radical polymerization technique. Such a technique enabled to obtain brushes with a very large range of densities up to a record of 1.1 chains/nm^2 [26]. NR was then used to demonstrate that all chains truly grow simultaneously in order to prove that the polymerization process was really controlled. To this aim, a brush was grown in two distinct steps: the brush was firstly initiated by deuterated monomers, stopped and then initiated again with hydrogenated monomers. Finally, although the brush was homogeneous from chemical point of view, it is an H/D di-bloc copolymer from the contrast point of view (figure 12.a).

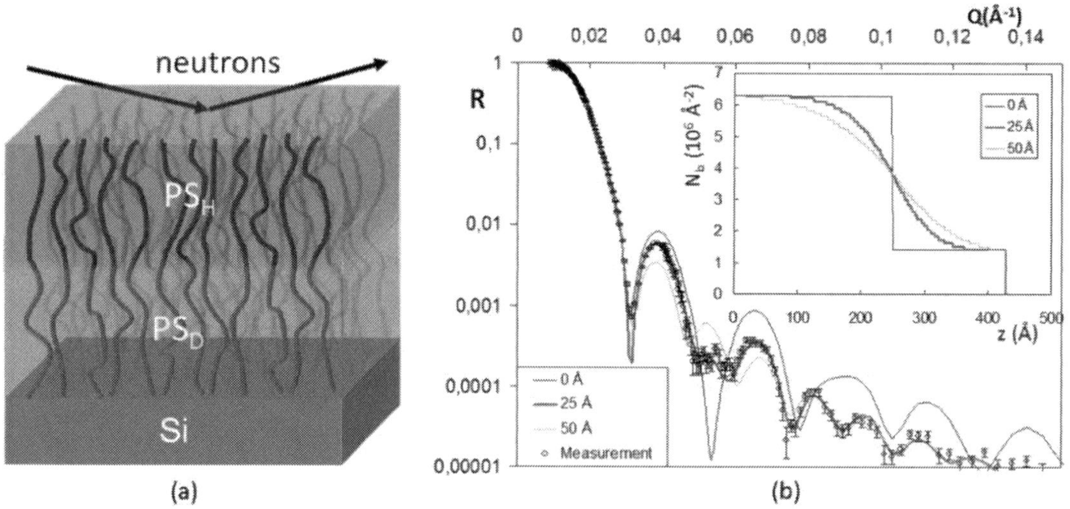

Figure 12. Dense brushes of Polystyrene grown initially by deuterated monomers and re-iniated by hydrogenated monomers [26]. (a) Sketch of the system from neutron contrast point of view at the end of the polymerization process; (b) Comparison between the NR curve obtained experimentally and the simulated curves corresponding to the different SLD profiles presented in inset.

The measurement of the interdiffusion layer between hydrogenated and deuterated parts enabled to probe the efficiency of the control of the polymerization. The NR reflectivity curve shows irregular interferences fringes (figure 12.b). This comes from the fact that there exist three distinct thickness within the system: those of the whole brush (420 Å), that can be measured independently by ellipsometry as the PS layer is homogeneous from optical refractive index point of view, those of the deuterated part (270 Å) and those of the hydrogenated part (150 Å). If there were no interdiffusion in the system, the NR signal would be completely dominated by the deuterated part since the PS_D/air contrast is much larger than the PS_H/air contrast. One would obtain almost regular fringes whose *d*-spacing would allow to recover the thickness of the deuterated part (simulation in blue in figure 12.b). Conversely, in case of a very large interdiffusion, one would only measure the typical reflectivity curve of the overall layer. The fringes would then again been regular but with a much larger *d*-spacing (simulation in green in figure 12.b). Between these two limit cases, a very weak variation of the interdiffusion layer strongly modify the reflectivity curve, especially when the interferences resulting from the deuterated layer and the whole layer are phase opposite, which induced the partial or total extinction of the fringes (see for example around 0.05 Å^{-1}). This enables a very precise determination of the thickness of the interdiffusion layer because some fringes appear or disappear upon variation of the thickness of a few Å. The obtained value from the best fit of the NR curve (25Å) is much smaller than the overall thickness, which proves that the polymerization approach was very efficient.

4.3 Behavior of thin polymer films in solvent

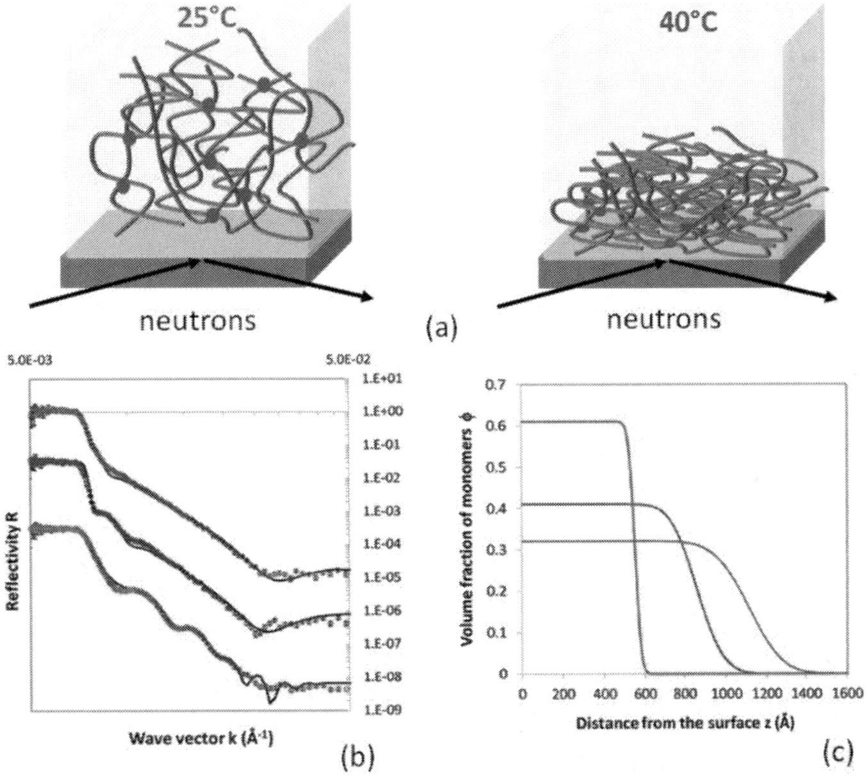

Figure 13: Aqueous behavior of a cross-linked hydrogel based on a thermo-sensible polymer (PNIPAM) tethered on a solid surface [27]. (a) Sketch of the system in two situations, respectively below the LCST (25°C, left) and upon the LCST (40°C, right); (b) NR curves corresponding respectively to 25°C (blue), 34°C (purple), and 40°C (red). The curves are shifted in intensity for clarity. (c) Volume fraction profiles corresponding to the best fits of the curves presented in (b).

One of the major actual trend in *Soft Matter* is the design of so-called *smart* materials, or more precisely stimuli-responsive materials, whose macroscopic properties can be tuned over a large range on demand in a controlled manner triggered either by a small change of environment (temperature, pH) or by an external stimulus (UV-light, magnetic field,…). The idea behind such a strategy is to play finely on the structure and/or interactions of the material at the nanometer scale to induce a large modification of the targeted macroscopic property to be controlled. Such an approach is especially suitable in the case of thin polymer films attached onto an interface in contact with a solution, owing to the possible applications in sensors, proteins repelling, adhesion, etc… The main question to answer when probing such systems is thus to describe the evolution of the conformation of the polymer thin layer as a function of the stimuli. To this aim, NR is then very powerful, simply because it enables to probe polymers at interfaces in solution on solid/liquid geometries with a good contrast, *e.g.* hydrogenated polymers in deuterated solvents. This is illustrated in Figure 13 representing the evolution on a cross-linked hydrogel of nanometric size of a thermo-sensible polymer (PNIPAM) tethered on a solid surface as function of temperature [27]. Such polymer has a Lower Critical Solubility Temperature in water T_{LCST} of 34°C, *i.e.* it collapses above T_{LCST}. Figure 13.b shows the evolution of the system as function of temperature. At low temperature, it shows damped fringes that are progressively more pronounced and shifted towards large q when increasing temperature. This qualitatively demonstrates that the system progressively collapses: the fringes amplitude increase owing to the increase of SLD' difference between the gel and the outside water due to progressive desolvation whereas the thickness reduces.

The volume fraction profiles extracted from the best fits confirm this qualitative approach (Figure 13.c). Below the LCST, the swelling of the gel depends on temperature and decreases when approaching T_{LCST}, whereas the gel is collapsed above T_{LCST}, even if expulsion of water is not complete ($\Phi(z=0)$ remains at 0.6 at 40°C). Please note that the mass of chains at the surface, that is the integral of the profile, is maintained at every concentration. Checking the mass conservation is indeed a good approach to constrain parameters during fitting.

4.4 Insight on specific part of a polymer chain: What is the distribution of chain ends in tethered chains?

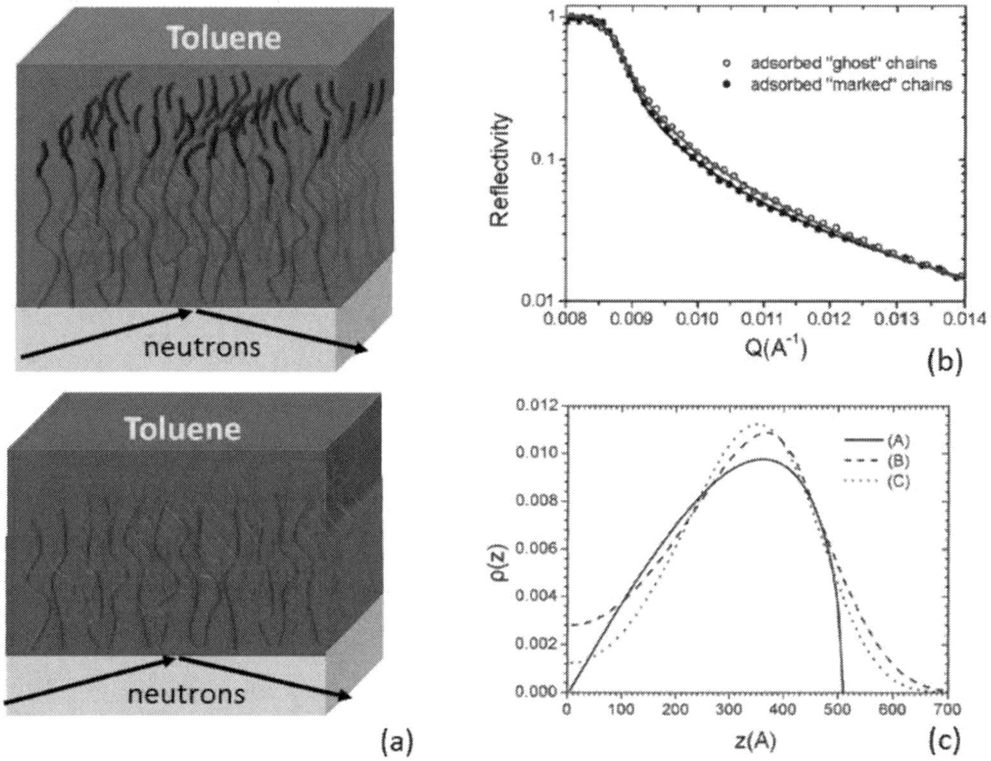

Figure 14: Distribution of chain ends in tethered brush from reference [28]. (a) Sketch of the system when the solvent (here toluene) has the same SLD as the deuterated PS chains in two situations: Measurement when the hydrogenated end chains are "marked" from the contrast points of view (top) and measurement with fully deuterated "ghost" chains. (b) Comparison of the experimental NR curves obtained respectively for ghost chains and marked chains. (c) Different profiles enabling to fit the NR data. Continuous line: Self-Consistent Theory; dashed line : modified analytical SCF theory; dotted line: Schulz function.

Contrast variation is also a powerful tool to probe some specific part of a polymeric chain. It has especially been used to measure the distribution of chain ends in tethered brushes in good solvent [28]. The principle is to achieve the synthesis of chains which are di-bloc polymers from the contrast point of view, namely chains for which the main part of the backbone is deuterated and the end is hydrogenated. A measurement in a solvent that matches the deuterated part of the chains allow then to probe only the hydrogenated chains ends, as shown in Figure 14.b that compares the NR curve measured in such contrast to those measured when the whole chains are deuterated and contrast-matched by solvent. Although the differences between these two NR curves are tiny, the determination of the specific contribution of the chain ends to the reflectivity is clearly achievable. The analysis of the experimental results was done both by simulation and by the Self-Consistent Theory. It supports the picture that free ends are not localized at the brush height, but are distributed throughout the whole brush.

4.5 What is the interdigitation kinetics between a polymer melt and a polymer brush ?

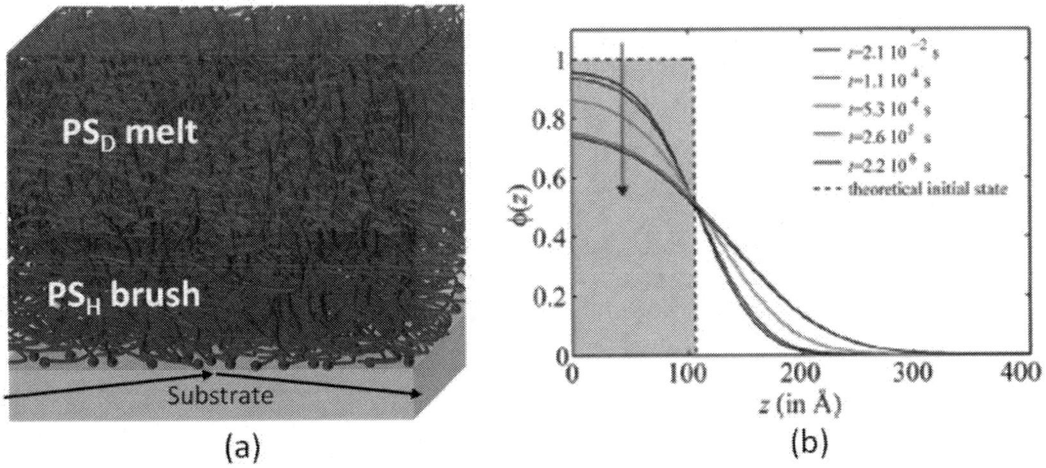

Figure 15: Probing the interdigitation kinetics between a polymer PS melt and a polymer PS brush [29]. (a) Sketch of the system from the contrast point of view; (b) Evolution of the segment density profile of the grafted chains for different annealing times for a given set of N_{brush}, σ and N_{melt}.

One of the case where NR is extremely powerful is for the study of interdiffusion processes between a thin polymer film in contact with a bulk polymer (brush/melt, melt/melt, brush/network, etc…). Indeed, this is the only one technique where it is possible to get some contrast between two melts of polymeric chains of the same chemical composition, provided that one is deuterated (see figure 15.a). The counterpart SANS experiment in bulk, *i.e.* the experimental verification of the Gaussian behavior of a chain within a melt, has been one of the milestone of neutron experiments for polymer science [30]. In NR, obviously only the interdiffusion SLD profile is probed but not the conformation. Since an NR measurement is an averaged static one, several measurements are then necessary to describe the kinetics. The strategy used is the following: The experimental system is let free to evolve at a given temperature above the T_g of the polymer for a given time t_1 outside the beam. It is then frozen below T_g and an NR measurement is done, which enables to capture a snapshot of the system at t_1. It is then possible to anneal the sample above T_g for a duration t_2 and to freeze the sample for a second measurement to capture the snapshot at t_2, and so on up to total equilibration of the system. Such an approach has been used in reference [29] to probe the kinetics of interdigitation of a PS brush in contact with a melt of the same homopolymer as function of the relevant molecular parameters of the system, namely the polymerization index of the chains from the brush N_{brush}, the grafting density of brushes σ and the polymerization index of the chains from the melt N_{melt}. This is illustrated in Figure 15.b that shows the segment density profile of the grafted chains for different annealing times for a given set of molecular parameters. It is then possible to extract a characteristic experimental time τ_{exp} from such series of profiles. The respective influences of N_{brush}, σ and N_{melt} on τ_{exp} can thus be finally probed in order to get a complete description of the interdigitation kinetics at the molecular level [29].

4.6 Ternary systems: behavior of a grafted brush in solution in presence of free chains

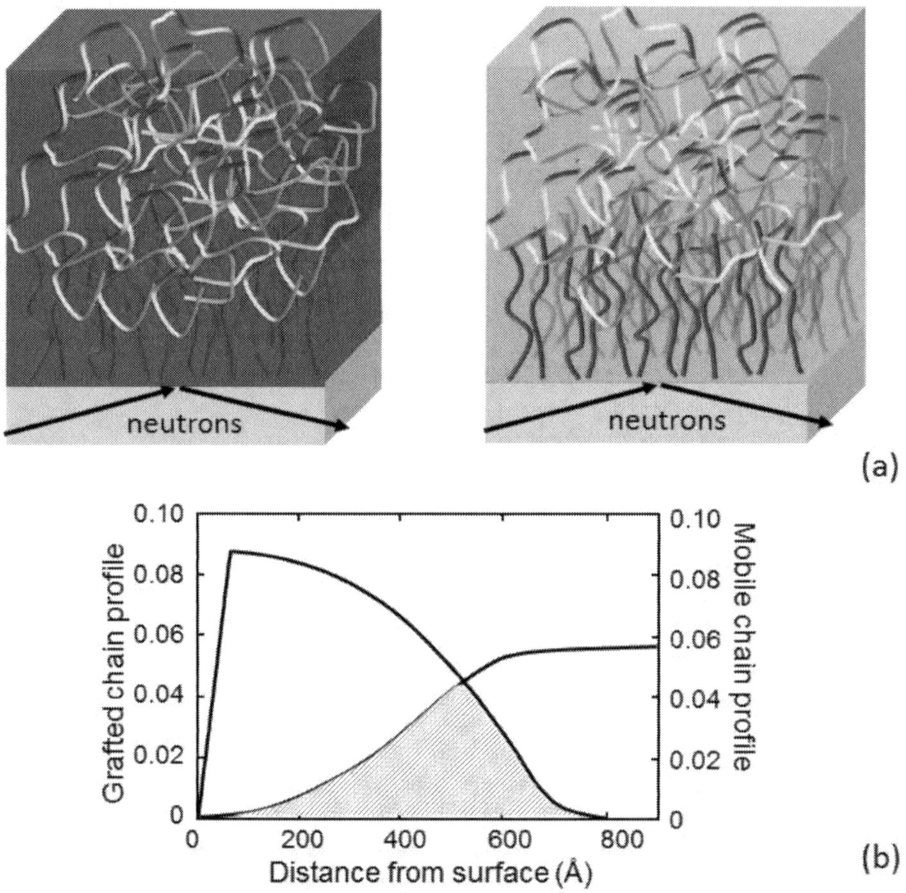

Figure 16: Contrast variation applied to the determination of the behavior of a grafted brush in solution in presence of free chains [31][32]. (a) Sketch of the system from the contrast point of view when the brush is contrast-matched by the solvent (left) or when the free chains are contrast-matched by the solvent (right); (b) Overview of the respective segment density profiles of the grafted chains and free chains, as obtained from the NR measurements in the contrast conditions depicted in (a).

The last example we present to depict the powerful possibilities of contrast variation is the case of a ternary system such as a mixture of two components in a solvent. In such a system, since the SLD of the solvent can be continuously tuned between those of the fully hydrogenated solvent and those of the fully deuterated one, it can be exactly adjusted to the SLD of one the component of the system, making it invisible to neutrons in order to probe specifically the structure of the other one, and *vice versa*. This strategy consisting in matching successively both components is routinely used in SANS but remains alas sparsely used in neutron reflectivity. This is illustrated for the case of the study of a tethered brush in a good solvent in contact with a semi-dilute solution of free chains of the same polymer species as the chains of the brush (Figure 16). In a first experiment, the SLD from the free chains was matched to determine the conformation of the chains of the brush. It appeared that the brush was contracted in presence of the free chains when $\Phi_{free} > \Phi_{brush}$, with respect to the case of the lone brush without free chains. This could either come from the equilibration of the osmotic pressure of the brush or from the penetration of the free chains within the brush that would screen the excluded-volume interactions [31]. In order to discriminate between these two hypotheses; it was mandatory to measure the SLD profile of the free chains in the vicinity of the surface by choosing a solvent matching the grafted chains. In such a contrast, the measurements revealed a depletion layer close to the surface due to the presence of the "ghost" brush [32]. By plotting the density profiles corresponding to these two contrast (Figure 16.b), on can show that free chains penetrate strongly within the brush. Measurements with various grafting densities σ of the grafted brush revealed that free chains are progressively expelled from the brush upon increasing σ [32].

5 Conclusion

Since the pioneering works in the early 80's [33], Neutron reflectivity has become a mature and widespread technique that is rather easily accessible owing to the numerous reflectometers existing in all centers devoted to neutron scattering throughout the world. Due to its unique properties (isotopic labelling, study of buried interfaces), it is a unique and invaluable tool for the determination of structures at the nanometer scale in Soft Matter. In particular, despite of the expanding growth of microscopy techniques, it remains the only technique that gives *at the same time* the structure and composition from a liquid surface [16].

Acknowledgements

The authors warmly thank Max Wolff for his careful reading of the manuscript and his very useful remarks and suggestions.

References

[1] J. Penfold and R. K. Thomas, "The application of the specular reflection of neutrons to the study of surfaces and interfaces," *J. Phys. Condens. Matter*, vol. 2, no. 6, pp. 1369–1412, 1990.

[2] T. P. Russell, "X-ray and neutron reflectivity for the investigation of polymers," *Mater. Sci. Reports*, vol. 5, no. 4, pp. 171–271, 1990.

[3] H. P. Wacklin, "Neutron reflection from supported lipid membranes," *Curr. Opin. Colloid Interface Sci.*, vol. 15, no. 6, pp. 445–454, 2010.

[4] G. Fragneto-cusani, "Neutron reflectivity at the solid/liquid interface : examples of applications in biophysics," *J. Phys. Condens. Matter*, vol. 13, pp. 4973–4989, 2001.

[5] C. Fermon, F. Ott, and A. Menelle, "Neutron reflectivity," in *Daillant, J., Gibaud, A. (Eds.), X-ray and Neutron Reflectivity: Principles and Applications, Lect. Notes Phys. 770 (Springer, Berlin Heidelberg)*, 2009, pp. 183–234.

[6] M. Wolff, "Grazing incidence scattering," *EPJ Web Conf.*, 2018.

[7] V. Sears, "Neutron scattering lengths and cross sections," *Neutron news*, vol. 3, no. 3, pp. 26–37, 1992.

[8] "http://www.neutron.anl.gov/reference.html." .

[9] L. G. Parratt, "Surface Studies of Solids," *Phys. Rev.*, vol. 95, no. 2, 1954.

[10] S. A. Holt, P. Le Brun, C. F. Majkrzak, D. J. Mcgillivray, F. Heinrich, and L. Mathias, "An ion-channel-containing model membrane : structural determination by magnetic contrast neutron reflectometry," *Soft Matter*, vol. 5, pp. 2576–2586, 2009.

[11] R. A. Campbell, H. P. Wacklin, I. Sutton, R. Cubitt, and G. Fragneto, "FIGARO: The new horizontal neutron reflectometer at the ILL," *Eur. Phys. J. Plus*, vol. 126, no. 11, p. 107, 2011.

[12] F. Cousin, F. Ott, F. Gibert, and A. Menelle, "EROS II: A boosted time-of-flight reflectometer for multi-purposes applications at the Laboratoire Léon Brillouin," *Eur. Phys. J. Plus*, vol. 126, no. 11, p. 109, 2011.

[13] L. T. Lee, D. Langevin, and B. Farnoux, "Neutron Reflectivity of an Oil-Water Interface," *Phys. Rev. Lett.*, vol. 67, no. 19, pp. 2678–2681, 1991.

[14] J. Bowers, A. Zarbakhsh, J. R. P. Webster, L. R. Hutchings, and R. W. Richards, "Neutron Reflectivity Studies at Liquid - Liquid Interfaces : Methodology and Analysis," *Langmuir*, vol. 17, no. 1, pp. 140–145, 2001.

[15] L. Besnard, M. Protat, F. Malloggi, J. Daillant, F. Cousin, N. Pantoustier, P. Guenoun, and P. Perrin, "Breaking of the Bancroft rule for multiple emulsions stabilized by a single stimulable polymer," *Soft Matter*, vol. 10, no. 36, pp. 7073–7087, 2014.

[16] G. Fragneto and A. Menelle, "Progress in neutron reflectometry instrumentation," vol. 126, no. 11, p. 106, 2011.

[17] "http://www.ill.eu/." .

[18] "http://www-llb.cea.fr/." .

[19] "https://europeanspallationsource.se/section/ess-instrument-suite." .

[20] G. Fragneto, "Neutrons and model membranes," *Eur. Phys. J. Spec. Top.*, vol. 213, no. 1, pp. 327–342, 2012.

[21] R. L. T. Narayanan, H. Wacklin, O. Konovalov, "Recent applications of synchrotron radiation and neutrons in the study of soft matter," *Crystallogr. Rev.*, vol. 23, no. 3, pp. 160–226, 2017.

[22] L. Braun, M. Uhlig, R. Von Klitzing, and R. A. Campbell, "Polymers and surfactants at fluid interfaces studied with specular neutron reflectometry," *Adv. Colloid Interface Sci.*, vol. 247, pp. 130–148, 2017.

[23] O. Félix, Z. Zheng, F. Cousin, and G. Decher, "Are sprayed LbL-films stratified? A first assessment of the nanostructure of spray-assembled multilayers by neutron reflectometry," *Comptes Rendus Chim.*, vol. 12, no. 1–2, pp. 225–234, 2009.

[24] G. Decher, "Fuzzy Nanoassemblies: Toward Layered Polymeric Multicomposites," *Science (80-.).*, vol. 277, no. 5330, pp. 1232–1237, 1997.

[25] C. Devaux and J. Chapel, "Controlled structure and density of 'living' polystyrene brushes on flat silica surfaces," *Eur. Phys. J. E*, vol. 7, no. 4, pp. 345–352, 2002.

[26] C. Devaux, F. Cousin, E. Beyou, and J. P. Chapel, "Low swelling capacity of highly stretched polystyrene brushes," *Macromolecules*, vol. 38, no. 10, pp. 4296–4300, 2005.

[27] M. Li, B. Bresson, F. Cousin, C. Fretigny, and Y. Tran, "Submicrometric Films of Surface-Attached Polymer Network with Temperature-Responsive Properties," *Langmuir*, vol. 31, no. 42, pp. 11516–11524, 2015.

[28] N. Spiliopoulos, A. G. Koutsioubas, D. L. Anastassopoulos, A. A. Vradis, C. Toprakcioglu, A. Menelle, and G. Mountrichas, "Neutron Reflectivity Study of Free-End Distribution in Polymer Brushes," *Macromolecules*, vol. 42, no. 16, pp. 6209–6214, 2009.

[29] A. Chennevière, E. Drockenmuller, D. Damiron, F. Cousin, F. Boué, F. Restagno, and L. Léger, "Quantitative analysis of interdigitation kinetics between a polymer melt and a polymer brush," *Macromolecules*, vol. 46, no. 17, pp. 6955–6962, 2013.

[30] J. Cotton, D. Decker, H. Benoit, B. Farnoux, J. Higgins, G. Jannink, R. Ober, R. Picot, and J. des Cloizeaux, "Conformation of polymer chain in the bulk," *Macromolecules*, vol. 7(6), pp. 863–872, 1974.

[31] L. Lee, B. J. Factor, F. Rondelez, and M. S. Kent, "Structure of Tethered Polymer Chains Immersed in a Solution of Mobile Polymer," *Faraday Discuss.*, vol. 98, pp. 139–147, 1994.

[32] L. Lee and M. S. Kent, "Direct Measurement of the Penetration of Free Chains into a Tethered Chain Layer," *Phys. Rev. Lett.*, vol. 79, no. 15, pp. 2899–2902, 1997.

[33] J. B. Hayter, R. R. Highfield, B. J. Pullman, R. K. Thomas, A. I. McMullen, and J. Penfold, "Critical Reflection of Neutrons," *J. Chem. Soc., Faraday Trans. I*, vol. 77, pp. 1437–1448, 1981.

[34] Q. Berrod, K. Lagrené, J. Ollivier, and J.-M. Zanotti, "Inelastic and Quasi-Elastic Neutron Scattering. Application to Soft-Matter.," *EPJ Web Conf.*, 2018.

Grazing incidence scattering

Max Wolff*
Department for Physics and Astronomy, Uppsala University, 75120 Uppsala, Sweden

Abstract. Reflectometry experiments probe the scattering length density along the normal of interfaces by analysing the specularly scattered intensity. Lateral fluctuations result in intensity scattered away from the specular condition. In this paper the principles and peculiarities of grazing incidence scattering experiments are explained. One specific example, the self assembly of polymer micelles close to interfaces, is taken as a show case in order to introduce the scattering geometry and accessible length scales. The basic idea of the distorted wave Born approximation is lined out and some scientific examples are summarized.

1 The importance of surface effects

With the development and miniaturization of modern technology the need for advanced experimental methods in surface science has increased tremendously. This fact is reflected, for example, by the development of atomic probes [1] or electron microscopy [2]. In the field of x-ray and neutron scattering the number of reflectometers, which are most relevant for surface science, is growing and with the years specular reflectometry measurements have become a routine tool to extract the density profiles of materials across interfaces with almost atomic resolution (see chapter 04001 by F. Cousin and A. Chenneviere). A list of neutron reflectometers as well as relevant citations describing those can be found in Ref. [3] and an introduction in reflectometry and grazing incidence is found in [4]. However, specular reflectivity measurements can only reveal the scattering length density (SLD) profile across an interface with the scattering potential averaged in the plane of the interface and over the coherence volume of the beam. To become more specific, inter-diffusion and roughness or even regular patterns on a surface can not be characterized (see Figure 1). In the case of roughness lateral correlations in the plane along the interface may be present. These give rise to interferences and a modulation of the scattered intensity. The lateral correlations can be extracted from so called off-specular (OFFSpec), grazing incidence small angle scattering (GISAS) or diffraction (GID). During the last 15 years most neutron reflectometers have become equipped with position sensitive detectors allowing an efficient collection of OFFSpec and grazing incidence scattering (GIS) data. Still the method is not exploited to its full potential, which results on the one hand from the fact that easy to use and robust fitting routines are not yet developed and on the other hand from the strongly limited flux and dynamic range on neutron instruments. Typically, OFFSpec and GIS intensities are three or more orders of magnitude lower in intensity than their specular counterparts.

*e-mail: max.wolff@physics.uu.se

 https://doi.org/10.1051/epjconf/201818804002

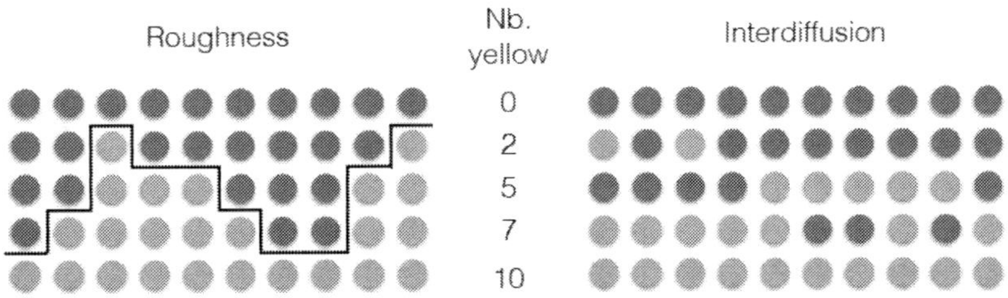

Figure 1. Roughness and inter-diffusion: In the left panel a continuous line can be drawn between the two materials, whereas in the right panel isolated islands exist. The integrated number of yellow and blue circles along the horizontal direction is the same. Specular reflectivity can not distinguish between the two scenarios.

This manuscript is structured as follows: First the scattering geometry and the peculiarities of OFF-Spec and GIS is explained on a specific example of polymer micelles self assembling at a solid boundary. Then the distorted wave Born approximation (DWBA) is outlined and finally some scientific examples are presented.

1.1 Scattering geometry

The scattering geometry in neutron reflectometry (NR) is usually defined in cartesian coordinates with the z direction along the normal of the interfaces. The x-direction is in the scattering plane and y is perpendicular to x and z (see Figure 2). In the case of a buried liquid interface a neutron beam may pass trough a silicon wafer, get scattered at the solid-liquid boundary and detected by an area detector. Note, since the absorption of neutrons in silicon is low and the scattering potential of deuterated liquids is large, such experiments can be performed in a straight forward way with a minimum loss in flux and a critical momentum transfer of total external reflection present.
The intensity scattered at the same incident and exit angles (orange arrows in Figure 2) is the specular reflectivity and contains the information about the SLD profile along the surface normal. Intensity scattered at different exit angles but still in the scattering plane (symbolized green) is usually called OFFSpec and intensity registered for exit angles out of the scattering plane is small angle scattering (SAS). The three components of the scattering vector Q, along the x, y and z-direction, are calculated

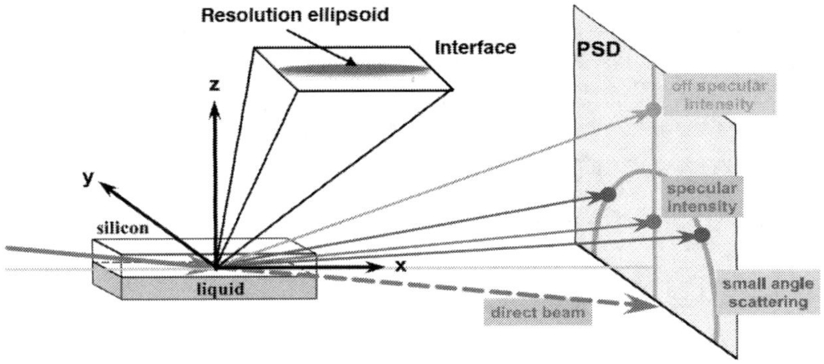

Figure 2. Scattering geometry for GIS experiments from the solid-liquid boundary [5].

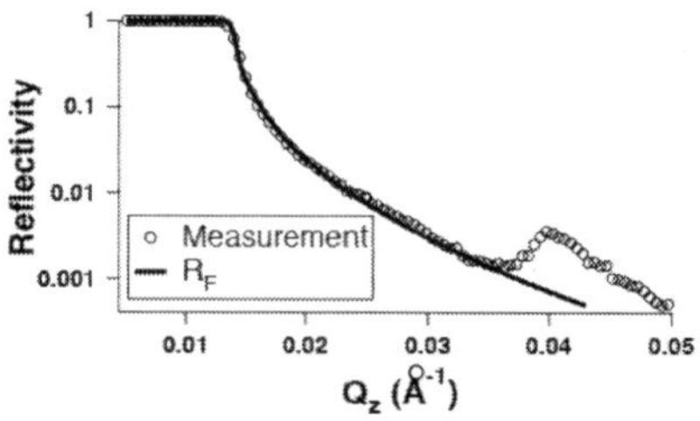

Figure 3. Reflectivity data taken with a micellar polymer solution in the liquid phase in contact with a silicon wafer compared to the Fresnel-reflectivity R_F.

as follows:

$$Q_x = \frac{4\pi}{\lambda}\left(\cos\alpha_f\cos\phi - \cos\alpha_i\right) \quad (1)$$

$$Q_y = \frac{4\pi}{\lambda}\sin\phi \quad (2)$$

$$Q_z = \frac{4\pi}{\lambda}\left(\sin\alpha_i + \sin\alpha_f\right) \quad (3)$$

with α_i and α_f the angle of the incident and exiting beam with respect to the sample's surface, λ the wavelength and ϕ the scattering angle out of the scattering plane.

2 Block-polymers and micelles

In the following the peculiarities of grazing incidence small angle neutron scattering (GISANS) and OFFSpec will be explained on one specific example, namely aqueous solutions of block polymers and their assembly at the solid-liquid boundary. This system offers a strong scattering signal on all length scales relevant for GIS experiments. In aqueous solution the molecules tend to form micelles (self-aggregartions of molecules). Depending on temperature and volume fraction these may form liquid or crystalline phases. The sample is a 18.5 % (in weight) solution of Pluronic F127 (EO_{99}-PO_{65}-EO_{99}) in deuterated water providing a good contrast for neutrons. The bulk properties of such samples are known in great detail [6, 7]. In the case of F127 face centred cubic (fcc) crystals are formed at elevated concentrations or temperatures.

Figure 3 shows the specular reflected intensity taken with the sample in a liquid phase at 295 K compared to the Fresnel-reflectivity. The data is corrected for the SAS by substracting the intensity for a detector off-set angle, from the specular condition, of 0.3°. At Q_z = 0.04 Å, the intensity is increased. This Bragg peak indicates several layers of adsorbed micelles at conditions, which are still in the liquid phase with no fcc structure present in the bulk.

3 Off-specular and GISANS

In the crystalline phase (temperature 298 K) a fcc dense packed structure is formed. Figure 4 (left panel) represents a GISANS pattern collected on an area detector for a monochromatic incident beam impinging the sample under an incident angle α_i close to the critical angle of total external reflection α_c. Specifically, α_i = 0.30° results in a penetration depth into the liquid of 40 μm (extinction length

https://doi.org/10.1051/epjconf/201818804002

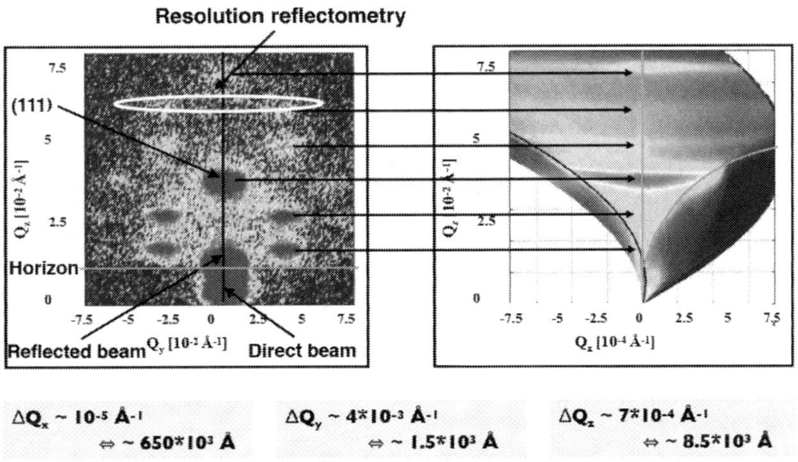

Figure 4. Scattering patterns for a micellar crystal in contact to a silicon wafer. The panel on the left hand side depicts the GISANS pattern form the cubic close packed structure. The panel on the right hand side depicts reflectometry data (specular and OFFSpec) from the same sample but with a relaxed resolution along the y-direction [5].

for a drop of 1/e of the incident intensity). α_c for a neutron wavelength of 0.44 nm is 0.25° and defined by the SLD of silicon and the polymer solution. For no momentum transfer $Q = 0$ the direct beam hits the detector at the very bottom in the middle. Slightly above the reflected beam is visible. Half way between these two peaks the sample horizon (brown line), dividing intensity transmitted by and reflected at the interface, is found. Clearly below the horizon the intensity is low (dark blue region below $Q_z = 0.01$ Å^{-1}) due to the enhanced absorption or incoherent scattering inside of the liquid sample as well as of the sealing of the liquid cell [10]. Note: the image shows raw data and Q_z was not corrected for refraction effects $Q_{z,in} = \sqrt{Q_z^2 - Q_c^2}$, with $Q_{z,in}$ the Q_z value inside the liquid. All other reflections are Bragg peaks resulting from a cubic dense packing of the polymer micelles close to the solid interface [11]. The projection of the resolution on the x, y and z axis for GIS is dependent on α_i and the exit angle α_f, since the coordinate system is fixed to the sample and not to the instrument [5]. Specifically, for this example and $Q = 0.04$ Å^{-1} the resolution $Q_x = 10^{-5}$ Å^{-1} is determined by the pixel size of the detector and the scattering angle. The resolution $Q_y = 4\times10^{-3}$ Å^{-1} and $Q_z = 7\times10^{-4}$ Å^{-1} is given by horizontal and vertical slits. In real space the resolution is defined by the coherence volume of the beam projected onto the respective axis of the coordinate system and is roughly 65 μm, 150 nm and 850 nm along the x, y and z-direction, respectively, for the incident beam angle of 0.3°.

On the monochromatic instrument α_i and α_f are changed and for each setting and an image, similar to the one shown in the panel on the left hand side, is detected. The panel on the right hand side (Fig. 4) shows the result of such a NR measurement, including the OFFSpec scattering. In this representation the horizon is again represented by a brown line. The two panels on the left and right (Figure 4) are orthogonal but intersect along the black lines. This line is almost identical with the sample horizon in the right panel. Due to the non-linear transformation of the coordinate Q_x the black line is curved in the panel on the right hand side.

For the NR measurement the resolution along y has been relaxed to 0.15 Å^{-1} (see white resolution ellipsoid in Fig. 4) in order to gain in flux. As a result a contribution from the Bragg reflections with

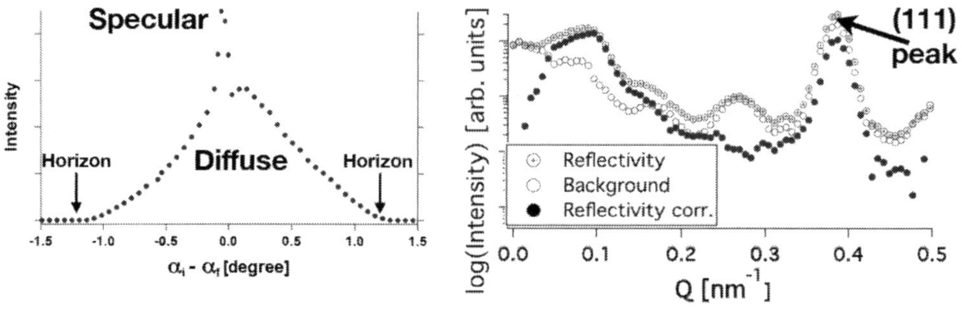

Figure 5. Left panel: Rocking scan for the sample in the crystalline phase at Q_z = 0.38 nm^{-1}. Right panel: Correction of the specular scattered intensity for SAS.

$Q_y \neq 0$ is detected in the Q_x, Q_z plane. In order to extract the true specular reflectivity this contribution has to be corrected for, since the intensity is not scattered specular but appears in the specular direction. This is a severe challenge for samples with strong SAS, when the SAS can be equal or stronger than the specular signal. Figure 5, right panel, demonstrates this effect [20]. The detected specular NR is plotted on a logarithmic scale versus Q_z (circles marked with a cross). The background, for an offset of α_i and α_f of 0.25°, is represented by open circles. For $Q_z > 0.2$ nm^{-1} the reflectivity is dominated by SAS and the true specular signal (black dots) can hardly be extracted with statistical significance.

The panels on the right and left hand side, Fig. 6, illustrate the unit cell of the cubic close packing and the GISANS plotted as color map onto the Ewald sphere, respectively. The detector plane represents a curved surface with Q=0 at the position of the direct beam. The second and only other point where the in-plane coordinates Q_x and Q_y are zero is the point where the specular scattered intensity is detected. All other points on the detector have none-zero Q_x and/or Q_y components.

Considering a densely packed cubic structure at the interface, the (111) lattice planes are parallel to the solid-liquid boundary and the Bragg reflections resulting from them should have no in-plane component ($Q_x = Q_y = 0$). This implies that the (111) reflection can only be detected in a GISANS

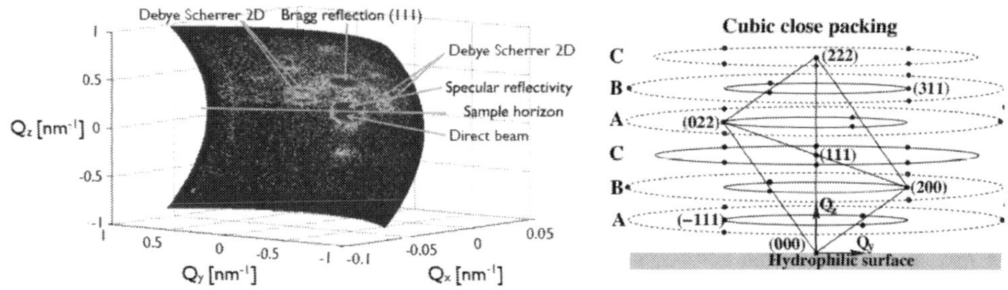

Figure 6. Left panel: GISANS scattering pattern plotted on the surface of the Ewald sphere in three dimensions [18]. Right panel: Reciprocal lattice for a cubic close packing (ccp). The dots mark the positions of reciprocal lattice points. For a two-dimensional powder with the unique axis along Q_z perpendicular to the interface these points are smeared out into circles.

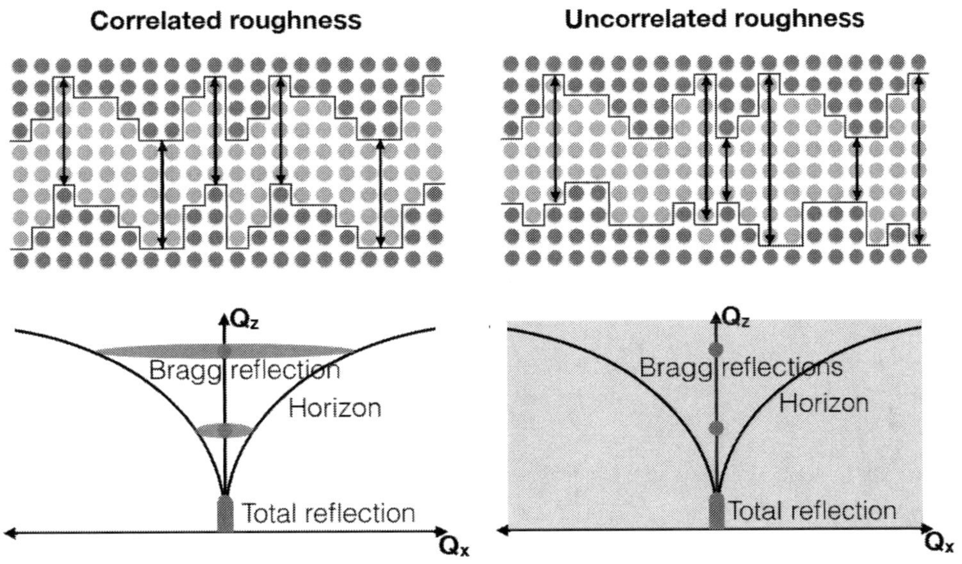

Figure 7. Structure and off-specular scattering from rough surfaces for perfectly correlated (left panels) and uncorrelated (right panels) interfaces.

experiment if it has a certain rocking width along Q_x or in other words correlated roughness of the micellar layers is present. All other four Bragg reflections are only visible if the crystalline structure is a two dimensional powder with respect to the sample interface [5], since the Q_y values are much larger than the Q_x values and the probability of a crystal orientation with the Bragg reflection on the Ewald sphere is unlikely.

The (111) reflection at $Q_z \approx 0.38$ nm^{-1} and $Q_y = 0$ is smeared out in the reflectivity measurement (right panel) along Q_x due to correlated roughness. Figure 5, left panel, represents a cut along this reflection along Q_x. Such a cut is called Rocking curve, the α_i and α_f are scanned for given scattering angle (see x-axis in Figure 5, left panel). The narrow component is the specular scattered intensity and reflects long range orientational correlations over the coherence volume of the neutron beam [21], indicated by the ellipsoid in Fig. 2 (resolution ellipsoid). The broad component is related to diffuse or OFFSpec scattering and results from the finite lateral correlations of the SLD fluctuations contributing to the Bragg peak along the plane of the interface. As mentioned the length scale probed along the x-direction is about two orders of magnitude larger than the one probed along y and z (see Figure 4). The GISANS together with an analysis of the width of the (111) reflection in the off-specular scattering regime can be used to analyse the texture as well as the size of the micellar crystallites [16]. Figure 7, left and right panel, depicts the extreme cases of perfectly correlated and uncorrelated interface roughness, respectively, and the resulting OFFSpec scattering. For a perfectly correlated roughness (left panels) between layers the OFFSpec scattering becomes concentrated along so called Bragg sheets, showing up as horizontal lines at a certain Q_z and running along Q_x. The width of the sheets is the reciprocal lateral correlation length. For uncorrelated layers the OFFSpec scattered intensity is smeared out over the whole Q space. Note, the discussion above is based on kinematic approximations and neglects the dynamic effects which can be important in reflectometry experiments. However, dynamical effects (apart from total reflection) are typically important for very well ordered samples, which are hardly found in soft matter applications.

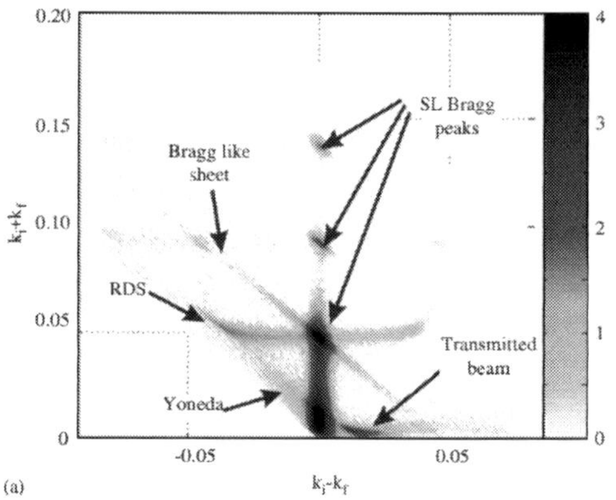

Figure 8. Two-dimensional intensity maps of the scattering from a [Ti(70 Å)/ Ni(70 Å)]$_{20}$ multilayer plotted as function of $k_i + k_f$ and $k_i - k_f$ [12].

Still dynamic effects resulting in so called Yoneda and resonant diffuse scattering (RDS) will be discussed briefly below. For illustration Fig. 8 depicts a scattering pattern collected with a TiNi multilayer with 20 repetitions [12]. The intensity is plotted over $k_i + k_f$ and $k_i - k_f$ instead of Q_x and Q_z to avoid the compression of the scattering pattern close to the critical momentum transfer, resulting from the non-linear transformation to Q [13]. Yoneda scattering [14] results from an constructive interference of the incoming and outgoing wave field at the critical momentum transfer of total external reflection and enhances scattering from roughness along so called Yoneda wings (sample horizon). In addition, since the incident or exiting wave impinges under the critical momentum the scattering pattern gets distorted close to the horizon. This results in an upturn (towards larger Q_z values) of the Bragg sheets in this region. For multilayers there is additional interference of the waves scattered at each interface and again for constructive interference roughness scattering is enhanced resulting in RDS or Holy scattering [15].

4 Modes of measurement

As described in the previous chapters neutron scattering experiments can be done in two modes. The classical way are monochromatic instruments with a continuous flux of neutrons. For each point in Q the scattering angle as well as the incident angle is chosen with respect to the sample, see Figure 9, bottom panel. For the time of flight (TOF) method a pulsed neutron beam is used with a polychromatic beam and the wavelength is extracted from the time of flight of the neutron from the source to the detector. With this method a range of momentum transfers is measured for one given setting of the instrument, see Figure 9, top panel.
As a result a whole range of Q values around Q_c and a range of penetration depth from the boundary can be probed. For a momentum smaller than Q_c the penetration depth is small, on the order of 20 nm (see Figure 10). Figure 10 shows the penetration depth of the neutrons into the liquid calculated according to Ref. [8, 9], using the sum of the absorption and the incoherent scattering cross sections. Note, since incoherent scattering is isotropic it only adds to the background but contains no information relevant for GISANS or OFFSpec scattering. Still the neutrons scattered incoherently are "lost" and can as such be treated as the absorbed ones. At the critical momentum the penetration depth has a distinct jump, in the case presented here by about four orders of magnitude from some nm straight to

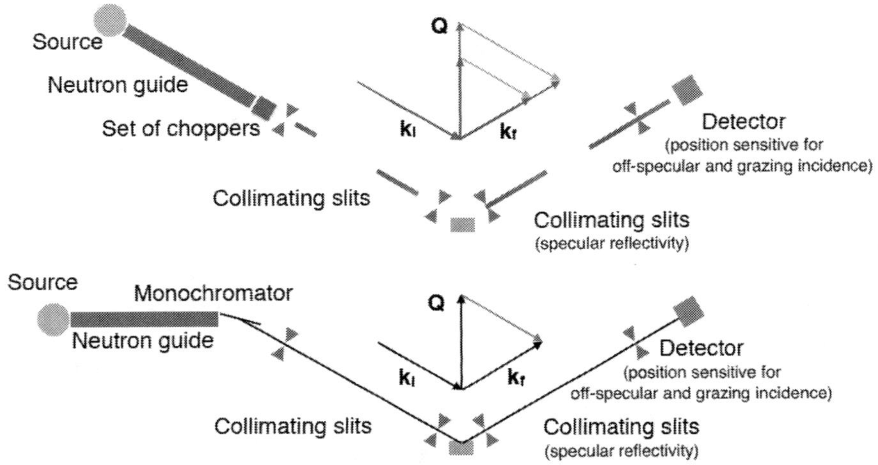

Figure 9. Modes for reflectometry experiments, the top panel shows time of flight or polychromatic instruments and the bottom panel monochromatic ones. On polychromatic instruments a range of Q values is probed for one setting of the instrument.

micrometers, which is typical for neutrons. The height of the jump at the critical momentum is related to the extinction properties of the material. The curve shown in Figure 10 is calculated for the silicon-polymer solution interface discussed above. The incoherent scattering cross section is included in the absorption and extinction related to scattering is neglected. It is seen that the penetration depth into the liquid changes by four orders of magnitude within 10 % above and below Q_c.
This implies that taking independent data in regions 1, 2 and 3 and achieving depth resolution is extremely challenging since an excellent $\frac{\Delta Q}{Q_c}$ is needed. In particular any tails of the resolution function

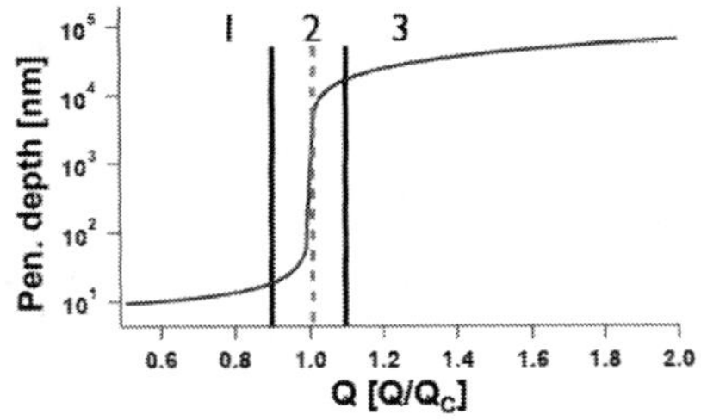

Figure 10. Penetration depth of a neutron into a micellar solution (20 % of protonated polymer solved in 80 % deuterated water) being reflected at the silicon polymer interface.

will immediately result in neutrons penetrating μm into the liquid and can dominate the scattering pattern [17]. $\frac{\Delta Q}{Q}$ is calculated by the following equation taking into account the wavelength distribution $\Delta\lambda$ and a angular divergence $\Delta\theta$:

$$\frac{\Delta Q}{Q} = \sqrt{\left(\frac{\Delta\lambda}{\lambda}\right)^2 + (\Delta\theta \cot\theta)^2} \tag{4}$$

For small angles θ this can be simplified:

$$\frac{\Delta Q}{Q} = \sqrt{\left(\frac{\Delta\lambda}{\lambda}\right)^2 + \left(\frac{\Delta\theta}{\theta}\right)^2} \tag{5}$$

The most efficient use of neutrons is achieved for equal relative resolutions regarding wave length and divergence. For a time of flight instrument the wavelength resolution is given by the length of the instrument, the time resolution of the detector as well as by the neutron pulse length and for a monochromatic instrument by the mosaic of the monochromator and the divergence in the neutron guide or the specifications of the velocity selector. The divergence of the beam can be adjusted on both types of instruments by slits. Let us assume a wavelength and angular resolution of 1 %. Regarding the wavelength this can easily be reached with time of flight methods or monochromators. For the angular resolution at a critical angle of 0.2° a 1 % resolution implies a divergence of 0.002° or two pinholes of 1 mm diameter at a distance of almost 30 m. In addition the footprint of e.g. 0.175 mm for a 5 cm sample under the angle of 0.2° is small.

Figure 11 depicts patterns of intensity measured on SANS-2D (Rutherford Institute, Didcot, England) for the polymer solution discussed above but now in contact to two silicon wafers with distinct surface energies [18]. The incident beam was collimated to provide a divergence of 0.29° and 0.04° along y and z direction, respectively. The incident beam angle was chosen to be 0.3° resulting in a critical wavelength for total external reflection of λ_c = 5.4 Å.

The three pictures for the hydrophilic silicon surface treated with piranha (top panels) and the hydrophobic one coated with octadecyltrichlorosilane (OTS) (lower panels) were taken simultaneously for one incident beam angle with a fixed detector angle. The left panels correspond to the integration of the detector images for wavelengths ranging from 6-15.6 Å resulting in a penetration depth of the neutron beam into the polymer solution of approximately 10 nm (region 1 in Figure 10). The intensity integrated for short wavelengths of 1.75-5 Å and thus a large penetration depth of about 30 μm is depicted in the right panels. The central column summarizes the intensities for wavelengths integrated around the critical wavelength, 5-6 Å, resulting in an intermediate penetration depth of about 10 μm. For the smallest penetration depth of about 10 nm a ring of scattered intensity is visible around the direct beam (Q=0) for the hydorphobic substrate. This ring is absent for the micelles in contact with the hydrophilic surface. Note, the presence of the 111 reflection for these wavelength demonstrates the limited depth sensitivity of the measurement, since the constructive interference arises from layers separated by more than 10 nm [17]. The difference in the intensity distribution for the sample in contact to one or the other solid substrate becomes more pronounced for larger penetration depths (middle panels). At least ten Bragg reflections can be distinguished with the sample in contact to the hydrophilic surface, whereas still the ring of scattered intensity is dominant for the hydrohpobic one. For the largest penetration depth (right panels) both detector images are almost identical, reflecting the bulk structure and a vanishing influence of the interface.

Another example of depth resolved time-of-flight GISANS addresses the morphology of defect rich deuterated poly(styrene-block-methyl methacrylate), P(S-b-MMAd), diblock copolymer films [19].

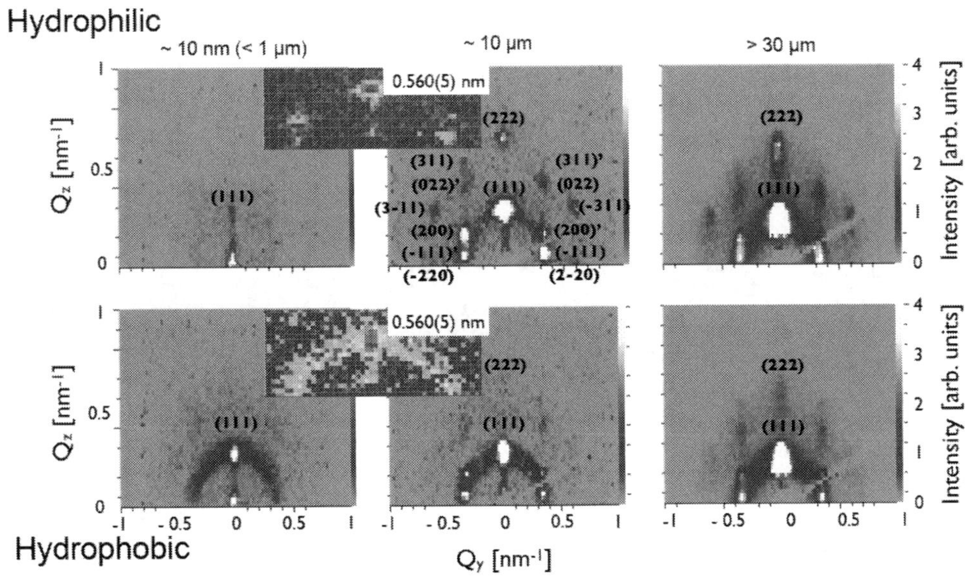

Figure 11. GISANS data taken for a solution of the polymer F127 in deuterated water in contact with a silicon surface treated with piranha (hydrophilic, top panels) and OTS (hydrophobic, lower panels). The penetration depths increases from left to right due to the different wavelengths in the incoming beam.

5 Reflection, refraction and scattering

Often the distorted wave Born approximation (DWBA) [22] is applied to describe OFFSpec and GIS. This method is a first order perturbation theory and requires that the SLD profile along the surface normal is known. This allows the calculation of the wave amplitude inside the sample. The scattering cross sections are then weighted with this wave amplitude in order to quantify the amount of OFFSpec and SAS. Figure 12 depicts the principle of the DWBA for a single interface. First the problem for a perfectly flat interfaces (plane waves) is solved and then the in-plane correlations are added as small perturbation (spherical waves). Luckily, the one-dimensional problem can be solved exactly but requires a specular reflectivity measurement. From that the SLD profile averaged over the coherence volume of the beam and the wave amplitude at different depth can be extracted. Note, since the coherence of the beam is strongly anisotropic it might be necessary to distinguish between a coherent (sum of wave amplitudes) and an incoherent (sum of intensities) summation along the Q_x and Q_y directions. This is particularly important for anisotropic samples, like e.g. stripe patterns [25] (see Fig. 13). The transmitted intensity below the critical momentum transfer is purely imaginary and becomes:

$$|\Psi(z)|^2 = t_r^2 e^{-\frac{z}{l}} \tag{6}$$

At each interface the neutron wave function has to be continuous and can be written as follows:

$$\Psi(z) = e^{ik_i z} + r_F e^{-ik_i z} \tag{7}$$

$$\Psi(z) = t_r e^{-ik_r z} \tag{8}$$

Equation 7 and 8 describe the wave function outside and inside the sample, respectively. r_F and t_r denoted the Fresnel and transmission coefficients and k_i and k_r the momentum of the incident and

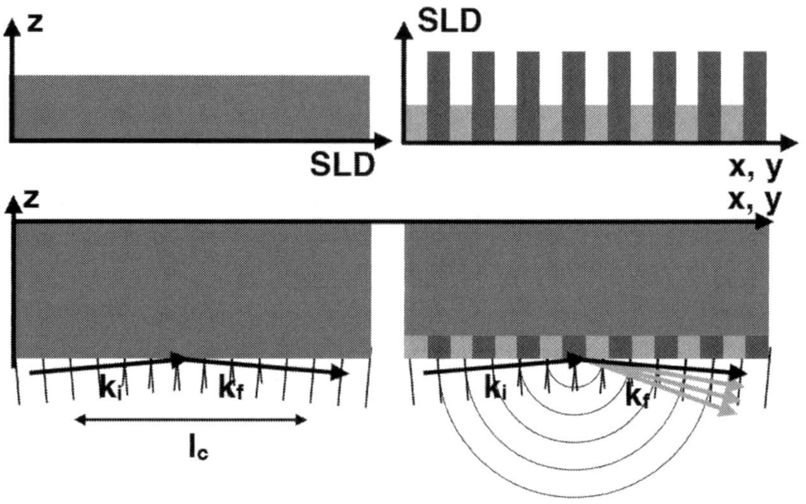

Figure 12. Scattering from lateral structures: The panels on the right hand side depict the scattering of plane waves from a sharp interface. On the left hand side lateral correlations are introduced. These give rise to spherical waves and OFFSpec scattering.

refracted wave. The intensity is calculated from the square of the wave function:

$$|\Psi(z)|^2 = 1 + r_F^2 + 2r_F \cos(k_i z) \tag{9}$$

The wave field oscillates as a function of z and forms a standing wave, with amplitude two, at the critical momentum transfer. This enhancement is called Yoneda scattering. Note, that this effect is similar to the enhancement of the wave field at Bragg reflections (constructive interference) leading to RDS for correlated roughness [15]. The enhancement of the wave field at the surface can be amplified further by using resonator layers [23, 24].

Once knowing the incident and outgoing wave fields for the mean potential the scattering cross section for off-specular or GIS can be calculated from lateral fluctuations $\Delta SLD = SLD - \langle SLD \rangle$.

$$\frac{d\sigma}{d\Omega} = \left\langle \left| f(\vec{q}_{\|}, p_{0f}, p_{0i}) \right|^2 \right\rangle \tag{10}$$

with

$$f(\vec{q}_{\|}, p_{0f}, p_{0i}) = -\int d\vec{r}\Psi_f(\vec{r})\Delta SLD(\vec{r})\Psi_i(\vec{r}) \tag{11}$$

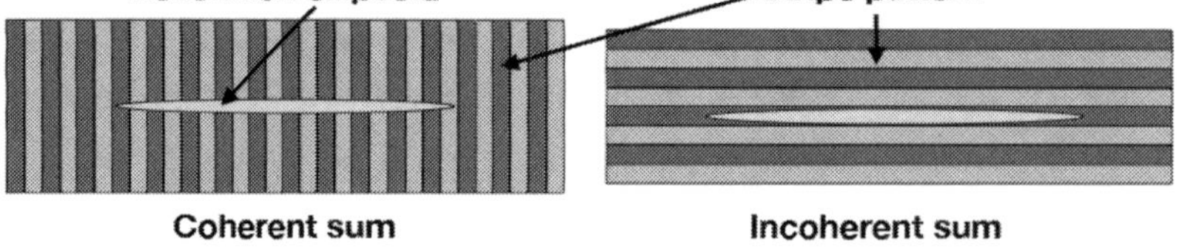

Figure 13. Coherent and incoherent summation: Depending on the orientation of the resolution ellipsoid with respect to in-plane correlations a coherent (left panel) or incoherent (right panel) summation has to be applied.

Using the fact that the normal and lateral components of the wave function are orthogonal and the components factorize we get the result for the scattering cross section:

$$\frac{d\sigma}{d\Omega} = \left\langle \left| f(\vec{q}_{\|}, p_{0f}, p_{0i}) \right|^2 \right\rangle \tag{12}$$

with:

$$f(\vec{q}_{\|}, p_{0f}, p_{0i}) = -\int dz \Psi_f(z) \Delta SLD(q_{\|}, z) \Psi_i(z) \tag{13}$$

and:

$$\Delta SLD(\vec{q}_{\|}, z) = \int d\vec{r}_{\|} \Delta SLD(\vec{r}_{\|}, z) e^{i\vec{q}_{\|}\vec{r}_{\|}} \tag{14}$$

This result is a Fourier transform as normally received in the Born approximation but weighted with the incident and outgoing wave amplitude. The formalism of the DWBA as well as examples from magnetic thin films and laterally patterned structures are summarized in Ref. [26]. While there is an agreement among software packages and fitting tools for specular reflectivity, currently programs to calculate diffuse scattering as well as GIS are sparse, e.g. the project BornAgain [27] allows the simulation of GISANS scattering patterns.
Since the DWBA is a first order perturbation theory the condition that the lateral fluctuations have to be small compared to the changes in SLD along the surface normal has to be fulfilled. For many experiments this may not be the case. As an example in the data set presented in Figure 3 right panel, the SAS intensity is almost equal to the reflected intensity. Concerning this fact a quantitative evaluation of GIS data is challenging and many studies restrict themselves to extract characteristic length scales or relative peak intensities.

6 Examples

The number studies OFFSpec and/or GIS using neutrons is limited. This fact results from the relatively low brilliance of the current neutron sources. Synchrotron x-ray sources offer orders of magnitude higher brilliance and GIS techniques are used there more routinely [28, 29].
Neutrons, due to their spin, high penetration power and isotope sensitivity, offer unique possibilities for the study of magnetism, buried interfaces and soft matter. In the case of GIS from soft matter, in particular buried interfaces, like the solid-liquid boundary discussed above are of high interest as well as the self organization and dewetting of polymers on solid substrates. Following along this line, in the end of the 80's, Anastasiadis et al. [30] demonstrated that flat multilayers of poly(styrene-b-deuterated-methylmethacrylate) are formed on a substrate. In the study they used Rocking scans, along the Q_x direction and show that the width of the specular line is resolution limited undermining the absence of correlated roughness. Some years later, in the beginning of the 90's, the first GISANS experiment using neutrons was done showing a hexagonal alignment of CTA35ClBz, CTABr thread-like micelles at the solid-liquid boundary under Poiseuille flow [34]. Later GISANS was used to study the lamellar orientation of supported thin films of poly(styrene-b-butadiene). It turns out that in low molar mass samples the lamellae are parallel to the substrate surface, whereas they are perpendicular for high molar masses [32]. Some of the earlier OFFSpec and GISANS experiments are summarized in the short review by Dalgliesh [33] and Hamilton [34], respectively. A more recent review discusses the use of GISAXS and GISANS for the study of block copolymer thin films [35] and a paper by Lauter et al. evaluates the power of so called complete reflectometry experiments evaluating NR, OFFSpec and GISANS [36, 37]
The low absorption of neutrons can be used to probe not only the reflected but also transmitted (Laue

scattering) intensity measuring below the samples horizon. This offers the opportunity to access a wide range of Q values [38], e.g. for the study of highly aligned phospholipid membranes. Transmitting the neutron beam trough a quartz crystal and reflecting from water polymer-supported single lipid bilayers allows the study of model cell membranes [39]. From off-specular neutron scattering the in-plane height-height correlations of interfacial fluctuations of such a lipid bilayer can be quantified. It turns out that with decreasing temperature the polymer swells and the polymer supported lipid membrane deviates from its initially nearly planar structure with a decrease of the correlation length characteristic for capillary waves. Ultra-thin deuterated poly(methyl methacrylate) films may spinodally dewet on polystyrene substrates as a result of dispersive force driven instabilities consisting of thermally excited capillary waves. The length scale and growth rate of this instability was studied by NR and OFFSpec measurements and are consistent with the predictions of a linear theory of spinodal dewetting at a liquid/liquid interface [40]. The kinetics during the dewetting process of blend thin films deuterated polystyrene and poly vinyl methyl ether was studied by time-resolved specular and OFFSpec NR measurements. It is reported that the droplets formation on the micrometer length scale occurs in the late stage of dewetting [41].

As indicated above the self assembly of three block polymer micelles at solid interfaces is a good showcase for GIS from solid-liquid boundaries and a range of molecules has been investigated. The alignment of the crystallites at a surface was found somewhat similar to shear alignment in a combined GISANS and transmission SANS experiment [42]. In addition the texture of the self assembly [11] as well as the reconstruction of the structure after the cession of shear [45] strongly depends on the surface energy of the solid substrate. For micellar solutions showing a phase transition from cubic to hexagonal ordering a hysteresis in the transition temperature has been reported [20, 43, 44].

Some neutron reflectometers offer the possibility to measure from horizontal samples and liquid surfaces. Such a geometry can be used to study surfactant layers at the air-water interface, like the ones used for deposition in the Langmuir-Blodgett technique. Following along this line it was shown from NR profiles that for surface pressures below the collapse point of a low molecular weight polystyrene film only specular reflection is present, but in the collapse region, some OFFSpec develops, indicating the presence of surface texture on the micron length scale in the plane of the interface [46]. The temperature effect on the conformal roughness in aerosol-OT lamellar adsorbed at the air-water and liquid-solid interface was studied by Li et al. [47]. They show that with increasing temperature the OFFSpec scattering becomes enhanced close to the specular line, indicating fluctuations of the membrane layers with defined correlation length and amplitude. In a combined grazing incidence neutron scattering and AFM study Gliss et al. have investigated the size and orientation of phospholipid bilayers [48]. They show that in the coexistence region of gel and fluid phase the domains grow in number rather than in size with decreasing temperature. Similar to the air-liquid interfaces liquid-liquid interfaces can be studied and some early experiments using x-ray and neutron GIS methods are reviewed by Schlossman [49].

In the field of magnetism OFFSpec and GIS offers unique opportunities to study the formation of magnetic domains [26]. The incorporation of magnetic polystyrene coated nano-particles into a lamellar polystyrene-poly(butyl methacrylate) P(Sd-b-BMA) film was studied and it was shown that an increasing particle concentration in the polystyrene layers results in a thickness expansion and decreasing correlated roughness [50]. The self-assembly of magnetic particles at a silicon surface as well as at the air interface was studied by OFFSpec neutron scattering [51] and from polarized GISANS experiments combined with reflectometry the magnetization and assembly of magnetic particles on a silicon substrate was studied [52].

7 Summary

GIS methods become a more readily accessible scientific tool for the study of surfaces and interfaces. With more sophisticated instrumentation and data evaluation tools one can expect a considerable increase in the number of scientific studies resulting from this approach. In this article the method of GIS is discussed, the peculiarities of the scattering geometry are highlighted and some recent results are reviewed. The key features of GIS and OFFSpec scattering are the wide range of length scales, from sub-nm to about 100 μm, being accessible along the different scattering directions. In the case of neutrons the depth sensitivity is limited due to the low brilliance of the source and weak absorption in most materials. The efforts of developing GIS methods is not restricted to elastic scattering experiments. During the last years several attempts have been made to study the near surface dynamics by grazing incidence neutron spin echo (GINSES) experiments [53–57].

References

[1] G. Binnig, C. F. Quate, Ch. Gerber, Phys. Rev. Lett., 56, 930 (1986).
[2] R. Reichelt, *Scanning Electron Microscopy, Science of microscopy* (Springer, New York, NY, 2007), 133-272.
[3] http.//www.reflectometry.net/biblio/Neutron_reflectometer_bibliography_instrument_sort.pdf
[4] J. Daillant, A. Gibaud (X-ray and neutron reflectivity: Principles and applications, Springer, Lecture notes in Physics 770, 2009)
[5] M. Wolff, A. Magerl and H. Zabel, Euro. Phys. J. E, **16(2)**, 141 (2005).
[6] G.Wanka, H.Hoffmann, W.Ulbricht, Macromolecules **27**, 4145 (1994).
[7] K. Mortensen, Polym. Adv. Technol. **12**, 2 (2001).
[8] L.G. Parratt, Phys. Rev. **95**, 359 (1954).
[9] H. Dosch, B. W. Batterman, D. C. Wack, Phys. Rev. Lett. **56**, 1144 (1986).
[10] D. van der Grinten, M. Wolff, H. Zabel, A. Magerl, Meas. Sci. Technol. **19**, 034016 (2008).
[11] M. Wolff, U. Scholz, R. Hock, A. Magerl and H. Zabel, Phys. Rev. Lett., **92**, 255501 (2004).
[12] M. Ay, C. Schanzer, M. Wolff, J. Stahn, Nuclear Instr. and Methods in Phys. Res. A **562**, 389 (2006).
[13] F. Adlmann, G. K. Palsson, J.-C. Bilheux, J. F. Ankner, P. Gutfreund, M. Kawecki, M. Wolff, J. Appl. Cryst. **49(6)**, 2091 (2016).
[14] Y. Yoneda, Phys. Rev. **131**, 2010 (1963).
[15] V. Holy, T. Baumbach, Phys. Rev. B **49**, 10668 (1994).
[16] M. Wolff, A. Magerl, H. Zabel, Langmuir (Letter) **25 (1)**, 64 (2009).
[17] F. Adlmann, G. K. Palsson, A. Korolkovas, J. Herbel, A. Korolkovas, B. Kitchen, A. Bliersbach, B. P. Toperverg, W. van Herck, M. Wolff, J. Phys.: Cond. matter (submitted).
[18] M. Wolff, J. Herbel, F. Adlmann, A. J. C. Dennison, G. Liesche, P. Gutfreund, S. Rogers, J. Appl. Cryst. **47**, 130 (2014).
[19] Müller-Buschbaum, G. Kaune, M. Haese-Seiller and J.-F. Moulin, J. Appl. Cryst. **47**, 1228 (2014).
[20] N. Wolff, S. Gerth, P Gutfreund, M. Wolff, Soft Matter **10(42)**, 8420 (2014).
[21] S. K. Sinha, E. B. Sirota, S. Garoff, H. B. Stanley, Phys. Rev. B **38**, 2297 (1988).
[22] George H. Vineyard, Phys. Rev. B **26**, 4146 (1982).
[23] F. Pfeiffer, V. Leiner, P. Hoghoj, I. Anderson, Phys. Rev. Lett. **88**, 055507 (2002).
[24] S. Jaksch, O. Holderer, M. Gvaramia, M. Ohl, M. Monkenbusch, H. Frielinghaus, Scientific Reports **7**, 4417 (2017).

[25] K. Theis-Bröhl, M. Wolff, A. Westphalen, H. Zabel, J. McCord, V. Höink, J. Schmalhorst, G. Reiss, T. Weis, D. Engel, A. Ehresmann, U. Rücker, B. P. Toperverg, Phys. Rev. B **73**, 174408 (2006).

[26] Hartmut Zabel, Katharina Theis-Bröhl, B. P. Toperverg, *Polarized Neutron Reflectivity and Scattering from Magnetic Nanostructures and Spintronic Materials in book: Handbook of Magnetism and Advanced Magnetic Materials*, (John Wiley & Sons, Ltd., 2007).

[27] http://bornagainproject.org/news

[28] M. Tolan, *X-ray Scattering from Soft Matter Thin Films* (Springer Tracts in Modern Physics vol. 148, Berlin, 1999).

[29] P. Müller-Buschbaum, Analytical and Bioanalytical Chemistry **376**, 3 (2003).

[30] S. H. Anastaisiadis, T. P. Russel, S. K. Satija, C. F. Majkrzak, Phys. Rev. Lett. **62**, 1852 (1989).

[31] W. A. Hamilton, P. D. Butler, S. M. Baker, G. S. Smith, John B. Hayter, L. J. Magid, and R. Pynn, Phys. Rev. Lett. **74**, 335 (1995).

[32] P. Busch, D. Posselt, D.-M. Smilgies, M. Rauscher, C. M. Papadakis, Macromolecules **40**, 630 (2007).

[33] R. Dalgliesh, Current Opinion in Colloid & Interface Science **7**, 244 (2002).

[34] W. A. Hamilton, Current Opinion in Colloid & Interface Science **9**, 390 (2005).

[35] Müller-Buschbaum, European Polymer Journal **81**, 470 (2016).

[36] H.J.C. Lauter, V. Lauter, B.P. Toperverg, Polymer Science: A Comprehensive Reference **2**, 411 (2012).

[37] V. Lauter, H.J.C. Lauter, A. G. Glavic, B.P. Toperverg, Reflectivity, *Off-Specular Scattering, and GISANS Neutrons, in book: Reference Module in Materials Science and Materials Engineering*, (Oxford: Elsevier, Editors: Saleem Hashmi, pp.1-27 (2016).

[38] C. Münster, T. Salditt, M. Vogel, R. Siebrecht, J. Peisl, Europhys. Lett. **46**, 486 (1999).

[39] M. S. Jablin, M. Zhernenkov, B. P. Toperverg, M. Dubey, H. L. Smith, A. Vidyasagar, R. Toomey, A. J. Hurd, J. Majewski, Phys. Rev. Lett. **106**, 138101 (2011).

[40] M. Sferrazza, M. Heppenstall-Butler, R. Cubitt, D. Bucknall, J. Webster, R. A. L. Jones, Phys. Rev. Lett. **81**, 5173 (2009).

[41] H. Ogawa, T. Kanaya, K. Nishida, G. Matsuba, J. P. Majewski, E. Watkins, J. Phys. Chem. **131**, 104907 (1994).

[42] M. Wolff, A. Magerl , H. Zabel, Thin solid films **515**, 5724 (2007).

[43] M. Wolff, A. Magerl, H. Zabel, J. Phys.: Cond. Matter **17**, S3645 (2005).

[44] M. Walz, M. Wolff, N. Voss, A. Magerl, H. Zabel, Langmuir **26**, 14391 (2010).

[45] M. Wolff, R. Steitz, P. Gutfreund, N. Voss, S. Gerth, M. Walz, A. Magerl, H. Zabel, Langmuir (Letter) **24**, 11331 (2008).

[46] P. M. Saville, I. R. Gentle, J. W. White, J. Penfold, J. R. P. Webster, J. Phys. Chem. **98**, 5935 (1994).

[47] Z.X. Li, J. R. Lu, R. K. Thomas, A. Weller, J. Penfold, R. P. Webster, D. S. Sivia, A. R. Rennie, Langmuir **17**, 5858 (2001).

[48] C. Gliss, H. Clausen-Schaumann, R. Günther, S. Odenbach, O. Randl, T.M. Bayerl, Biophysical Journal **74** 2443 (1998).

[49] M. L. Schlossman, Current Opinion in Colloid & Interface Science **7**, 235 (2002).

[50] V. Lauter-Pasyuk, H. J. Lauter, G. P. Gordeev, P. Muüller-Buschbaum, B. P. Toperverg, M. Jernenkov, W. Petry, Langmuir **19**, 7783 (2003).

[51] A. Vorobiev, G. Gordeev, W. Donner, H. Dosch, B. Nickel, B.P. Toperverg, Physica B **297** 194 (2001).

[52] K. Theis-Bröhl, M. Wolff, I. Ennen, C. Dewhurst, A. Hütten, B. P. Toperverg, Phys. Rev. B **78**, 134426 (2008).

[53] H.Frielinghaus, O.Holderer, F.Lipfert, M.Monkenbusch, N.Arend, D.Richter, Nuclear Inst. and Methods in Physics Research, A **686**, 71 (2012).

[54] F. Lipfert, H. Frielinghaus, O. Holderer, S. Mattauch, M. Monkenbusch, N. Arend, and D. Richter, Phys. Rev. E **89**, 042303 (2014).

[55] K. Gawlitza, O. Ivanova, A. Radulescu, O. Holderer, R. von Klitzing, S. Wellert, Macromolecules **48(16)**, 5807 (2015).

[56] S. Jaksch, O. Holderer, M. Gvaramia, M. Ohl, M. Monkenbusch, H. Frielinghaus, Sci Rep. **7**, 4417 (2017).

[57] H. Frielinghaus, M. Gvaramia, G. Mangiapia, S. Jaksch, M. Ganeva, A. Koutsioubas, S. Mattauch, M. Ohl, M. Monkenbusch, O. Holderer, Nuclear Inst. and Methods in Physics Research A **871**, 72 (2017).

EPJ Web of Conferences **188**, 05001 (2018) https://doi.org/10.1051/epjconf/201818805001
JDN 23

Inelastic and quasi-elastic neutron scattering. Application to soft-matter

Quentin Berrod[1], *Karine* Lagrené[1], *Jacques* Ollivier[2], and *Jean-Marc* Zanotti[1,*]

[1]*Laboratoire Léon Brillouin (CEA-CNRS), Université Paris-Saclay, CEA Saclay, 91191 Gif-sur-Yvette Cedex, France*

[2]*Institut Laue-Langevin, 38042 Grenoble Cedex, France*

Abstract. Microscopic dynamical events control many of the physical processes at play in condensed matter: transport, magnetism, catalysis and even function of biological assemblies. Inelastic (INS) and Quasi-Elastic Neutron Scattering (QENS) are irreplaceable probes of these phenomena. These experimental techniques reveal the displacements of atoms and molecules over distances spanning from angstroms to a few tens of nanometers, on time scales ranging from a fraction of picoseconds to microseconds. In this context, the different INS and QENS machines (Time-of-Flight (ToF), Backscattering (BS) and Neutron spin-echo (NSE)) stand at a central position. After introducing an underlying basic theoretical toolbox for neutron scattering, the principles and key elements of a ToF measurement are described. While, here, we mainly focus on disk choppers spectrometers, all the INS/QENS instruments share a common ground: they directly and simultaneously probe correlation functions in both time and space, so that the scattering vector (Q) dependence of the systems characteristic time(s) can be measured. To illustrate, the potentialities of the technique in the field of soft-matter, we show a multiscale approach of the dynamics of a polymer melt. The system is probed from the molecular to the mesoscopic scale (1 ps to 0.6 μs and 0.1 to 40 nm), in bulk and under nanometric confinement. We address the different dynamical modes of a high mass entangled polymer chain: local monomer dynamics, Rouse modes up to the reptation process. This study exemplifies that, used in conjunction with hydrogen/deuterium isotopic effects, high resolution QENS can be bridged to the Zero Average Contrast (ZAC) method to probe, in a non destructive way, the dynamics of a single polymer chain in bulk but also under severe nanometric confinement. Connection and complementarity of the neutron derived analysis with Pulsed-Field Gradient and Relaxation NMR techniques are discussed.

*e-mail: jmzanotti@cea.fr

Contents

1 Introduction

Within the framework of the wave-particle duality, a wave can be associated to a neutron. The thermal neutron beams[1] made available to the scientific community are designed to provide a wavelength range of the order of magnitude of interatomic distances. The interaction of a neutron beam with matter thus leads to a scattering phenomenon, and its analysis gives access to the local structure of the sample. A second particularly interesting feature of thermal neutrons is that their energy is of the order of magnitude of many excitations at play in condensed matter. The analysis of the energy exchanged between a neutron and the sample thus also makes it possible to access the dynamics of the system. A key point to keep in mind is that the underlying physical mechanism leading to the scattering phenomenon is a nuclear process driven by a spin-spin interaction between the spin of the incident neutron and the nuclear spin of the target nucleus. Neutron scattering then offers the interesting feature to play with strong isotopic effects. In the field of soft-matter, the Hydrogen/Deuterium (H/D) isotopic effect can be advantageously considered to *i)* mask a dynamical contribution of the system (solvent, side-chains...) and *ii)* finely balance the incoherent and coherent scattering to assess individual (self-diffusion coefficient for example) or collective (phonons) dynamical information. Also, the scattering vector ($\vec{Q}$) dependence can be used to access the geometry of the atomic motions (jump distances in a jump-diffusion process for example). The first sections of the manuscript are devoted to provide more detailed information of the points/properties described above. For the sake of consistency, throughout this manuscript, all the specific example, illustrating this lecture will be related to neutron experiments on a single polymer system: Poly(Etylene Oxide) (PEO[2], $-[CH_2-CH_2-O]_n-$, 35 kg.mol^{-1} *i.e.* above the critical entanglement mass M_e = 3.5 kg.mol^{-1}).

1.1 Phenomena accessible by neutron scattering

The time-of-flight (ToF) technique allows the study of dynamical phenomena of nuclear or magnetic origin. The quasi-elastic region, which extends up to a few meV around the elastic peak (zero energy transfer point: $\hbar\omega$ = 0 meV) is characteristic of phenomena and/or relaxations processes in the system. Over the whole energy range, from 0 to several hundreds meV, the signal can also corresponds to the exchange of quanta of energy between the neutron and the sample (Fig.1). The energy transfer range above 1 meV is often referred to as *spectroscopic* (neutron spectroscopy).

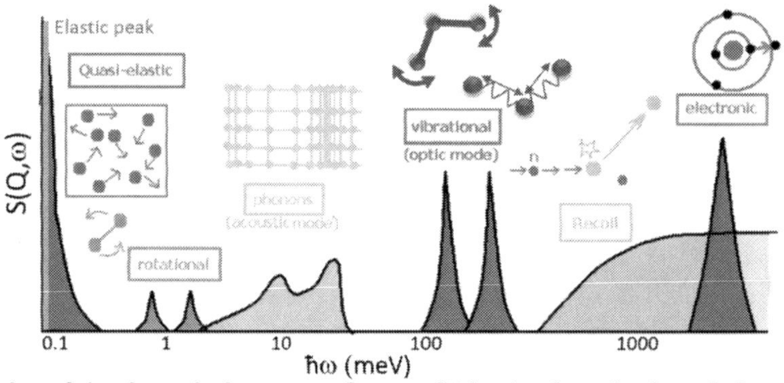

Figure 1. Schematics of the dynamical structure factors, $S(Q, \omega)$, of a selection of phenomena accessible by quasi-elastic and inelastic neutron scattering (from [1]).

[1]**Thermal neutrons:** neutrons of 1.8 Å wavelength. The associated energy is in the order of magnitude of the thermal agitation at room temperature *i.e.* 293 K, 25.25 meV, 203.65 cm^{-1}, 2200 m.s^{-1} or 6.11 Thz.

[2]Fully hydrogenated and fully deuterated PEO will be respectively designed by ^{h}PEO and ^{d}PEO.

https://doi.org/10.1051/epjconf/201818805001

1.2 Spatial and temporal ranges probed by neutron time-of-flight

The Fig.2 specifies the field of action of time-of-flight spectrometers in the neutron techniques instrumental scenery. They allow to cover the wide range of the (Q, ω) domain, intermediate between the filters instruments (*triple axis*) and the high resolution instruments like the backscattering or spin-echo spectrometers. This representation, however, disregards the counting times required for measuring an exploitable signal and certain technical limitations: the sample quantity and/or the low scattering may, for example, be an obstacle to backscattering studies. In spin-echo, the absence of polarization - induced by a coherent/incoherent scattering competition - can make it impossible to measure the dynamics of the system in certain domains of scattering vector.

The temporal range probed by time-of-flight calls upon a fundamental instrumental notion which

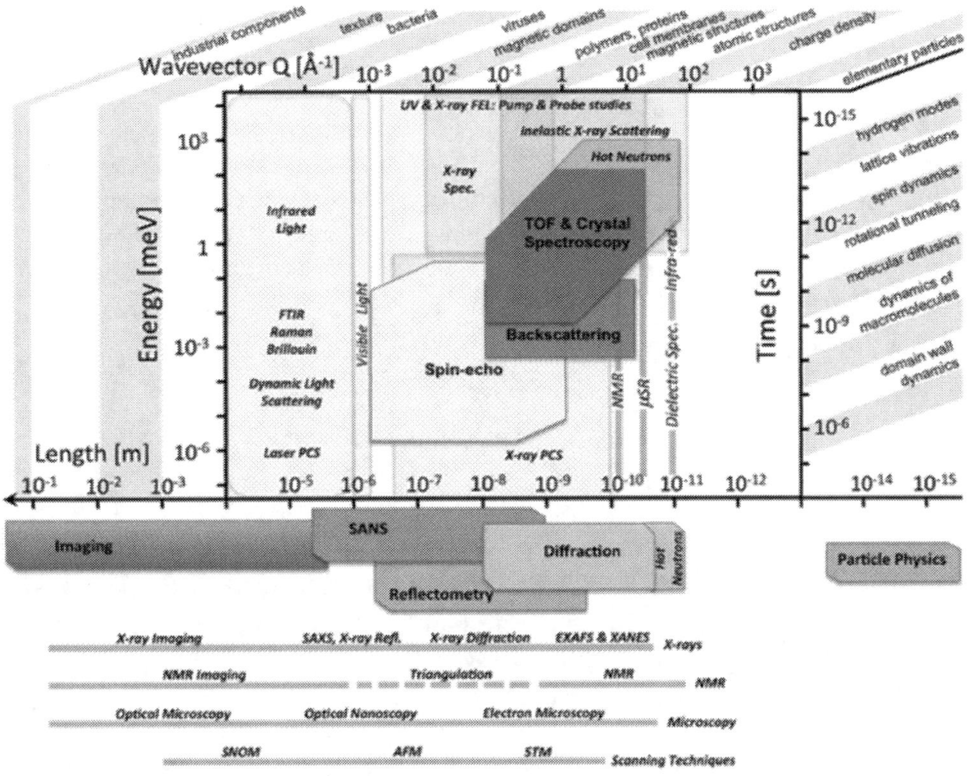

Figure 2. a) Complementarity of inelastic neutron scattering (triple axis also called "filter techniques"), time-of-flight, backscattering and neutron spin-echo) with other spectroscopic techniques. The spatial (r [Å] $\approx 2\pi/Q$ [Å^{-1}]) and temporal (t [ps] $\approx 0.658/\hbar\omega$ [meV]) regions associated with the main physical phenomena potentially observable are shown. Connection, not shown on this graph, to thermodynamic quantities, like the specific-heat is also possible (see Eq.25). One can note that neutron scattering has the capability to simultaneously probe a significant region of this spatial and temporal map. Also, the spatial dependence of the correlation time is directly accessible. From the European Spallation Source Web site (https://europeanspallationsource.se/science-using-neutrons).

will be discussed in greater detail hereafter: the energy resolution. It results from instrumental imperfections. This energy resolution limits not only the precision of the measurement but also defines the upper limit to the (longest) accessible correlation times (see section 3.3).

2 Neutron scattering: Few theoretical elements

2.1 Basic principles

As in all radiation scattering experiments, a neutron scattering experiment consists of *illuminating* a sample with a neutron beam of wave vector $\vec{k_0}$, wavelength λ_0 and energy E_0. Under the effect of the interaction $V(\vec{r}, t)$ between the neutron and the system, downstream of the sample, the radiation is characterized by a wave vector $\vec{k_1}$ and an energy E_1 (Fig.3). The diffusion angle, 2θ, is defined as the angle between the vectors $\vec{k_1}$ and $\vec{k_0}$. The energy of a neutron with wavevector $\vec{k}$ is:

$$E = \hbar\omega = \frac{\hbar^2|\vec{k}|^2}{2m_n}, k = \frac{2\pi}{\lambda} \quad (1)$$

where m_n is the neutron mass and $\hbar$, the Planck constant divided by 2π. The momentum and energy conservation rules lead to:

$$\begin{aligned} \vec{Q} &= \vec{k_1} - \vec{k_0} \\ \hbar\omega &= \hbar(\omega_0 - \omega_1) \end{aligned} \quad (2)$$

where $\vec{Q}$ [3] is the scattering vector and $\hbar\omega$ is the energy gained or lost by the neutron. Below, by convention, the negative values of ω will correspond to an energy loss of the sample, therefore an energy gain of the neutron. Similar to a classical collision (e.g. like at the pool game), the conservation law of momentum and energy imposes the kinematic condition:

$$\frac{Q^2}{k_0^2} = 2 - \frac{\omega}{\omega_0} - 2cos(2\theta)\sqrt{1 - \frac{\omega}{\omega_0}} \quad (3)$$

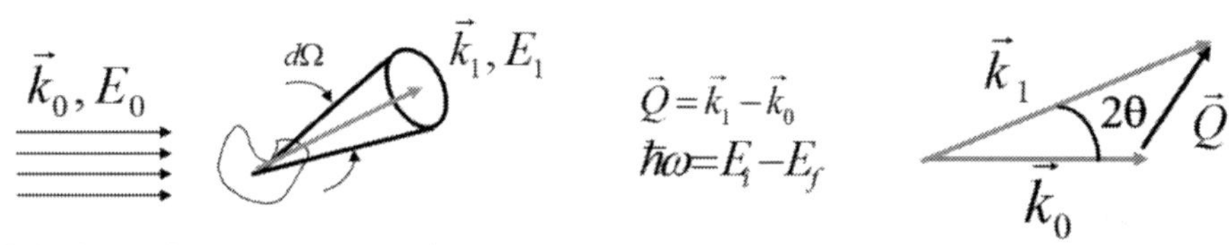

Figure 3. Principle of a neutron scattering experiment. As a result of interaction with the sample, incident neutrons ($\vec{k_0}$ and ω_0) can undergo a wave vector and energy changes: respectively $\vec{Q} = \vec{k_1} - \vec{k_0}$ and $\hbar\omega = \hbar(\omega_0 - \omega_1)$. In this general case of an inelastic interaction ($\hbar\omega \neq 0$), Q is a function of θ and ω (Eq.3). For an elastic scattering process $\vec{k_1} = \vec{k_0}$ (*i.e.* $\omega_0 = \omega_1$), the elastic scattering vector is $\vec{Q}_{el} = \frac{4\pi}{\lambda} sin(\theta).\vec{u}$, where $\vec{u}$ denotes a unit vector (not shown). In the reciprocal domain, *i.e.* the domain where the neutron detector directly operates, the two working quantities are Q and ω. The two dual quantities in the real space, *i.e.* the domain of the sample, are the time, t, and space, r. One can switch from one domain to the other by Fourier Transform over r and t (Eq.4).

Experimentally, one measures [2] the differential scattering cross-section *i.e.* the number of neutrons scattered over a solid-angle $d\Omega$ and by energy transfer $d\omega$. This quantity is ruled by the time dependence of the positions $\vec{r_i}(t)$ of the atoms of the system.[4]

$$\frac{d^2\sigma}{d\Omega d\omega} = \frac{1}{2\pi}\frac{k_1}{k_0}\sum_{i,j}\left\langle b_i b_j\right\rangle \int_{-\infty}^{+\infty} \left\langle e^{i\vec{Q}\vec{r_i}(0)} e^{-i\vec{Q}\vec{r_j}(t)}\right\rangle e^{-i\omega t}.dt \quad (4)$$

[3]From now on, the scattering vector $\vec{Q}$ will be simply written Q.
[4]From now on, the time dependence of the position $\vec{r_i}(t)$ will be written $r_i(t)$

2.2 Coherent and incoherent scattering

The neutron scattering process is the result of a nuclear interaction depending on the spins of both the neutron and the target nucleus. This process can be taken into account by a single quantity, the scattering length b_i. The development of the $\langle b_i b_j \rangle$ term of Eq.4 leads to an expression where the scattered intensity is expressed in terms of coherent and incoherent contributions (Fig.4):

$$\left(\frac{d^2\sigma}{d\Omega d\omega}\right) = \left(\frac{d^2\sigma}{d\Omega d\omega}\right)_{coh} + \left(\frac{d^2\sigma}{d\Omega d\omega}\right)_{inc} \tag{5}$$

The terms of this equation are weighted by the neutron scattering σ_{coh} et σ_{incoh} cross-sections:

$$\left(\frac{d^2\sigma}{d\Omega d\omega}\right)_{coh} = \frac{k_1}{k_0}\frac{\sigma_{coh}}{4\pi} S(Q,\omega) \tag{6}$$

$$\left(\frac{d^2\sigma}{d\Omega d\omega}\right)_{inc} = \frac{k_1}{k_0}\frac{\sigma_{inc}}{4\pi} S_{inc}(Q,\omega) \tag{7}$$

In these expressions, as they carry the physics of the system, the important terms are the coherent and incoherent dynamical structure factors, $S(Q,\omega)$[5] and $S_{inc}(Q,\omega)$, respectively. The cross-sections prefactors only indicate the fact that the technique used to assess them is neutron scattering. $S(Q,\omega)$ and $S_{inc}(Q,\omega)$ obey the so-called *sum rules*:

$$\int_{-\infty}^{+\infty} S(Q,\omega).d\omega = S(Q) \tag{8}$$

$$\int_{-\infty}^{+\infty} S_{inc}(Q,\omega).d\omega = 1 \tag{9}$$

The coherent structure factor $S(Q)$ (Eq.8) is usually a Q dependent quantity showing for example Bragg peaks for crystalline materials or halos for liquids. After integration over ω, the incoherent dynamical structure factor is a Q-independent signal (Eq.9). In Small Angle Neutron Scattering (SANS) or diffraction experiments, it is therefore often labeled as a "background" and one tends to minimize its contribution. In what follows, we will show that this incoherent contribution is much more than a background: its dependence in energy carries key informations on the dynamics of the system. This feature is made crystal-clear by the use of the van Hove formalism [3].

2.3 The van Hove formalism

Let's consider a particles assembly and $G(r,t)$ a correlation function indicating the probability to observe a particle at a position r at time t, provided that a particle P stands at a position $r = 0$ at time $t = 0$. One can split $G(r,t)$ in a *self*, $G_s(r,t)$, and a *distinct*, $G_d(r,t)$ contributions:

$$G(r,t) = G_s(r,t) + G_d(r,t)$$

$G_s(r,t)$ is the probability that the initial particle P stands at r at time t, while $G_d(r,t)$ is the probability that it is a different particle. The van Hove formalism [3] establishes a direct connection between the correlation functions above and the scattered intensity:

$$S(Q,\omega) = \frac{1}{2\pi}\int G(r,t)e^{i(Qr-\omega t)}.dr.dt \tag{10}$$

$$S_{inc}(Q,\omega) = \frac{1}{2\pi}\int G_s(r,t)e^{i(Qr-\omega t)}.dr.dt \tag{11}$$

[5]Note that, by convention, the coherent dynamical structure factor does write $S(Q,\omega)$ and not $S_{coh}(Q,\omega)$

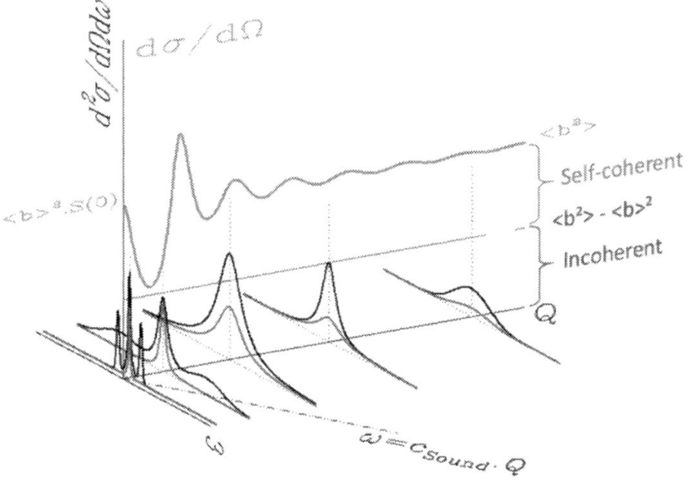

Figure 4. Schematic representation of the contributions to the neutron intensity scattered by a liquid sample in the quasi-elastic region. The coherent and incoherent inelastic contributions are shown as black and red lines on the basal (Q, ω) plane. After integration over energy, the incoherent inelastic contribution reduces to a flat background of intensity $< b^2 > - < b >^2$ (Eq.9). As, here, for the sake of clarity, we neglect the strong inelastic processes, the structure factor $S(Q)$ (Eq.6, green line), oscillates on top of this incoherent contribution, around the level of the self-coherent contribution value $< b^2 >$. From [4].

• **The coherent intensity** is partly related to the "distinct" $G_d(r,t)$ term. It therefore accounts for a collective information. As it appears from equation 8, for a coherent system, after integration in energy, the analysis of the scattered intensity provides informations on the structure of the system. One should note, that the "self" term also contributes to Eq.10 (terms $i = j$ of the array Eq.4). This so-called "self-coherent" contribution is a Q independent signal, adding up to the incoherent scattering to define the high Q limit of diffracted signals (see Fig.4). $S(Q)$ is the integral of the coherent dynamical structure factor over the energy, and the Q Fourier transform of the total van Hove correlation function at $t = 0$:

$$S(Q) = \int_{-\infty}^{+\infty} S(Q,\omega).d\omega = \int G(r, t = 0).e^{-iQr}.dr$$

If one now consider the dependence in energy of the coherent intensity, it is possible to assess collective dynamical processes like phonons for example.

• **The incoherent intensity** is only connected to the self contribution "self" $G_s(r,t)$. As this signal only carries an individual information, it does not give access to any direct information regarding the structure (Eq.9) of the system. Depending on the energy associated with the incoherent process, one defines:

• ***the elastic***(*i.e.* $\hbar\omega = 0$) **incoherent scattering**: it can be used to access indirect dynamical information (mean-square displacements, confinement volume of a particle).

• ***the quasi-elastic*** (*i.e.* energy exchange of few meV) **incoherent scattering**: it can be used to measure self-diffusion coefficients, residence times in jump diffusion processes.

• ***the inelastic incoherent scattering***: it can be used to perform vibrational spectroscopy. It can be used to extract the vibrational density of states *i.e.* the distribution in energy of the vibrational modes of the system.

Coherence or incoherence of the scattered intensity are driven by the isotopic composition of the system. Specific devices (polarizers, analyzers) can be used to specifically access the coherent or the incoherent contribution of the scattered intensity. This can be performed, for example, on NSE instruments.

2.4 Taking advantage of isotopic effects

Experimentally, coherent and incoherent scattering can be used to distinguish individual and collective processes. The table 1 shows the values of the coherent and incoherent neutron cross-sections of most of the isotopes used in soft-matter experiments. The hydrogen atom shows an incoherent scattering

Element	b_{coh} (fm)	σ_{coh} (barn)	σ_{inc} (barn)	σ_{abs} (barn)
C	6.64	5.55	0.001	0.0035
N	9.36	11.01	0.50	1.90
O	5.803	4.232	0	0.000191
S	2.847	1.0188	0.007	0.53
H	-3.7406	1.7583	80.27	0.3326
D	6.671	5.592	2.05	0.000519

Table 1. Coherent and incoherent scattering cross-sections of different elements and isotopes. The absorption cross-sections, σ_{abs}, are indicated for an incident wavelength of 1.8 Å [5].

cross-section 40 times higher than the one of the deuterium, its isotope. Also, by comparing the numerical values, it is clear that in hydrogenated compounds, the incoherent contribution of hydrogen largely dominates all the scattering contributions from the other atoms. The incoherent dynamic structure factor of such compounds can therefore be directly related to the individual dynamics of the sample's protons.
In the soft-matter and biology fields, specific deuteration can be used to minimize the detrimental contributions of specific moieties: typically the solvent and/or side-chains can be deuterated. In section 4.4, we also show that a mixture of hydrogenated and deuterated polymer chains can be used to access the dynamical contribution of a "single" chain.

2.5 Derivation of $S_{inc}(Q,\omega)$ for selected examples of dynamical modes

In this section, we provide an overview, in both the time and energy domains, of how key typical dynamical modes can be apprehended in inelastic and quasi-elastic neutron scattering. This is also summarized on Fig.5.

2.5.1 Long range translational and jump diffusion

• *Random walk diffusion (Fick's law)*

The isotropic diffusion mechanism of a particle can be described as a cyclic phenomenon: following an interaction, with another particle by example, the particle is set in motion and travels a path r_{elem} in space over a time τ. Then, a new event changes the speed and direction of the particle so that the initial history of the particle (position, speed, direction) is lost. One can show that [7], if we observe

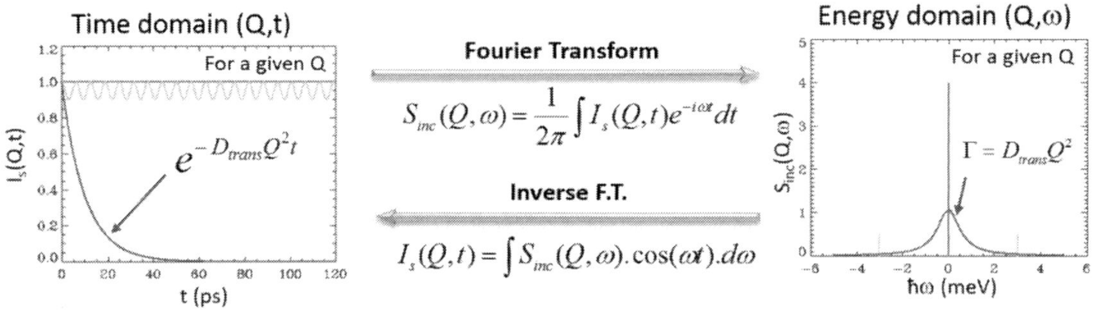

Figure 5. Schematic representation of the inelastic spectra of different dynamical modes. We consider the autocorrelation function $I_s(Q,t)$ of a particle in the (Q,t) domain. The switch to the (Q,ω) domain is achieved by time Fourier Transform of $I_s(Q,t)$. By definition of an autocorrelation function, whatever the dynamics of the particle, its position at $t=0$ is known *i.e.* $I_s(Q,t=0)=1$. If the particle is immobile, whatever t, its position is also known *i.e.* $I_s(Q,t)=1$ (see red curve). The Fourier transform in time of this horizontal straight line is a dirac function in energy positioned at $\hbar\omega=0$. This physically corresponds to a situation where neutrons do not exchange any energy with the particle. One talks of "elastic scattering". If now a particle experiences a vibrational mode at frequency $\hbar\omega_0$, the autocorrelation function $I_s(Q,t)$ is a sinusoidal function of time. Its Fourier transform materializes as an elastic dirac along with two diracs centered at $\hbar\omega_0$ and $-\hbar\omega_0$. These two diracs of the inelastic zone of the spectrum ($\hbar\omega \neq 0$) represents the simpler expression of $g(\omega)$, the vibrational density of states (see section 2.5.3). In the case of a particle experiencing a long-range translational diffusion (*c.f.* section 2.5.1), $I_s(Q,t)$ takes an exponential form (Eq.14). The corresponding dynamical structure factor is a Lorentzian line centered around $\hbar\omega=0$ with HWHM, Γ (Eq.15). From [6].

the system at a scale larger than r_{elem} and a long time compared to τ, the probability of finding the particle at a position r at time t obeys the relation:

$$\frac{\partial G_s(r,t)}{\partial t} = D_t \nabla^2 G_s(r,t) \tag{12}$$

where D_t is the long range diffusion coefficient. With the boundary conditions $G_s(Q,0)=\delta(r)$ and $G_s(r,\infty)=0$, the solution of this differential equation is:

$$G_s(r,t) = \frac{1}{(4\pi D_t t)^{\frac{3}{2}}} e^{-\frac{r^2}{4D_t t}} \tag{13}$$

Physically, this relationship means that the probability of knowing the position of the particle at a distance r from its initial position is a Gaussian function of space (r). The square of the standard deviation (quantifying "the spreading" of the curve) is directly proportional to time. The passage in the reciprocal space is made by space Fourier Transform:

$$I_s(Q,t) = \frac{1}{2\pi}\int G_s(r,t).e^{i(Qr)}.dr = e^{-D_t.Q^2.t} \tag{14}$$

The corresponding incoherent dynamic structure factor is obtained by time Fourier transform:

$$S_{inc}(Q,\omega) = \frac{1}{\pi}\frac{D_t Q^2}{(D_t Q^2)^2 + \omega^2} \tag{15}$$

$S_{inc}(Q,\omega)$ is a Lorentzian line in energy with integrated intensity 1 (*i.e.* $\int_{-\infty}^{+\infty} S_{inc}(Q,\omega).d\omega = 1$), of Half-Width at Half Maximum (HWHM) $\Gamma = D_t Q^2$. Γ shows a linear dependence vs Q^2 with a slope D_t. It is referred as the so-called "DQ^2 law" (Fig.6).

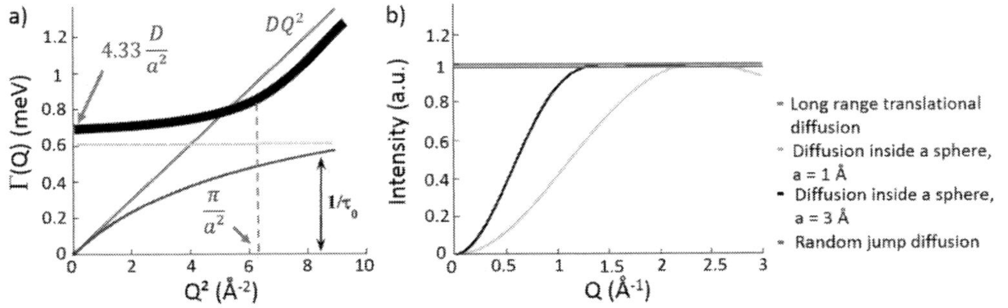

Figure 6. Q dependence of the quasi-elastic HWHM and intensity (integrated in ω) of selected dynamical processes. In red: long range translational diffusion (Fick's law or DQ^2 law, Eq.15). Blue: jump diffusion (Eq. 16). Green: reorientational diffusion (Eq.22). Black: diffusion within a confinement volume [7, 8].

• *Jump diffusion.*

In the situation where the (Q,ω) range used allows to access phenomena with spatial and temporal characteristics of the same order of magnitude than those of the elementary process leading to the diffusion, the conditions that led us to write Eq.12 is no longer valid. It is the case when working at large vector of diffusion. In the model of jump diffusion [7], it is assumed that between two events leading to diffusion, the particle remains on a given site for a time τ_0, very long in front of the jump duration τ. In this case, the dynamic structure factor writes:

$$S_{inc}(Q,\omega) = \frac{1}{\pi}\frac{f(Q)}{f(Q)^2+\omega^2} \tag{16}$$

$S_{inc}(Q,\omega)$ is therefore a Lorentzian line with a HWHM $f(Q)$:

$$f(Q) = \frac{D_t Q^2}{1+D_t Q^2\tau_0} \tag{17}$$

At small Q, the Q dependence of $f(Q)$ is similar to the one of the DQ^2 law. At large Q, $f(Q)$ leaves linear behavior to tend towards the asymptotic value of the reverse of the residence time τ_0 (Fig.6).

2.5.2 Diffusion in restricted geometry. The notion of EISF

• *The concept of EISF.*

The autocorrelation function of the individual dynamics of a particle is described by the van Hove autocorrelation function.

At t = 0, $G_s(r,0) = \delta(r)$. The "self" contribution is obtained by inverse Fourier transform: $I_s(Q,t) = \int G_s(r,t).e^{iQr}.dr$.
Let's consider that the particle is experiencing a perturbation at t = 0. After relaxation, at $t = \infty$, the

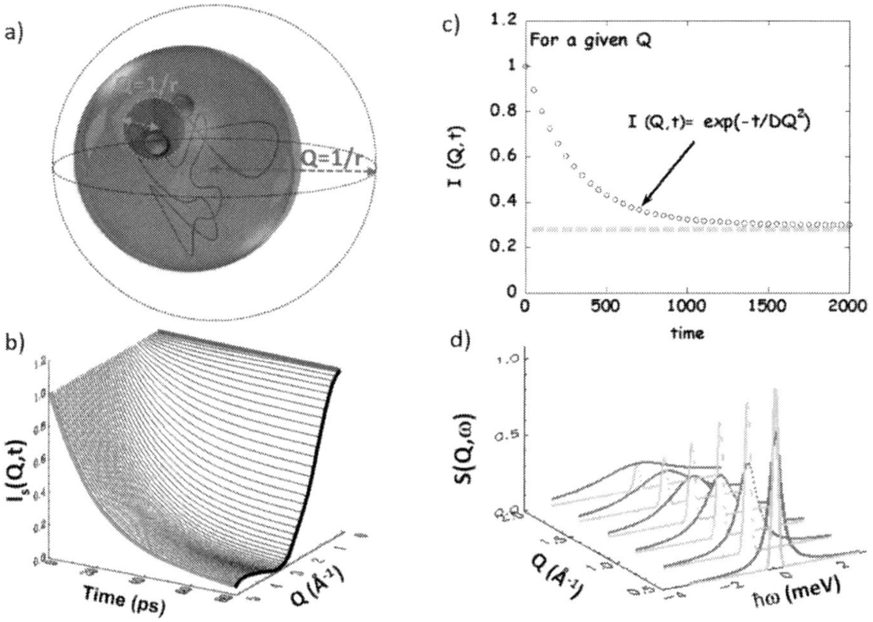

Figure 7. The Incoherent Elastic Structure Factor: EISF. a) Consider a particle (red dot) experiencing a diffusion process (trajectory in red) in a confinement volume (brown sphere). b) 3D representation of the evolution of the corresponding intermediate self scattering function $I_s(Q,t)$. Each curve represents the evolution as a function of time of $I_s(Q,t)$, at a given Q value, *i.e.* at a scale $r \approx 2\pi/Q$ of observation of the system. Whatever this scale, at $t = 0$, by definition of an autocorrelation function, the position of the particle is known: $I_s(Q, t = 0) = 1$. At longer time, *(i)* if we look at the system on a very local scale (light blue dotted sphere in (a)) *i.e* at large Q, the particle diffuses and informations about its position are quickly lost: we end-up at $I_s(Q, t = 0) \approx 0$ (light blue line in (b)). *(ii)* If we observe the system from afar, *i.e.* on a much larger scale than the size of the confinement cage (blue dotted sphere in (a)), $1/r$ is large so Q is small), at any moment the position of the particle is known: it sits somewhere inside the confinement cage so $I_s(Q,t) \approx 1$ (blue line in (b)). c) Between these two extreme scales of observation, $I_s(Q,t)$ adopts an intermediate behavior, characterized by the presence of a plateau (in green). The curved defined by $I_s(Q, t = \infty)$ is a function of a single parameter Q: this is nothing else than the form factor (*Structure Factor*) of the confinement volume (in black in (b)). Since the particle is subject to a diffusion process, next to the plateau, the time dependent contribution takes the form of Eq.14. It carries the information on the dynamics of the system. d) 3D representation of the evolution of the dynamic structure factor, $S_{inc}(Q,\omega)$, recorded on a ToF spectrometer. In the (Q,ω) domain, for each Q-value, the plateau and exponential of the (Q,t) domain of (b) leads to an elastic peak of intensity $I_{El} = I_s(Q, t = \infty)$ and a quasi-elastic intensity ($I_{Quasi} = 1 - I_s(Q, t = \infty)$), respectively (cf. Fig.5). In the (Q,t) domain, for each Q-value, the plateau is characterized by the elastic fraction of the signal. It is the same in the (Q,ω) domain: $EISF = I_{El}/(I_{El} + I_{Quasi})$. As it is derived from a plateau in the time domain of an autocorrelation function, hence the elastic intensity of an incoherent scattering, this form factor is called an Elastic Incoherent Structure Factor (EISF). From [6].

system is back to equilibrium, and the correlation with the position at t = 0 has been lost [7]:

$$\begin{aligned} I_s(Q, t = +\infty) &= \left\langle e^{-iQr(0)} e^{-iQr(t=+\infty)} \right\rangle \\ &= \left\langle e^{-iQr(0)} \right\rangle \left\langle e^{-iQr(t=+\infty)} \right\rangle \\ &= \left| \left\langle e^{-iQr(t=+\infty)} \right\rangle \right|^2 \\ &= \int G_s(r, t = +\infty) e^{iQr} .dr \end{aligned} \tag{18}$$

$I_s(Q, t = +\infty)$ is therefore the form factor of the region of space accessible to the particle (Fig.7). The dynamic incoherent structure factor can be decomposed in a time dependent quantity and a constant value with time, corresponding in the time domain to: $I_s(Q, t)$ and $I_s(Q, t = +\infty)$, respectively:

$$S_{inc}(Q, \omega) = I_s(Q, t = +\infty).\delta(\omega) + S^{qe}_{inc}(Q, \omega) \tag{19}$$

where $S^{qe}_{inc}(Q, \omega)$ accounts for the quasi-elastic intensity noticeable around the elastic peak $\delta(\omega)$. The **elastic** peak intensity of an **incoherent** scattering spectrum of a particle subjected to a confined diffusion is therefore closely related to the form factor (**structure factor**) of the space of confinement of the particle. We talk about Elastic Incoherent Structure Factor (EISF). A more phenomenological description is given on Fig.7.

• *Experimental determination of the EISF.*

By integration of the dynamical structure factor at constant Q:

$$\begin{aligned}\int_{-\infty}^{+\infty} S_{inc}(Q, \omega)|_{Q\ constant}\, d\omega &= \int_{-\infty}^{+\infty} \left(\int_{-\infty}^{+\infty} I_s(Q, t) e^{-i\omega t}.dt \right).d\omega \\ &= \int_{-\infty}^{+\infty} I_s(Q, t).\delta(t).dt \\ &= I_s(Q, 0) = 1\end{aligned} \tag{20}$$

According to Eq.19 and Eq.20, $I(Q, \infty)$ is the elastic contribution of the total scattered intensity (elastic + quasi-elastic). Therefore, after fitting the total intensity with a model accounting for both the elastic and quasi-elastic contributions, the EISF is calculated according to:

$$EISF(Q) = \frac{I^{el}(Q)}{I^{el}(Q) + I^{qe}(Q)} \tag{21}$$

where $I^{el}(Q)$ and $I^{qe}(Q)$ are the elastic and quasi-elastic intensity, respectively. The comparison of this experimental value with a theoretical EISF derived from a model makes it possible to access to the geometry of a particle motion.

• *Rotational diffusion*

In the case of a particle diffusing freely over a sphere of radius b, with a good approximation [9], the incoherent dynamical structure factor writes

$$S_{inc}(Q, \omega) = B_0(Q)\delta(\omega) + [1 - B_0(Q)]\, L(\omega, \Gamma_r) \tag{22}$$

where D_R is the rotational diffusion coefficient, $B_0(Q) = j_0(Qb)^2$ and $L(\omega, \Gamma_r)$ stands for a Lorentzian line with a HWHM $\Gamma_r = 2D_R$. At the opposite of the free (non-confined) diffusion: *i)* the particle being confined in a restricted portion of the space induces the existence of an EISF and hence the presence of an elastic peak in the expression of the dynamical structure factor, *ii)* Γ_r is Q independent. This is a property common to all the reorientational processes.

2.5.3 *Vibrational dynamics*

• *Vibrational density of states and mean-square displacements.*

Unlike infra-red or raman spectroscopy, there is no selection rules for inelastic neutron scattering. Any accessible transitions in the energy range where the experiment is carried out, are allowed. We can therefore access a true weighted density of state. Note however that the intensity excitations strongly depend on the temperature of the system. The frequency distribution function of the vibrations (eigen modes, resonant frequencies ...) of a system is the so-called **vibrational density of states** (VDOS), $g(\omega)$. At a temperature T, the intensity scattered elastically by an assembly of N harmonic oscillators, of mass M, experiencing various isotropic inelastic modes in three dimensions (of density of states $g(\omega)$) is proportional to:

$$S(Q,\omega=0) = e^{-Q^2\langle u^2\rangle/3}.\delta(\omega) \quad with \quad \left\langle u^2\right\rangle = \frac{\hbar}{6MN}\int \frac{g(\omega)}{\omega} coth(\frac{\hbar\omega}{2k_BT}).d\omega \tag{23}$$

For $\hbar\omega \gg 2k_BT$, at high enough temperatures, $coth(x) \simeq \frac{1}{x}$ Eq.23 simplifies:

$$\left\langle u^2\right\rangle = \frac{3k_bT}{M}\int \frac{g(\omega)}{w^2}.d\omega \tag{24}$$

In purely harmonic systems, the frequency of the vibrational modes are strictly independent of the temperature and therefore of the *shape* of $g(\omega)$. However, as the intensity is proportional to the population of the different modes, $g(\omega)$ is controlled by the temperature of the sample *via* the Bose population factor $n(\omega)$ (equivalent to the Maxwell-Boltzmann statistics at high temperature). Eq.23 shows that the vibrational modes (vibration of atoms around their position of equilibrium) induce a Gaussian loss of intensity, clearly observable at large Q: this is the famous Debye-Waller effect existing also in X-ray diffraction. The scattered intensity in the inelastic zone of the spectrum gives access to $g(\omega)$. On the other hand, according to Eq.23, a fit of $ln(S(Q,\omega\approx 0))$ *vs* Q^2 is an estimate of the total spatial extension induced by the different mode, $\left\langle u^2\right\rangle$ (Fig.8a). However, it should be noted that for a solid, essentially static, the neutrons mainly undergo an elastic diffusion and that the inelastic bands are much less intense than the elastic peak (ratio of the order of one to hundred). Fine measurement of the density of vibrational states, $g(\omega)$, demands an excellent signal-to-noise ratio. **A deviation from the linearity of $\left\langle u^2\right\rangle$ *vs* T is characteristic of the appearance/disappearance of dynamical (diffusional for example) and/or vibrational modes**. In practice, the temperature where the quasi-elastic signal effectively is determined by performing a so-called "Elastic Scan": the sample temperature is submitted to a temperature ramp (typically 0.5 K/min) and at each temperature the scattering is recorded for a short time (typically 5 minutes). As the scattering is generally intense around the elastic peak, the uncertainty of the slope of the $ln(S(Q,\omega\approx 0))$ *vs* Q^2 linear fit is moderate and a curve showing the temperature evolution of of $\left\langle u^2\right\rangle$ with several hundreds of point can be plotted (see Fig.8b). Once the temperature range of the quasi-elastic signal has been determined, one can perform long scans at selected temperature to record the quasi-elastic signal with a good statistics (see section 4.2).

• *Connection with thermodynamics*

According to Klug *et al.* [11] the vibrational heat capacity can be accurately estimated below 100 K from the inelastic neutron spectra (INS) using the harmonic oscillator model

$$C_v = R\int \frac{g(\omega)\,(\hbar\omega/k_BT)^2\,exp(\hbar\omega/k_BT)}{(exp(\hbar\omega/k_BT)-1)^2} \tag{25}$$

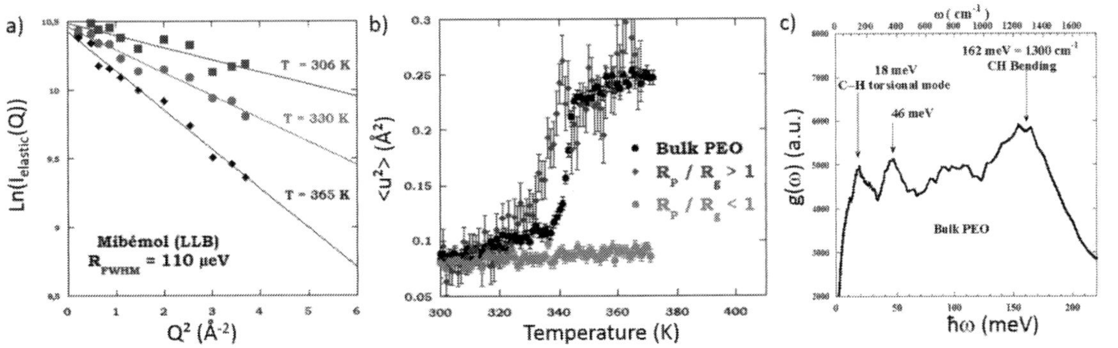

Figure 8. a) Q dependent Gaussian loss of elastic intensity I_{el} of 35 kg/mol ^{h}PEO at selected temperature. b) Practical result of an "elastic scan" (see text) showing the temperature evolution of $< u^2 >$ obtained from a linear fit of $ln(I_{el})$ vs Q^2, in the [300-373 K] r ange. For bulk PEO, the clear jump of $< u^2 >$ at 331 K, indicates the sample melting. Above this temperature, diffusive motions come into play, and are detected as a broadening of the the quasi-elastic line of the $S_{inc}(Q,\omega)$ spectrum. The plot b) illustrate the influence of confinement on PEO dynamics. The confining material is an Aluminum Oxyde Membrane (AAO, see Fig.16). On can define two characteristic lengths in the system: R_P, the AAO pore diameter, and R_g, the radius of gyration of the polymer chain (for a 35 kg/mol PEO chain, R_g =69 ±1 Å). Depending on the ratio of R_P/R_g, one can detect *i)* a clear downshift of the melting point (Gibbs-Thomson effect [10]) for $R_P/R_G > 1$ and *ii)* a glass-like arrest of the polymer dynamics for $R_P/R_G < 1$. c) Vibrational density of states, $g(\omega)$ of a bulk PEO measured on a spallation source (inverted geometry spectrometer Fig.14). Note that the high incident energy of incoming neutrons allows to reach high neutron energy loss.

where Cv stands for the isochoric molar heat capacity, R is the gas constant, and k_B is the Boltzmann's constant. The anharmonic part of the specific heat associated with the thermal expansion $CpCv = TV\alpha^2/\kappa_T$ (where V is the molecular volume, α is the volume thermal expansion, and κ_T is the isothermal compressibility) is estimated to be negligible (for ice: 0.025 J mol^{-1} K^{-1} at 100 K [12]). Thus the isobaric heat capacity, Cp, is virtually indistinguishable from the isochoric specific heat Cv. Above 100 K, the vibrational heat capacity of ice becomes noticeably anharmonic [11] and deviations of the calculated heat capacity from the experimental values are expected. The theory is exact for simple crystals and valid in a good approximation for amorphous systems with dominant incoherent scattering. In biomolecules, on average, half of the atoms are protons and their roughly homogeneous distribution reflects the collective dynamics. Also as neutron scattering originate from nuclear interactions, no selection rule applies, so that $g(\omega)$ can be considered as a genuine VDOS [13].

2.5.4 The total dynamical structure factor

The total dynamical structure factor is the convolution of the different elemental structure factors described in the previous sections:

$$S_{inc}(Q,\omega) = S_{inc}^{trans}(Q,\omega) \otimes S_{inc}^{rot}(Q,\omega) \otimes S_{inc}^{vib}(Q,\omega) \qquad (26)$$

The vibrational contribution is the sum of an elastic and an inelastic contribution due to the internal vibrations of the molecule ($S_{inc}^{V}(Q,\omega)$) and the lattice vibrational modes ($S_{inc}^{L}(Q,\omega)$):

$$S_{inc}^{vib}(Q,\omega) = e^{-2W(Q)}\left(\delta(\omega) + S_{inc}^{V}(Q,\omega) + S_{inc}^{L}(Q,\omega)\right) \qquad (27)$$

The inelastic part $S^{V}_{inc}(Q, \omega)$ is composed of a collection of δ functions each centered at frequencies corresponding to each molecular vibrational modes. The contribution $S^{L}_{inc}(Q, \omega)$ can be more complex, but can usually be taken into account by a flat background in the quasi-elastic region of the spectrum. In isotropic samples, it is possible to perform a powder average and the Debye-Waller factor writes:

$$e^{-2W(Q)} = e^{-Q^2\langle u^2\rangle/3} \tag{28}$$

$\langle u\rangle^2$ is the mean-square displacement of the atoms undergoing vibrational modes. The convolution product in Eq.26 simplifies:

$$S_{inc}(Q, \omega) = e^{-Q^2\langle u\rangle^2/3}\left(S^{trans}_{inc}(Q, \omega) \otimes S^{rot}_{inc}(Q, \omega) + S^{I}_{inc}(Q, \omega)\right) \tag{29}$$

where $S^{I}_{inc}(Q, \omega)$ is an inelastic term due to the convolution of $S^{V}_{inc}(Q, \omega)$ and $S^{L}_{inc}(Q, \omega)$ by $S^{rot}_{inc}(Q, \omega) \otimes S^{trans}_{inc}(Q, \omega)$. A detailed discussion of the approximations leading to this results can be found at page 66 of the reference [7]. In the quasi-elastic region of the spectrum, this equations reduces to:

$$S_{inc}(Q, \omega) = e^{-Q^2\langle u^2\rangle/3}\left(S^{trans}_{inc}(Q, \omega) \otimes S^{rot}_{inc}(Q, \omega)\right) + \mathcal{B}(Q) \tag{30}$$

In all cases, the time reversibility is not considered *i.e.* $I(Q, t) = I(Q, |t|)$, and the auto-correlation functions are considered as even and the corresponding Fourier Transforms are real.

2.6 Aspects related to the sample temperature

Let's consider a neutron with incident energy $\hbar\omega_0$ interacting with an assembly of harmonic oscillators at temperature T, *i.e.* with an average energy $E = k_B.T$:

- If $k_B T >> \hbar\omega_0$: this is a situation where the incident neutron is a *cold* particle coming in interaction with a *hot* medium. A significant fraction of the oscillators are in an excited state so the incident neutron has a large probability to gain energy upon interaction with the sample. The probability that the neutron loses energy is not zero, however, it can not lose more energy than it carries. The maximum energy transfer is then limited to $+\hbar\omega_0$.

- If $k_B T << \hbar\omega_0$: this is the reverse situation, where the incident neutron is a *hot* particle coming in interaction with a *cold* medium. The sample excited levels are poorly or not populated. It is thanks to the energy of the neutron that they will be populated (hence the neutron will be *colder* after the interaction). In this case, it is the initial energy of the neutron that limits the maximum observable energy transfer and not the temperature of the sample.

The considerations above explains that at a given scattering vector Q, the observed intensity is a function of ω but also the temperature, T, of the sample. This can be formalized using the Bose population factor: $n = 1/(e^{-\hbar\omega/k_B T} - 1)$:

$$\left(\frac{d^2\sigma}{d\Omega d\omega}\right) = \frac{k_1}{k_0}\frac{1}{4\pi}.n_B(\omega, T).\left[\sigma_{coh}.S(Q, \omega) + \sigma_{inc}.S_{inc}(Q, \omega)\right] \tag{31}$$

with $n_B(\omega, T) = n + 1$ if $\omega < 0$, and $n_B(\omega, T) = n$ if $\omega > 0$. One can note that $2n + 1 = coth(\hbar\omega/2k_B T)$. These considerations on the $\hbar\omega_0/k_B T$ ratio have important consequences on the choice of the experimental conditions and on the type of spectrometer, to optimize the observation of an excitation (see sections 3.2.2 and 3.2.3).

3 Instrumentation

Two types of neutron sources exist: spallation sources that deliver a pulsed beam and continuous sources (reactors). In a schematic way, in both cases, the wavelength distribution $\Phi(\lambda)$ *vs* λ of the beams delivered is characterized by a maxwellian contribution of wavelength $\lambda > 0.5$ Å. On pulsed sources, $\Phi(\lambda)$ shows, with respect to continuous sources, an important additional contribution (from order of 50 % of the total flux) of the form $\Phi(\lambda) \approx 1/\lambda$ so-called ephithermal contribution. This strong profile of very energetic neutrons allows (i) to access the intramolecular vibration dynamics at high energy ($\hbar\omega > 1000\ cm^{-1}$) and at very low temperature, and (ii) to measure concomitantly dynamical (quasi-elastic) events and structural properties (small angle scattering and diffraction).

All types of inelastic neutron spectrometers have been used to study the dynamics of macromolecules (polymers and proteins). Next to successful studies of the collective dynamics of lipidic membranes [14] one can hardly cite studies taking advantage of the potentialities of three-axis spectrometers in the field of soft-matter. We have therefore made the choice to focus here on inelastic spectrometers using the time-of-flight technique. In all cases the central part of the instrument consists of the sample. We then distinguish the primary spectrometer (before the sample) and the secondary (after the sample). The functions of primary and secondary spectrometers vary by type of technology used: direct or inverted geometry. In both cases, the use of speed diagrams allows to easily understand the operation of a spectrometer (Fig.12-14).

3.1 Principle of the time-of-flight technique

Fig.9 describes the principle of a neutron inelastic time-of-flight measurement. The incident neutrons of a monochromatic beam with energy E_i, are scattered by the sample with an energy E_f (Eq.1) in a direction oriented at an angle $\phi = 2\theta$ of the incident beam. Over this process the neutrons experience a momentum transfer, Q, and an energy transfer, $\hbar\omega$ (Eq.2) with the sample[6]. Eq.2 shows the conservation laws of momentum and energy.

The key step in the determination of $S(Q,\omega)$ by the time-of-flight methods is to measure the energy of each scattered neutron, by evaluating its travel time (time of flight), t, over a known distance. In the context of the example presented here, this is a simple measurement of v_f, the velocity[7] of the scattered neutron, over the sample-detector distance L_{SD}.

$$v_f = \frac{L_{SD}}{t} \tag{32}$$

in other terms, the ToF spectrum:

$$\frac{d^2\sigma}{d\Omega dt} \tag{33}$$

is the measurement, by unit of solid angle $d\Omega$ of the arrival time distribution function of the arrival times $t = n\Delta\tau$ ($\Delta\tau$ is the sampling time unit) of the neutrons onto the detector (Fig.9). One can show [2] that the differential cross-section by energy $d\omega$ and solid angle $d\Omega$ units is directly connected [8] to the dynamical structure factor $S(Q,\omega)$ [2]:

$$\frac{d^2\sigma}{d\Omega d\omega} = N\frac{k_f}{k_i}\frac{\sigma}{4\pi}S(Q,\omega) \tag{34}$$

[6]The signs of Q et $\hbar\omega$ are fixed by convention.

[7]**Key formulas and orders of magnitudes:** the energy of neutrons of wavelength λ is $E\ [meV] = 81.79/(\lambda\ [\text{Å}])^2$. The time-of-flight, τ, over a distance D, for neutrons of wavelength λ is $\tau\ [\mu s] = 252.8.D\ [m].\lambda$ [Å]: neutrons of 5 Å spend 1.26 ms to fly over a distance of 1 meter.

[8]Here one disregards the existence of coherent and incoherent cross section to consider a single global cross section σ.

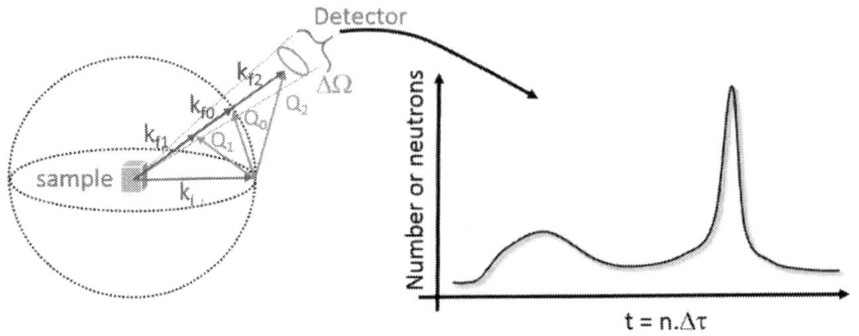

Figure 9. Schematic diagram of an inelastic measurement by time-of-flight: at an arbitrary origin of the time, a sample is briefly *illuminated* by a neutron burst. The distribution of the arrival time of the scattered neutrons at the detector equipped with an internal clock is then recorded. The so-called time-of-flight spectrum $\frac{d^2\sigma}{d\Omega dt}$ is shown here in black. The time scale is discretized into a succession of n constant time channels $t = n.\Delta\tau$. As the neutron-sample interaction is generally inelastic, spatial and energy quantities are linked by Eq.3. The three possible configurations are represented: $\|\vec{k}_{f_1}\| < \|\vec{k}_i\|$: inelastic scattering with loss of energy of the neutrons (Stokes), $\|\vec{k}_{f_0}\| = \|\vec{k}_i\|$: elastic scattering, $\|\vec{k}_{f_2}\| > \|\vec{k}_i\|$: inelastic scattering with energy gain of the neutron (anti-Stokes). Adapted from [15].

By writing the expression 33 in the form:

$$\frac{d^2\sigma}{d\Omega dt} = \frac{d\omega}{dt}\frac{d^2\sigma}{d\Omega d\omega} \tag{35}$$

and considering that $E_f = \frac{1}{2}m_n v_f^2$, and by combining Eq.1, 32 and 34, one can writes:

$$\frac{d^2\sigma}{d\Omega dt} = \frac{N\sigma}{h\sqrt{E_i}}\left(\frac{m_n}{2}\right)^{\frac{3}{2}}\frac{L_{SD}^3}{t^4}S(Q,\omega) \tag{36}$$

The ToF spectrum is therefore directly related to the dynamical structure factor, but on the *raw* data it appears "deformed" by the term in time of flight to the power 4. This can have important consequences on the appearance of the raw spectra (Fig.10). Since the weight of this term in $1/t^4$ becomes significant when t is large, the very precise subtraction of the electronic background, clearly detected on the first time channels (n from 1 to about 10), is essential.

3.2 Implementation

The spectrometers of the so-called "ToF-ToF" technology are the ones which make it possible to illustrate the implementation of an inelastic neutron scattering experiment in the simplest way. The two essential steps of the measurement are the monochromatization of the incident beam and the energy analysis of the scattered beam. Both steps use the same principle: the energy discrimination of neutrons according to the time spent to travel a known distance (generally in the order of a few meters).

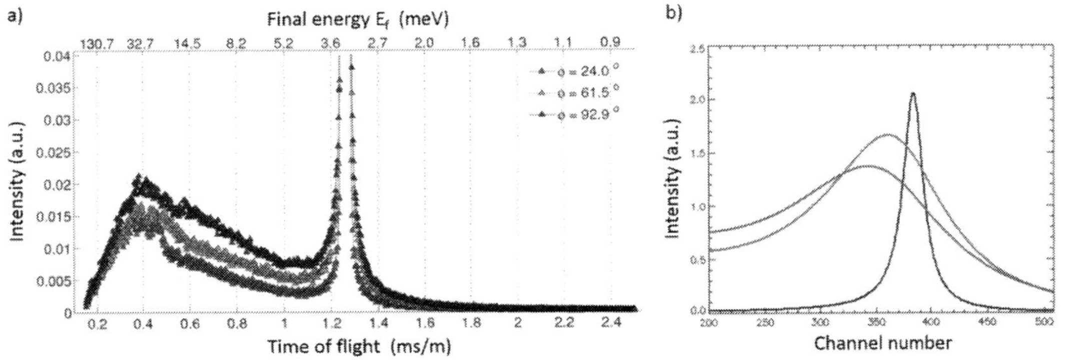

Figure 10. a) Example of a "raw" ToF spectrum. The abscissa axis is in ToF (the ToF scale, t, is proportional to the channel-width Δ: $t\ [\mu s] = n.\Delta\tau\ [\mu s]$, where n is a so-called channel number). Additionally, this scale has been here divided by the sample to detector distance (4 m). The corresponding energy scale is also shown (5 Å incident neutrons) on top of the plot. Note that the scale is linear in time but not in energy. b) Several Lorentzian dynamical structure factor $S(Q,\omega) = \frac{\Gamma}{\pi}/(\Gamma^2 + \omega^2)$ plotted in a channel number scale. All these quasi-elastic signals are centered at zero energy transfer ($\hbar\omega = 0$), corresponding here to the elastic channel $n_{Elastic} = 385$. Three Lorentzians with Γ values 0.05, 0.3, and 0.4 meV are shown in black, red and blue respectively. Based on its HWHM, Γ, the raw spectrum may appear to be an inelastic signal, *i.e.* showing an offset with respect to the elastic channel. This is a consequence of the t^4 term of Eq.36.

3.2.1 Monochromatisation

A first chopper[9] equipped with a single slot, rotating at the speed Ω_c, "chops" the white incident beam[10] in short-time *pulses* with a frequency $\nu[Hz] = \Omega_c[rpm]/60$ (Fig.11). At a distance L_{12}, a second chopper, identical to the first one, but out of phase by an Φ angle, reproduces this chopping operation of the beam with a time offset $\delta[s] = \Phi/\nu[Hz]$ modulo $1/\nu$. Choosing the value of Φ allows to select the desired wavelength. As they are the ones that travel the distance L_{SD} in a time δ, only the neutrons with wavelength

$$\lambda_i = \frac{h}{m_n}\frac{1}{L_{SD}}\frac{\Phi}{2\pi\nu_c} \tag{37}$$

are transmitted by the second chopper. The others ones are absorbed.

• *Analysis:*

In practice, the instant, t_0, the beam reaches the second chopper, defines the origin of the measurement times (Fig.11): the internal clock of the detector is started. The neutrons scattered by the sample are recorded according to their arrival time on the detector. It is the probability of an energy exchange between the neutron and the sample, described by the dynamic structure factor $S(Q,\omega)$, that spread, over the sample-detector distance L_{SD}, the neutron arrival time on the detectors. The neutrons that have gained energy arrive first at the detector, followed by elastically distributed neutrons arriving in the elastic channel (Fig.12). Finally, the neutrons having lost energy to the sample arrive last.

[9]A chopper is a neutron absorbing disk rotating in a continuous beam in which a or multiple slits have been made. The incident neutrons are absorbed by the chopper except when the slits passes in front of the beam. This system induces a "chopped" beam.

[10]By analogy with light, we denote by *white beam* a beam presenting an extended spectrum in wavelengths and therefore a distribution of energies. A *monochromatic* beam is composed of a single wavelength. See the section 3.3 for consideration on uncertainty on the wavelength determination and the underlying notion of resolution in energy of a spectrometer.

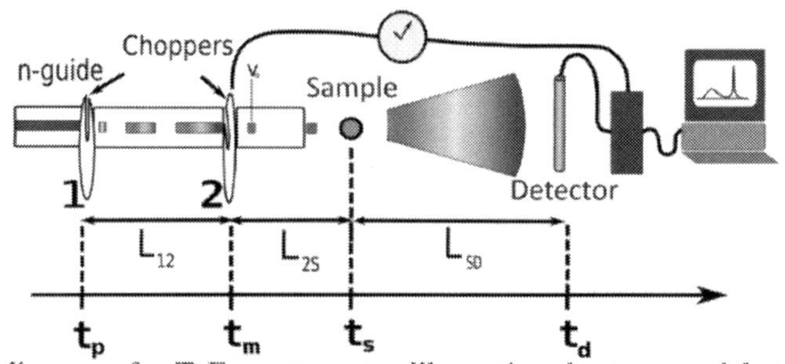

Figure 11. Schematic diagram of a ToF spectrometer illustrating the two crucial steps of the measurement: the monochromatization of the white (see 3.2.1) incident beam, then the time of flight analysis at the detector (details in section 3.2). For the sake of simplicity a single detector is represented. On a real spectrometer, the measurement in time of flight is performed simultaneously on several hundred detectors. From [15].

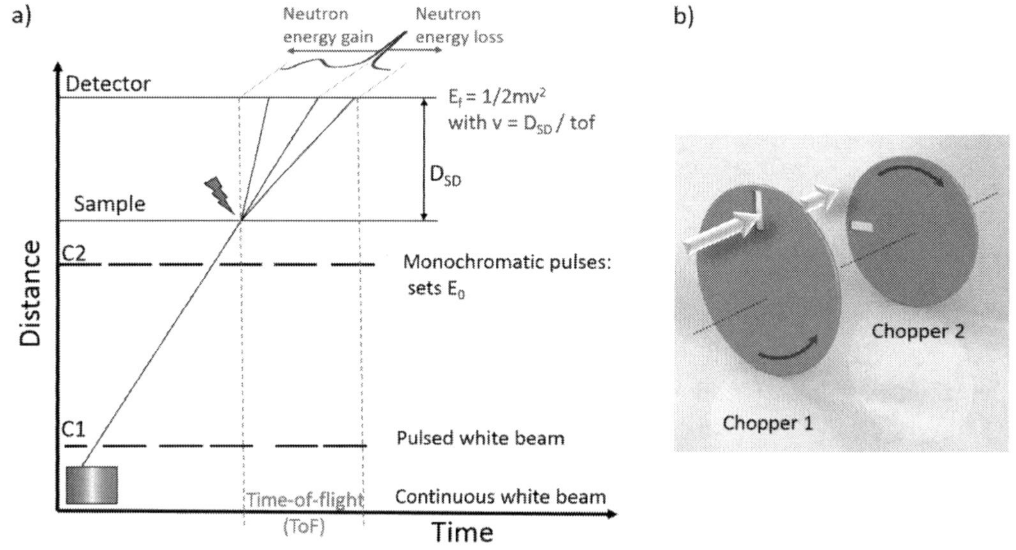

Figure 12. Time-distance diagram. This figure implements the so-called representation "Time-distance diagram". The time is plotted on the abscissa and the distances on the y-axis. A neutron of energy E *i.e.* with speed $v = \sqrt{2E/m_n}$, is represented by a line with a slope *1/v*. a) General principle: case of the direct geometry (the inverted geometry principle is shown on Fig.14. To perform a direct geometry ToF measurement, it is necessary to extract pulses of monokinetic neutrons, from a white beam (see 3.2.1). In principle, three devices are necessary: two choppers and a detector with an internal clock. From the continuous white beam (multicolored square) the first chopper (C1) produces a pulsed white beam. A phase adjustment φ of the second chopper (C2) extracts from the white beam issued from C1 the neutrons having the desired wavelength (energy E_0). When these neutrons reach the sample, the detector clock is triggered. By interaction with the sample, the neutrons gain energy (speed increase, so also the slope on the graph), lose energy (decrease of speed, so decrease of the slope) or undergo an elastic interaction ($\hbar\omega = 0$, no change of slope). Then, they continue their trajectory until they reach the detector D located at a distance D_{SD} from the sample. At every ToF moment (in each time channel), the number of neutrons arriving on the detector is recorded. The energy of each neutron having experienced a "time of flight", ToF, over the SD distance is $E_f = 1/2.m_n(D_{SD}/ToF)^2$. They have undergone an exchange of energy, $\hbar\omega = E_i - E_f$. An elementary derivation allows to switch from the ToF spectrum recorded on the detector to $S(Q, \omega)$ (Eq.36).

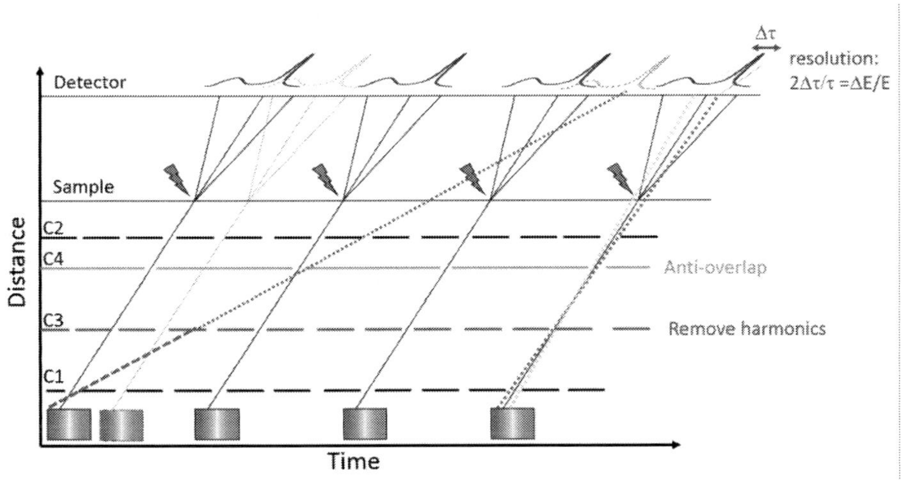

Figure 13. Time-distance diagram: direct geometry. In practice, to be usable, the system requires a little sophistication. A third chopper C3 acts as an anti-harmonic filter: it prevents neutrons having an integer sub-multiple speed not to reach the sample (red dots). A fourth and last chopper (called anti-overlap chopper), C4, having an integer sub-multiple rotational speed of the other choppers allows to space-out in time the pulses arriving on the sample. This ensures that the fastest neutrons of the $n+1$ pulse scattered by the sample can not be cought-up with the slowest neutrons of the pulse n. The window opening time (in front of the neutron guide) being a defined value (not infinitely small), there is an uncertainty of the order of a few percent on the neutron energy selected by the choppers system. Suppose a purely elastic sample-neutron interaction $\hbar\omega \approx 0$. The uncertainty on the speed of the incident neutrons produces an uncertainty about their arrival time (green/blue curves). The elastic peak is not infinitely thin, but has an intrinsic width $\Delta\tau$: this is the energy resolution of the instrument. The higher speed the choppers, the shorter the window opening time in front of the guide. The uncertainty about the neutron departure time decreases but also the flux. One have always to compromise between flux and resolution.

3.2.2 Direct geometry time-of-flight instruments: The chopper spectrometers

In this type of technology, both the energies of the incident and of the scattered neutrons are selected by ToF.

- *Compromise between flux and resolution:*

One of the causes of the finite energy resolution, ΔE, of the ToF spectrometers is the existence of a wavelength distribution, $\Delta\lambda$, at the detectors (Fig.13): $\Delta E/E = 2\Delta\lambda/\lambda = 2\Delta\tau/\tau$ where τ is the time of flight of neutrons and $\Delta\tau$, the associated time distribution. The most natural solution for reaching a resolution in energy of a few percent is to limit the distribution in incident neutron energy. This is the strategy adopted in the design of spectrometers with 2 choppers. The price to be paid for the gain in resolution is the low flux at the sample level. Also, these spectrometers can only occupy a position at the end of a guide.

3.2.3 Inverted geometry time-of-flight instrument

There is an alternative to choppers instruments technology: the so-called "inverted" geometry (Fig.14). The principle is to illuminate the sample with a white beam, and to detect by Bragg reflection on a crystal, the neutrons scattered with a specific energy. In Eq.3, the fixed parameter is $\hbar\omega_1$ and $\hbar\omega_0$ takes different values. This technology is particularly suited for pulsed sources that have a very wide spectrum of incident wavelengths, from 0.01 to 20 Å. With this type of spectrometer, it is possible to measure excitations of several hundred meV at very low temperatures (see discussion 2.6).

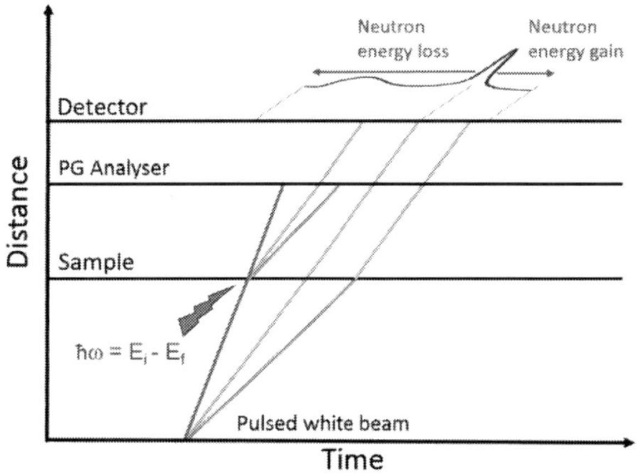

Figure 14. Time-distance diagram: inverted geometry. The sample is *illuminated* with a white (see 3.2.1) neutron beam. Only the neutrons scattered with a selected final energy (filtered by Bragg reflection on a monochromator) are sent to the detectors. While the experimental strategy is different (white *vs* monochromatic incident beam) for both the direct (Fig.12) and indirect geometry, the recorded information is the probability for a neutron to be scattered at a specific Q value with an energy exchange ω: this is the dynamical structure factor $S(Q, \omega)$.

•*An advantage: The simultaneous measurement of the structure and of the dynamics*

It is possible to have in the ToF base of the spectrometer, a detector without interposing the analyzer that selects the final energy $\hbar\omega_1$. The neutrons scattered by the sample arriving in this detector, located at a diffusion angle 2θ, can be used to perform a diffraction experiment *i.e.* to measure [11] $I(Q)$ *vs* Q. Some detectors can therefore be used in diffraction simultaneously with detectors that retain the classical operation of energy analysis.

3.3 Experimental energy resolution effects and consequences

As a first approximation,[12] the signal $S(Q, \omega)_{exp}$ measured by a spectrometer is the convolution product of the theoretical signal $S(Q, \omega)_{theo}$ (which would be measured if the device was infinitely resolutive) by an experimental resolution $R(Q, \omega)$ (Fig.15):

$$S(Q, \omega)_{exp} = S(Q, \omega)_{th} \otimes R(Q, \omega) \tag{38}$$

Therefore, by inverse Fourier Transform and applying the convolution theorem:

$$I(Q, t)_{exp} = I(Q, t)_{theo}.R(Q, t) \tag{39}$$

[11] $Q = \frac{4\pi}{\lambda} sin(\theta)$: contrary to the "standard case", here θ is fixed and λ varies.

[12] When one talks about energy resolution of a spectrometer this is the "elastic" resolution. In reality the resolution is a function of the energy transfer. On a ToF chopper spectrometer, the resolution is proportional to the scattered neutron energy. Eq.38 is only a valid approximation in the quasi-elastic region.

where $R(Q,t)$ is the inverse Fourier Transform of $R(Q,\omega)$.
Generally, $R(Q,\omega)$ is experimentally determined (Fig.13) by the measurement of the scattering spectrum of a vanadium sample (this is a metal, hence a purely elastic scatterer below its melting point of 1910°C) having the same geometry than the sample. The vanadium being an incoherent scatterer, it also provides information on the detector efficiency. For relaxational or Gaussian processes, the time-energy equivalence[13] writes:

$$\Delta E\ [meV] = \frac{0.658}{\Delta t\ [ps]} \tag{40}$$

These resolution effects imply that in neutron scattering, the observed physical phenomena and their associated quantities (distances, correlation times) are only valid at the resolution of the spectrometer (*i.e.* within the time (energy) / space (Q) domains accessible experimentally).

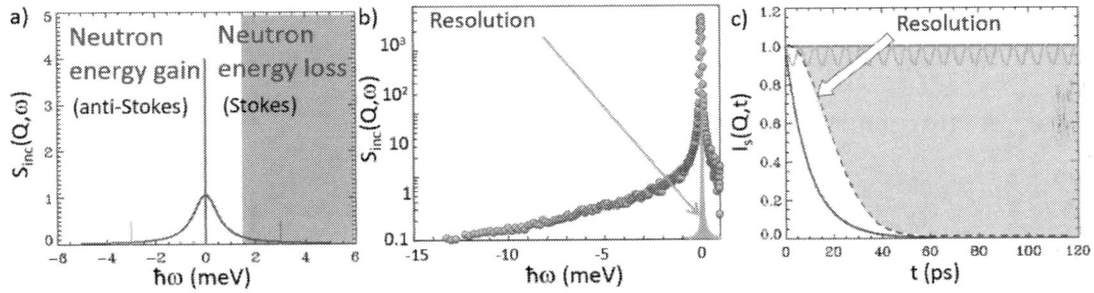

Figure 15. Theoretical neutron scattering spectrum and experimental limitations. a) The maximum energy a neutron can provide to the sample is limited by its own incident energy $\hbar\omega_0$. The red part of the neutron energy loss side is therefore inaccessible to neutrons with energy $\hbar\omega_0$. b) Experimental $S(Q,\omega)$ spectrum of water confined in a polymer membrane (Nafion). The elastic, quasi-elastic and inelastic zones of the spectrum are clearly visible. One can see that the reality of a measurement is far from the idyllic vision presented in a) where the spectral bands are fine and the excitations clearly defined. The complexity of the experimental spectrum is partly due to the sample dynamics but also to the intrinsic energy resolution of the spectrometer (gray curve, see also Fig.13). c) The considerations exposed in Fig.13 showed that the energy resolution is Gaussian in the (Q, ω) domain. Fourier transform of a Gaussian (of HWHM $\Delta\omega$) and a Lorentzian (of HWHM Γ) in the energy domain are, in the time domain, a Gaussian (HWHM: $1/\Delta\omega$) and an exponential decay (with correlation time $\tau = \hbar/\Gamma$), respectively. The point to keep in mind is that in both cases, the broader the curve in energy, the narrower the curve in time. Thus, the Fourier transform of the quasi-elastic line, wider than the energy resolution (curve b), is a temporal signal (purple exponential) narrower than the resolution (gray dotted curve in c). We can see on c) that to measure long times, the limiting factor is the maximum time corresponding to the resolution. Graphically, this translates into the fact that it is impossible to measure correlation time corresponding to exponentials located in the region above the gray dotted line. This area is "forbidden" as symbolizes the light gray area. **The only way to measure longer correlation times is to increase the resolution in energy: the narrower the Gaussian elastic peak (energy domain) the broader its Fourier transform (time domain).** However, the price to pay to increase the resolution is usually a decrease in flux.

[13] A decaying exponential with a correlation time of 10 ps is, in the energy domain, a Lorentzian line with HWHM $\Gamma = 0.658/10 = 65.8\,\mu eV$.

4 The tools at work: A multiscale study of a polymer melt in bulk and under nanometric confinement. Connection to NMR

After decades of developments, very high-achieving theoretical tools are now on hand to finely interpret both the structural and dynamical behaviors of polymeric systems. The theoretical models describing the physics of bulk polymer melts have, for example, made possible to establish a connection between the rheological properties and the dynamics of the chains at the microscopic scale. For long entangling chains, the diameter of the reptation tube, d_{Rep}, controls the level of the plateau of the stress relaxation modulus $G_e \approx 1/d_{Rep}^2$ and, in turn, the polymer viscosity. Polymer physics is therefore a canonic example of a successful multiscale approach of a complex system in bulk. Nevertheless, despite the key practical and industrial relevance of such situations, the peculiar properties of polymers in interfacial situations or in strong confinement are far from being fully understood.
In the present section, we address the different dynamical modes of a high-mass entangled polymer chain: the local monomer dynamics, the Rouse modes and the reptation process. This study exemplifies that, used in conjunction with hydrogen/deuterium isotopic effects, high resolution QENS can be bridged to the Zero Average Contrast (ZAC) method to probe, in a non destructive way, the dynamics of a single polymer chain in bulk but also under severe nanometric confinement. Connection and complementarity of the neutron derived analysis with Pulsed-Field Gradient and Relaxation NMR techniques are also discussed.

4.1 Polymer dynamics in a nutshell

A polymer chain experiences different dynamical regimes [16, 17], spread over broad time and length scales. At short time and at the local scale (1 ns, 50 Å), the chain just experiences random entropic forces and freely fluctuates in an isotropic way. This is the Rouse regime. At longer times and larger scale (100 ns, 500 Å), the neighboring chains induce a topological constraint and the chain fluctuates alongside a fictive tube [18]: the so-called reptation tube (regime of the local reptation). The tube diameter is similar to the distance between two entanglements *i.e.* around 50 Å. Then, at even longer times (1000 *ns*), in the so-called full reptation regime, the chain can finally escape the tube. Such a microscopic model is extremely useful and efficient, since with very few parameters (a monomeric friction coefficient and the length of a chain segment), it is possible to bridge the local dynamics to macroscopic properties like, for example, the polymer viscosity.
In numerous scientific fields, reducing the dimensions of a physical system down to a scale that matches the characteristic sizes of its natural fluctuations in bulk, usually leads to unexpected and surprising new physical behaviors [19]. In soft-matter, despite the key practical and industrial relevance of such situations, the peculiar properties of polymers in interfacial situations [20] or strong confinement [21] are far from being fully understood. Several theoretical approaches have addressed the issue of the polymer conformation, when a melt is confined inside cylindrical pores [22]. It has been proposed that, under confinement, a modification of the monomer-monomer excluded volume interaction could lead to an increase of the longest polymer relaxation time and therefore of the confined polymer viscosity [23].

As a matter of fact, even in the recent past, confinement was achieved in media with no macroscopic orientation: "powders" in the crystallographic sense of this word. This has two consequences when probing the physical properties of confined fluids:

- first, all the retrieved structural and dynamical information are blurred by isotropic average: key points like the relative directions of the motions of a molecule with respect to the pore (radial or longitudinal) are basically lost.

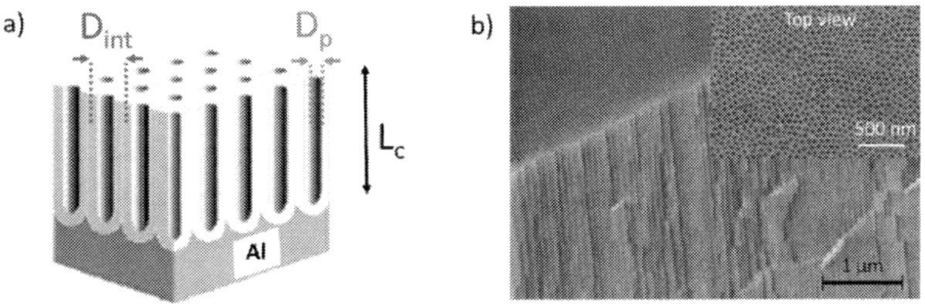

Figure 16. a) Schematic structure of an anodized alumina membrane (AAO). From [24]. The membrane topology can be described by the pore diameter, D_p, the inter-pores distance, D_{int}, and the channels length, L_c. b) Scanning Electron Microscope image of a typical AAO membrane. The highly oriented array of cylindrical pores (showing an imperfect hexagonal arrangement, see inset) goes from the top down to the bulk of the membrane. We illustrate in section 4.3 how the macroscopic orientation of such a pore network can be used to lift the limitations of the powder average. From [25].

- at a second level, the tortuosity (imposed by the pore morphology and/or the random spatial arrangement of the powder grains) is making it extremely challenging to link the local transport properties (few Å or nm), like the self-diffusion coefficient, to the macroscopic transport flow of the material.

Following recent progresses in inorganic chemistry, porous media with pores showing macroscopic orientation are now available (Fig.16). The "powder average" limitation can therefore be easily overcome by simply orienting the porous sample in a measuring device (confocal microscope, NMR, infra-red, neutron spectrometers ...). It is then relatively easy to firmly disentangle dynamical phenomena in the parallel or radial direction of the pore axis.

The Rouse model successfully describes the physics of short-chain polymer melts [26]. This model ignores the atomic structure, to assimilate the polymer to an *ideal* (or *Gaussian*) chain made by a linear succession of beads (with a diameter σ, called the Kuhn length[14]) interacting through a harmonic potential. The conformational entropy is the restoring force that prevents large excursions from the average equilibrium conformation. The full mathematical development of the model leads to the following intermediate scattering function:

$$I^{Rouse}(Q,t) = e^{-\frac{2}{\sqrt{\pi}}\left(\frac{3k_BT}{\sigma^2\zeta_0}Q^4 t\right)^{\frac{1}{2}}} \tag{41}$$

where k_B is the Boltzmann constant, T the temperature, ζ_0 the monomeric friction coefficient and σ the length of a chain segment. $W = 3.k_BT.\zeta_0^{-1}\sigma^{-2}$, the elemental Rouse rate, is the quantity generally used to characterize the correlation time associated to the polymer dynamics. This expression also writes:

$$I^{Rouse}(Q,t) = e^{-\left(\frac{t}{\tau}\right)^{\beta}} \tag{42}$$

One now clearly sees that Eq.41 is a *stretched exponential*, with the stretched exponent β *exactly* equal to $1/2$. Also, within the framework of the Rouse model, the system is characterized by a single correlation time τ and τ follows a Q^{-4} power law. The Rouse model is able to describe the polymer

[14] A real polymer chain can be described as an equivalent so-called *Kuhn chain* made of a succession of N connected *Kuhn segments* (or beads) of *Kuhn length* σ. This theoretical treatment introduced by Kuhn, makes it possible to assimilate a real complex polymer chain to a random coil *i.e.* a random walk of the Kuhn segments [17].

dynamics up to Q values around 0.3 $Å^{-1}$ with a remarkable accuracy. At higher Q, the detailed local structure and dynamics of the chain under scrutiny give strong contribution and Eq.41 fails to account for the data. Despite significant deviations from the ideal form of the Rouse model and therefore a lack of clear physical significance, stretched exponentials are often used to account for QENS and NMR relaxation data from polymer melts. In many cases, β is found to differ from its ideal 0.5 value and becomes Q dependent, τ does not follow a power law in Q^{-4} and these deviations change from polymer to polymer.

4.2 Local polymer dynamics: A neutron-derived physical model

4.2.1 QENS derived dynamical susceptibility as a tool to determine the actual number of dynamical contributions

This section is devoted to the definition of a model to account for the dynamics of bulk PEO as sensed on the typical structural and dynamical ranges of a ToF experiment (few Å and few tens of ps), *i.e.* in the situation where the pure stretched exponential description of a polymer dynamics fails.
As this has been shown in section 2.5.3, mean-square displacements as measured by incoherent neutron scattering (elastic scans) can be informative and provide a global view of the dynamics of a system. But we need to clearly state that, when the instrumental resolution (in energy) of the spectrometer matches the relaxation characteristic times of the system under investigation, the way to take the best of a QENS experiment, is to undertake a full measurement of the inelastic and/or quasi-elastic signal. Then, fitting the experimental data with a meaningful model can lead to more detailed insights in the physics of the system.
To illustrate this strategy, we start with an elastic-scan measurement (Fig.8) to obtain a global view of the dynamics of the confined polymer. Then, we attempt to derive a more physically meaningful model of the local (0.1 nm), short time (10 ps) dynamics of a linear polymer like PEO. We consider *i)* orientational diffusion along the polymer chain, *ii)* local conformational transitions and *iii)* long-time, large-scale motions. The model takes into account the spatial component of the local dynamics, described in terms of the scattering vector Q.
PEO can be confined within the porous structure of an AAO membrane by directly melting the polymer under vacuum onto the membrane. The filling process is completed after a few days. Under confinement the melting occurs at a significantly lower temperature (T_M = 317 K) than in bulk (T_{MBulk} = 333 K): a clear Gibbs-Thomson effect is detected. This strong thermodynamical effect has also a signature, in terms of local mobility, on the polymer mean-square displacement measured by incoherent neutron scattering (Fig.8).

In order to unveil the various dynamical processes involved, full QENS measurements are shown in Fig.17 in terms of the dynamical susceptibility function:

$$\chi''(Q,\omega) = S(Q,\omega)/n_B(\omega,T) \tag{43}$$

where $n_B(\omega,T) = \left(1 - exp(-\frac{\hbar\omega}{k_B T})\right)^{-1}$ is the Bose population factor, T the temperature in Kelvin, k_B the Boltzmann constant and $\hbar$ the Planck constant divided by 2π. In quasi-elastic scattering, a relaxation mechanism with characteristic time τ appears, in $S(Q,\omega)$, as a quasi-elastic broadening of HWHM $\Gamma = \frac{\hbar}{\tau}$ and, in $\chi''(Q,\omega)$, as a band centered at $\omega \approx \Gamma$. Two dynamical regions can be clearly identified in the bulk polymer (Fig.17a). A low-energy dispersive band ($E \approx 0.07$ meV at $Q_0 = 0.55$ $Å^{-1}$) merges at higher Q with a non-dispersive band ($E \approx 5$ meV), giving rise to a single broad band for Q values above 1.7 $Å^{-1}$. This purely qualitative analysis leads to the conclusion that the dynamics of the systems can be accounted for by a system with at least two distinct correlations times: one being

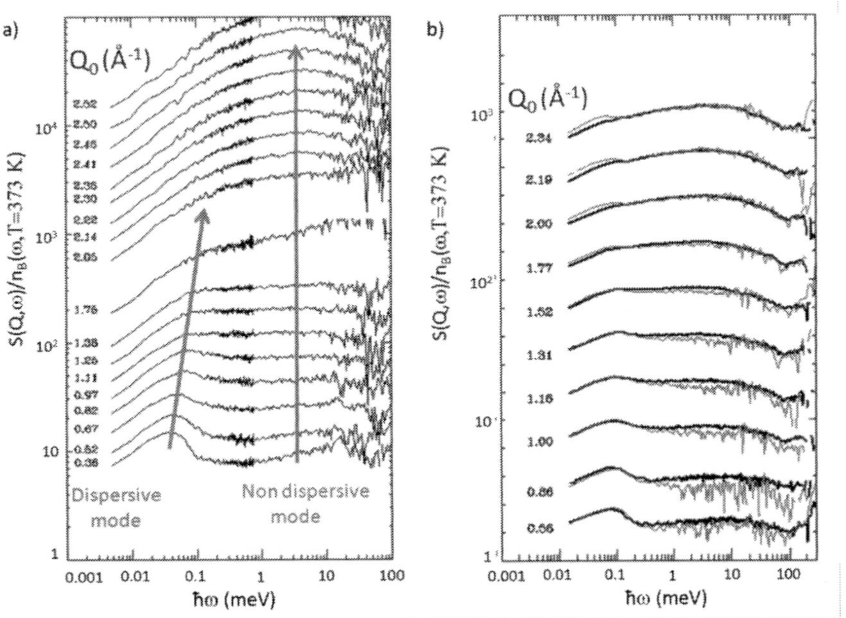

Figure 17. Q dependence of the dynamical susceptibility spectra, $\chi''(Q,\omega)$ *vs* $\hbar\omega$, calculated from the QENS data of: a) bulk PEO (100 kg/mol, QENS spectrometer at ANL/IPNS) and b) PEO confined in AAO with 35 nm diameter pores (red line, Mibémol spectrometer at LLB). Curves are shifted for clarity and values of Q_0, the elastic scattering vector is indicated. For the confined sample, the corresponding signal of bulk PEO is shown (black line). In all cases, T = 373 K. From [27].

dispersive, the other being Q independent. The same conclusion applies for the confined polymer (Fig.17b).

4.2.2 A physical appealing model of the local polymer dynamics: The building blocks

• *Local conformational transitions.*

If the conformational motions experienced by a monomer are characterized by a single correlation time τ_0, the corresponding scattering function is:

$$S_{inc}^{Conf}(Q,\omega) = A_0(Q).\delta(\omega) + \frac{1 - A_0(Q)}{\pi}.L_{inc}^{Conf}(Q,\omega) \tag{44}$$

where $L_{inc}^{Conf}(Q,\omega)$ is a Lorentzian function of HWHM $\hbar/\tau_0(Q)$ and $A_0(Q)$ is the Elastic Incoherent Structure Factor (EISF) associated with the geometry of the motion. For a purely incoherent sample, as the one we are dealing with here, the EISF is the infinite time limit of the intermediate scattering function $I_{Self}(Q,t)$. The Q dependence of $I_{Self}(Q,t)$ at $t = +\infty$, is the form factor of the localization volume of the particle (see section 2.5.2 and Fig.7).

•*Local PEO dynamics: A mechanism unveiled by 2D* ^{13}C *solid-state NMR. Extrapolation to S* (Q,ω).

In the case of crystalline PEO, few 2*D* ^{13}C solid-state NMR experiments have shown [28] that the local chain conformational transitions (few Å) can be described as an *helical jump* (HJ) motion. This process has been detected at 225 K, *i.e.* just few kelvins above the glass transition and is therefore

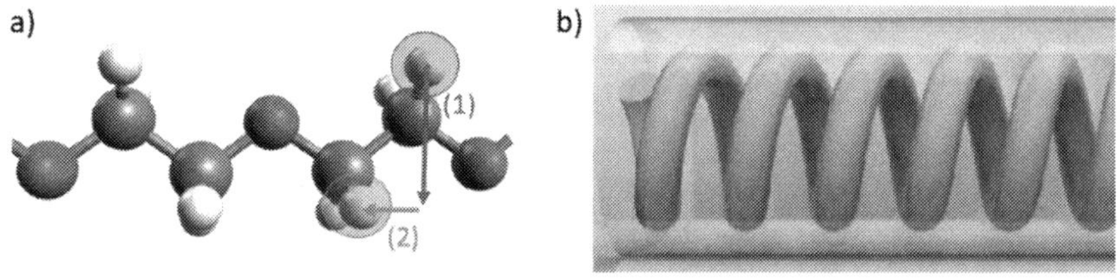

Figure 18. a) Schematic representation of a PEO chain ($-[CH_2 - CH_2 - O-]_n$) with the two fundamental motions involved described in this section. A local "jump over two sites" conformational change (noted 1) with non dispersive correlation time τ_0 and EISF $A_0(Q)$, is coupled to a translation of the whole polymer chain (noted 2) with a dispersive (in Q^{-2}) correlation time τ_1 and EISF $A_1(Q)$. Altogether, those two motions give rise to a helical jump (HJ) motion. Its envelope is represented in green on sketch b). The jump distance derived from $A_0(Q)$ is about 2 Å. This HJ motion extends up to 26 Å. $A_1(Q)$ has therefore a form factor of a cylinder of 2 Å in radius and 26 Å in length. As a summary, we reach the conclusion that in bulk the whole PEO polymer chain is sliding, as a whole, along its own physical contour.

very slow. It involves a rotation of a monomer by 180° with a concomitant chain translation of one monomer unit (Fig.18).
This translation can be described as a Random Walk in Continuous Time (RWCT), a generalization of the discrete random walk in which the waiting time probability density shows a time dependence $\Psi(t)$, instead of being a Dirac function corresponding to a single residence time (like in jump diffusion, see section 2.5.1). If $\Psi(t)$ is Poissonian, which is the case for conformational changes in PEO melt [29], the autocorrelation function of the motion is given by:

$$G^{HJ}(t) = e^{-t/\tau_1} I_0(t/\tau_1) \tag{45}$$

where τ_1 is the rate of the conformational change and I_0 is the zero-order modified Bessel function of the first kind. The long time-scales over which the HJ motions take place at 225 K are inaccessible by QENS. However, we can expect these motions to be still present above the melting point, but several orders of magnitudes faster, and thus within the dynamical range of QENS. Due to the effect of damping, originating from both the averaged dynamics of the neighboring chains and the presence of entanglements, one can expect that the diffusion of the polymer chains involved in the HJ motions should not extend beyond a few nanometers. These diffusive motions being localized, they give rise to a scattering function of the form:

$$S_{inc}^{Diff}(Q,\omega) = A_1(Q).\delta(\omega) + \frac{1 - A_1(Q)}{\pi} S_{inc}^{HJ}(Q,\omega) \tag{46}$$

where $A_1(Q)$ is the form factor (EISF) of the volume over which the diffusion takes place and $S_{inc}^{HJ}(Q,\omega)$ is the Fourier transform of Eq.45 with τ_1 being Q-dependent.

• *Long time and large scale rheology-related motions.*

At times beyond those where this atomic description applies, larger scale viscoelastic properties have to be taken into account. The present work focuses on high Q (> 0.3 Å^{-1}) data in which these larger scale motions are not likely to be accurately described. For the present purposes, they will be accounted for by a single relaxation time $\tau_2(Q)$, leading to the scattering function:

$$S_{inc}^{Glob}(Q,\omega) = \frac{1}{\pi} \int e^{-t/\tau_2}.e^{-i(Qr-\omega t)}.dt \tag{47}$$

Assuming that the three different dynamical mechanisms described above are uncorrelated, the final scattering function $S_{inc}^{PEO}(Q,\omega)$, describing in full the PEO dynamics, is a convolution of EQ.44, 46 and 47. The correlation time τ_0, representing the local reorientational dynamics, is expected to be much shorter than τ_1 and τ_2 so that, to a good approximation:

$$S_{inc}^{Conf}(Q,\omega) \otimes S_{inc}^{HJ}(Q,\omega) \approx S_{inc}^{Conf}(Q,\omega) \tag{48}$$

$$S_{inc}^{Conf}(Q,\omega) \otimes S_{inc}^{Glob}(Q,\omega) \approx S_{inc}^{Conf}(Q,\omega) \tag{49}$$

and $S_{inc}^{PEO}(Q,\omega)$ simplifies [27] to:

$$S_{inc}^{PEO}(Q,\omega) = B_0(Q).\delta(\omega) + B_1(Q).S_{inc}^{Glob}(Q,\omega) \otimes S_{inc}^{HJ}(Q,\omega) + B_2(Q).S_{inc}^{Conf}(Q,\omega) \tag{50}$$

where $B_0(Q)$, $B_1(Q)$ and $B_2(Q)$ are respectively the intensities of the elastic peak, a narrow and a broad quasi-elastic contribution. It is important to note that by fitting the QENS data with Eq.50, we impose the time (or equivalently the energy) dependence of the quasi-elastic signal (*i.e.* the shape of the contributions), but their relative intensities are free parameters. For each Q value, from $B_0(Q)$, $B_1(Q)$ and $B_2(Q)$ we can derive $A_0(Q)$ and $A_1(Q)$, the form factors of the conformational changes and of the extension of the HJ motion (Fig.19).

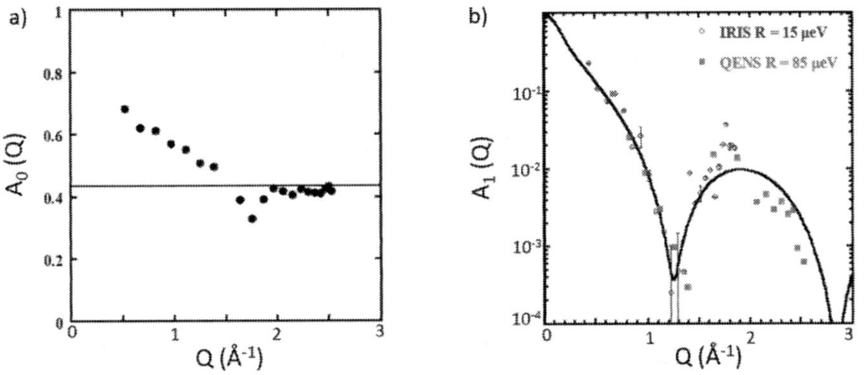

Figure 19. a) $A_0(Q)$, form factor (EISF) for the conformational changes, derived from the $B_0(Q)$, $B_1(Q)$ and $B_2(Q)$ intensities of Eq.50. The full horizontal line indicates the level $(1-a)$ of the weighting factor balancing the short and long correlation time contributions to the ^{13}C spin-lattice relaxation NMR spectral density (see Eq.51). b): Semi-log plot of $A_1(Q)$ deduced from the fit of IRIS data (RAL/ISIS, UK, high resolution, low-Q, in blue) and QENS data (ANL/IPNS, USA, low-resolution, high-Q data, in red), with Eq.50. The full line represents the form factor of a cylinder of 2 Å radius and 26 Å long. From [27].

We find the Q dependencies of the three correlation times of the model τ_0, τ_1 and τ_2, to be respectively consistent with Q^0, Q^{-2} and Q^{-4} power laws (Fig.20). At 373 K, the polymer dynamics is described in terms of transverse motions of the chain segments over a distance of few nm, with a local monomeric diffusion coefficient derived from the Q^{-2} dependence of τ_1: $D_{Mono} = 1.78 \times 10^{-9}\ m^2.s^{-1}$. From this value, a monomeric friction coefficient $\zeta_0 = k_B T / D_{Mono} = 2.89 \times 10^{-12}\ N.s.m^{-1}$ can be derived and used as numerical input to the Doi-Edwards theory [16]. This leads [27] to a chain center-of-mass diffusion coefficient $D_{cm} = 9.4 \times 10^{-15}\ m^2.s^{-1}$. This value is in very good agreement with Pulsed Field Gradient NMR (PFG-NMR) data [30] (Fig.20).

Within the framework of this model, the two distinct dynamical modes clearly detected in the dynamical susceptibility plots of the bulk polymer (Fig.17) can be attributed to the helical jump motion

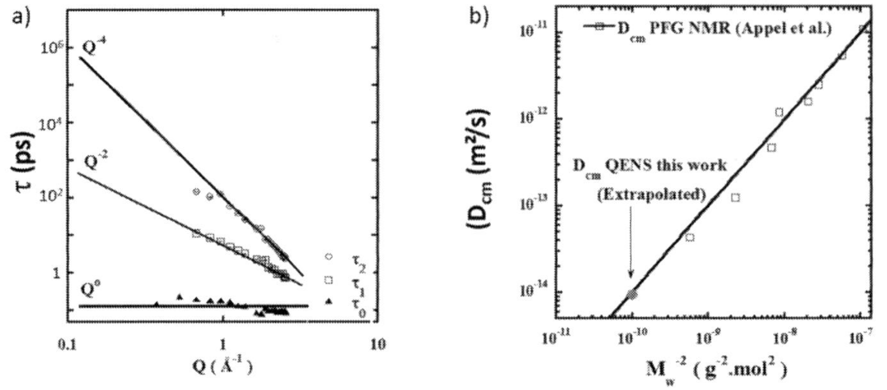

Figure 20. a) Correlation times τ_0, τ_1 and τ_2 derived from the fits of the QENS data with Eq.50. The straight lines indicate the power-law behaviors. From the Q^{-2} dependence of τ_1, we derive a a monomeric friction coefficient $\zeta_0 = 2.89 \times 10^{-12} N.s.m^{-1}$. From this value, we can derive the polymer chain center-of-mass diffusion coefficient: $D_{cm} = k_B.T.M_e/(3.\zeta_0.M_W)$ with $M_W = 100$ kg/mol. b) Molecular weight dependence of the PEO chain center-of-mass self-diffusion coefficient D_{cm} (PFG-NMR data from [30]). The D_{cm} value derived from the neutron data in a), is shown as a red point. From [27].

(Q^{-2} dependent correlation time τ_1, narrow dispersive band) and to the non dispersive conformational jump with correlation time τ_0 (broad dispersive band around 5 meV). Since no significant difference is detected at 373 K in the susceptibility plot of the PEO confined in 35 nm pores, such an interpretation seems to hold on that case also.

The high Q limit of $A_0(Q)$ around 0.5 is characteristic of a jump motion over two sites (*i.e.* the probability to be on one site is 1/2). From the Q dependence of $A_0(Q)$ we can estimate a jump distance of about 2 Å. The experimental values of $A_1(Q)$ are consistent with a form factor of a cylinder of 2 Å in radius and 26 Å in length. From this geometrical information and the Q dependence of the correlation times of the model (Fig.20), we propose the microscopic mechanism of a PEO polymer chain dynamics shown in Fig.18: the whole PEO polymer chain is sliding as a whole along its own physical contour.

4.2.3 Bridging the QENS and the NMR relaxation models

Detailed molecular dynamics can be also assessed by NMR. Hall and Helfand (HH) [31] derived a bimodal autocorrelation function to describe damped conformational changes along a polymer chain. Subsequently, Dejean de la Batie *et al.* [32] showed that this autocorrelation function could not account for the high value of the spin-lattice relaxation time T_1 observed in ^{13}C spin-lattice relaxation time experiments. To explain their data, they complemented the HH model with an additional fast, small-amplitude anisotropic motion that they identified as librations of C-H vectors. They propose the so-called Dejean-Laupretre-Monnerie (DLM) model:

$$G(t) = (1-a).e^{-t/\tau_1}.e^{-t/\tau_2}.I_0(t/\tau_1) + a.e^{-t/\tau_0} \tag{51}$$

In the first term, following Hall and Helfand [31], τ_1 is the correlation time associated with the conformational jumps responsible for orientational diffusion along the polymer chain, while τ_2 represents a damping mechanism consisting of either non-propagating isolated motions or distortions of the chain with respect to its stable local conformation. The fast relaxation τ_0 corresponds to the correlation time

of librations of C-H vectors inside a cone of half-angle θ and a is a geometrical factor depending on $cos(\theta)$.
In the case of PEO with $M_w = 9200$ g/mol at 373 K, Dejean de la Batie *et al.*, [32] report $\tau_1 \approx 4$ ps, $\tau_1/\tau_0 = 200$, $\tau_1/\tau_2 = 200$, and a = 0.56, corresponding to the libration of a C-H bond in a cone of half-angle $\theta = 40°C$.
There is clearly a formal similarity between the DLM and the QENS model. For the sake of comparison, we express the latter in the variables Q and t:

$$I(Q,t) = A_0(Q).A_1(Q) + A_0(Q)(1 - A_1(Q)).e^{-t/\tau_1}.e^{-t/\tau_2}.I_0(t/\tau_1) + (1 - A_0(Q)).e^{-t/\tau_2} \quad (52)$$

As immediately apparent from a comparison of Eq.51 and 52, the presence of the variable Q could lead to the impression that it is difficult to directly compare models derived from QENS and NMR data. Based on the simple qualitative argument that the dipole-dipole interactions giving rise to the NMR T_1 relaxation is short-ranged, we can argue that NMR assesses the atomic scale dynamics (two or three covalent bonds). This corresponds to a Q range of $Q \approx 2\pi/3 \approx 2$ Å^{-1}. Replacing $A_0(Q)$ and $A_1(Q)$ by 0.44 and 0, their numerical values at $Q \approx 2$ Å^{-1} (Fig.19), Eq.52 exactly reduces to Eq.51. Therefore, starting from a $2D$ ^{13}C solid-state derived molecular motion (HJ), we have proposed a detailed model of the PEO motion at the molecular level. The correlation times and characteristic distances of the model are respectively consistent with PFG-NMR and T_1 NMR relaxation data.

4.3 The Rouse regime

Neutron Spin-Echo (NSE) is in principle a suitable technique to match the time and spatial scales relevant to the Rouse regime. Moreover, by specific orientation of the macroscopically oriented AAO membranes with respect to the incident neutron beam (see Fig.21) of a NSE spectrometer, it is possible to probe the intermediate self-correlation function perpendicular (radial dynamics) and parallel (longitudinal dynamics) to the axis of the cylindrical AAO pores.

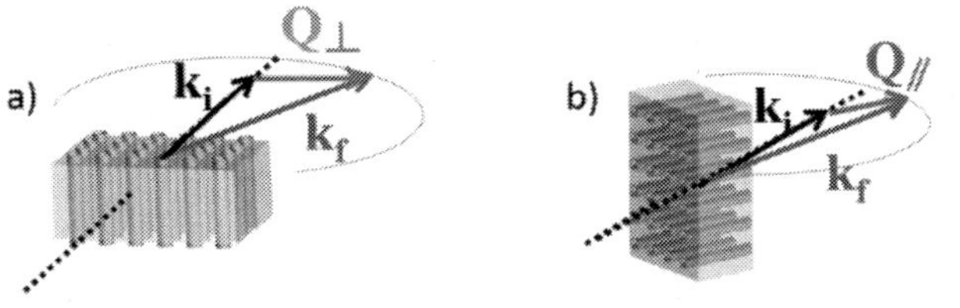

Figure 21. By orienting macroscopically oriented porous samples with respect to the neutron beam, the dynamics perpendicular/radial (a) or parallel (b) to the pores can be independently retrieved. From [33].

Severe restriction applies to NSE measurements: the dynamics is assessed through the polarization, $< P >$, of the neutron beam after interaction with the sample. $< P >$ is a function of I_{coh} and I_{inc} the coherent and incoherent scattered intensities: $< P >= I_{coh} - I_{inc}/3$. Therefore, no observable are available (or with very poor statistics) in Q ranges where $I_{coh} \approx I_{inc}/3$. The experimental polarization of bulk and AAO confined PEO is shown in Fig.22a. As this has been stated earlier, the Rouse model applies for Q below 0.3 Å^{-1}. For $Q < 0.3$ Å^{-1}, the polarization is good for both samples, but a full $I(Q,t)$ measurement shows that the signal is purely static (coherent elastic scattering of the aluminum cell and/or the alumina matrix). In the 0.3 Å^{-1} region, the polarization of the bulk makes the $I(Q,t)$ accessible, but the polarization of the confined sample is almost null. Altogether, the smallest Q

available where the sample dynamics is in the NSE time window and the polarization is not null for both the bulk and the confined polymer is 0.62 Å^{-1}. The corresponding data are shown on Fig.22b.

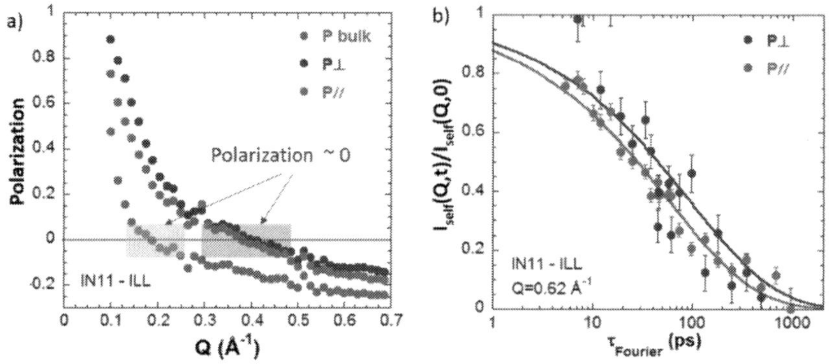

Figure 22. NSE results. The sample is ^{h}PEO (M_w =35000 g.mol^{-1}) confined within the macroscopically oriented structure of an AAO (24 nm in diameter), at 373 K. AAO membranes being macroscopically oriented, specific orientation of the sample in the neutron beam makes it possible to access the dynamics parallel and perpendicular to the pores. a) Polarization of the neutron beam for bulk PEO, and for the two orientations of the AAO confined sample. b) Neutron spin-echo measurements (IN11-ILL) of $I_{Self}(Q,t)$ (Eq.42), the self intermediate scattering function, at Q = 0.62 Å^{-1}. No differences are detected between the polymer dynamics perpendicular (blue symbols) and parallel (red symbols) to the axis of the cylindrical AAO pores. From [34].

Even at this rather large Q value the Rouse model Eq.41 holds (we do observe a master curve in Fig.23). $W.\sigma^4$ values of 11 ± 2, 3.3 ± 0.7, and 2.0 ± 0.5 Å^4.ps^{-1} have been measured for the bulk PEO, the PEO confined in the parallel (alongside the AAO pore axis) and perpendicular direction, respectively. Within the statistics, at this rather local scale, no dynamical difference can be detected between the radial and longitudinal dynamics. However, under confinement the polymer dynamics decreases by a factor of about four.

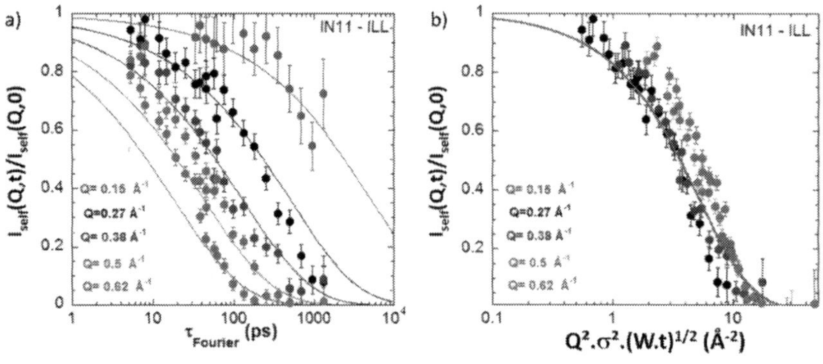

Figure 23. a) NSE spectra of bulk ^{h}PEO (M_w =35000 g.mol^{-1}) for various Q. b) Same data as in a) fitted with Eq.41. When plotted as a function of the Rouse variable $u = Q^2\sigma^2.(W.t)^{1/2}$, the data fall on a master curve for $W.\sigma^4$ values of 11 ± 2 Å^4.ps^{-1}. From [34].

4.4 The reptation regime: Observing a *single* polymer chain (Zero Average Contrast)

Since porous materials are usually very good scatterers, Small Angle Neutron Scattering (SANS) is a key technique for accessing their structure. But this advantage turns to a strong handicap when it comes to access the dynamics under confinement in the SANS regime: the fraction of intensity scattered inelastically by the confined material is several orders of magnitude smaller than the elastic contribution due to the static confining matrix. Then within the typical statistics of a neutron experiment (few %) the dynamical information cannot be recovered. In the following section, we detail a general way to lift this difficulty.
As we have stated at the beginning of this section, the relevant time scale to probe polymer dynamics is at least one order of magnitude larger than the one assessed on the previous NSE experiment. A Q range in the 10^{-2} Å^{-1} region (now fully experimentally accessible on spin-echo spectrometers), is indeed much more appropriate. However, inelastic experiments in such a SANS region in situations of hard confinement are almost impossible: *(i)* the scattering of the confining porous matrix is usually extremely intense and *(ii)* since the porous material has no dynamics, this signal is purely elastic (Fig.24). As a consequence, the relevant information *i.e.* the inelastic scattering of the sole confined material is usually a negligible contribution to the total scattering, so that the related error bars prevent any practical use of the signal. The Zero Average Contrast (ZAC) technique makes it possible to overcome this difficulty (Fig.24).

The ZAC technique has been used to access the conformation of a polymer in semi-dilute solution, confined in the porous structure of Vycor [35]. A solution of hydrogenated (H) and deuterated (D) polymer chains in a solvent (S) have been confined in a porous medium (Vycor). If the solvent scattering length densities matches the scattering length densities of the confining matrix, the SANS intensity $I(Q)$, can be written as a sum of only three partial structure factors:

$$I(Q) = (n_D - n_0)^2 S_{DD} + (n_H - n_0)^2 S_{HH} + 2(n_D - n_0)(n_H - n_0) S_{HD} \tag{53}$$

where, n_H, n_D and n_0 are the neutron coherent scattering length densities of the hydrogenated polymer, the deuterated polymer and the solvent. In our case, the confining matrix can be assimilated to the solvent. If one makes the assumption that the H and D polymer chains are identical, Eq.53 becomes:

$$I(Q) = (n_D - n_H)^2 x_d (1 - x_d) \upsilon \Phi N P(Q) + (x_d n_D + (1 - x_d) n_H - n_0)^2 [\upsilon \Phi N P(Q) + \upsilon \Phi^2 Q(Q)] \tag{54}$$

where $P(Q)$ and $Q(Q)$ are respectively the intra-chain and the inter-chain contributions, x_d is the proportion of deuterated chains, v the molecular volume of a monomer and Φ the polymer volume fraction. If x_d can be set to meet the condition:

$$x_d . n_D + (1 - x_d) . n_H - n_0 = 0 \tag{55}$$

then, $I(Q)$ is only proportional to $P(Q)$, the form factor of a single chain. The coherent scattering length density of an AAO membrane can then be derived upon membrane impregnation by an isotopic mixture of water (x_d = 70 vol%): $\rho_{AAO} = 4.21 \times 10^{10}$ cm^{-2}. The coherent scattering length density of H and D PEO being $\rho_{hPEO} = 6.20 \times 10^9$ cm^{-2} and $\rho_{dPEO} = 6.88 \times 10^{10}$ cm^{-2} respectively, the condition Eq.55 is fulfilled for $x_d = 0.56$.

As shown in Fig.24a) by the perfect canceling out of the intense AAO peak, the H/D PEO mixture perfectly matches the AAO membrane. This is a direct experimental evidence that the polymer is fully and homogeneously confined within the AAO porous network. As shown by the strong similarity of the form factor, $P(Q)$, of a single confined polymer chain and bulk polymer chain form factor, no large scale significant structural change of the confined chain is detected by respect to its bulk analogue. In

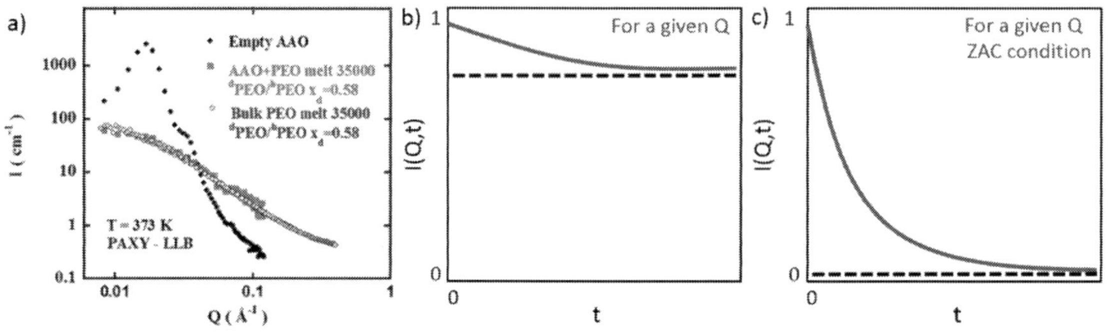

Figure 24. a) Relative intensities of an empty AAO (pore diameter 18 nm) membrane and the same membrane filled with a $^dPEO/^hPEO$ mixture with a deuterated volume fraction $x_d = 0.58$. The form factor of a single polymer chain measured on the same $^dPEO/^hPEO$ mixture in bulk (*i.e.* a typical bulk ZAC measurement) is shown for comparison. b) Due to the intense "parasitic" SANS elastic contribution (black dotted line) of the confining material, large scale NSE experiments to assess the dynamics of the confined material are, in practice, impossible (see text): the relevant inelastic contribution related to the confined polymer dynamics (full red line) has a negligible contribution. c) A way through is to combine ZAC conditions (to assess the signal of a single chain) with a matching condition of the confining material. Once matched, the SANS elastic intensity contribution of the membrane to the NSE $I(Q,t)$ signal becomes negligible (black dotted line), making the polymer correlation function (full red line) experimentally fully accessible (experimental data shown on Fig.25). Fig.a from [25].

particular, the radius of gyration $R_G = 6.7 \pm 0.1$ nm is not affected by the confinement.
As far as the dynamical aspect is concerned, the reptation mechanism in a tube with diameter d_{Rep} leads to a transient localization of the chain. This is accounted for in $I_{Chain}(Q,t)$, the intermediate scattering function of a single polymer chain [26]:

$$\frac{I_{Chain}(Q,t)}{I_{Chain}(Q,t=0)} = [1 - p(Q)].F(Q,t) + p(Q) \tag{56}$$

with $p(Q) = e^{-\frac{Q^2 d_{Rep}^2}{36}}$, the form factor of the reptation tube. Note that $p(Q)$ shows a Q dependent but time independent plateau. F(Q,t) writes:

$$F(Q,t) = e^{-\frac{u^2}{36}}.erfc(-\frac{u}{6}) \tag{57}$$

where $u = Q^2\sqrt{W\sigma^4 t}$ is the Rouse variable, σ is the length of a chain segment and $W = 3.k_BT.\zeta_0^{-1}\sigma^{-2}$ is the elemental Rouse rate. W is a function of ζ_0, the monomeric friction coefficient, k_B the Boltzmann constant and the temperature T.

The NSE data of the bulk and confined polymer (Fig.25) follow the model Eq.56. A fit of the $I_{Chain}(Q,t)$ data set provides a very precise determination of both the temporal and structural parameters of Eq.56. The description of the confined PEO data with the fitted parameters $W\sigma^4 = 1778 \pm 63$ Å^4 ns^{-1} and $d_{Rep} = 67 \pm 3$ Å is very satisfactory (Fig.25). The same data treatment applied to the bulk signal leads to $W\sigma^4 = 2066 \pm 60$ Å^{4}ns^{-1} and $d_{Rep} = 65 \pm 2$ Å. Both the dynamics and the reptation diameter of the bulk and confined polymer are indistinguishable (Fig.25). This conclusion is supported by several recent independent results obtained by quasi-elastic neutron scattering (NSE included) [25, 36, 37].
We should nevertheless stress-out that quasi-elastic neutron scattering is sensitive to both translational

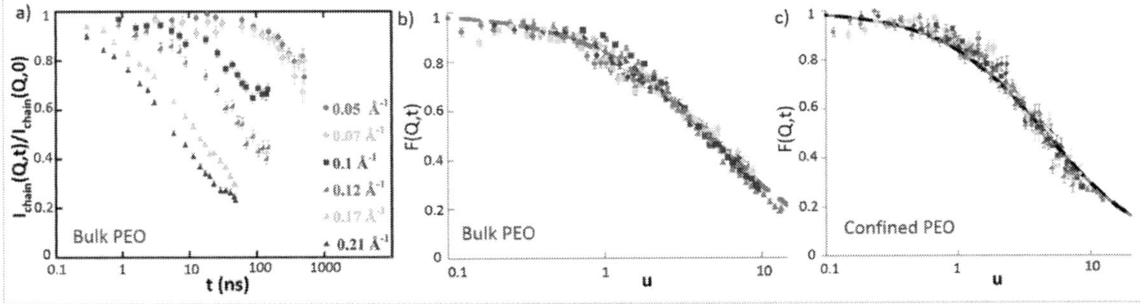

Figure 25. a) NSE intermediate scattering function of a single chain 35 kg.mol^{-1} PEO confined within an AAO membrane. The polymer is a mixture of deuterated and hydrogenated PEO (x_D = 56 vol%). This isotopic composition matching the AAO membrane, the intense elastic contribution due to the AAO membrane is canceled out and the scattered intensity is purely inelastic. $[I_{Chain}(Q,t)/I_{Chain}(Q,t=0) - p(Q)]/[(1-p(Q)]$ for the bulk (b) and confined polymer (c). The symbols used refer to the same Q values than in a). The full and dotted lines stand for the master curve $F(Q,t)$ (Eq.56) with the fitted parameters $[W\sigma^4 = 1778 \pm 63$ Å^4 ns^{-1}; $d_{Rep} = 67 \pm 3$ Å], $[W\sigma^4 = 2066 \pm 60$ Å^4 ns^{-1}; $d_{Rep} = 65 \pm 2$ Å] for the confined and bulk polymer, respectively. From [25].

and rotational fluctuations. We observe in Fig.25, that the curvilinear Rouse motion of the local reptation regime (Doi-Edwards Regime II) satisfactorily describes the experimental $F(Q,t)$ of both the bulk and confined samples. Within the statistics of our NSE data, there is therefore no need to invoke any rotational contribution to describe the coherent intermediate scattering function of a single polymer chain.

4.5 Multiscale study of a polymer melt: Conclusion

Starting from a $2D$ ^{13}C solid-state NMR derived molecular motion (HJ), we have proposed a detailed model of the PEO motion at the molecular level. The correlation times and characteristic distances of the model are respectively consistent with PFG-NMR and T_1 NMR relaxation data. This physically meaningful model of the local (Å), short time (0.1-10 ps) dynamics of highly entangled ^{h}PEO melt takes into account *i)* a helical jump motion along the polymer chain, *ii)* a local conformational transitions and *iii)* a long-time, large-scale motions. The model also accounts for the spatial dependence and geometry of these local motions via the Elastic Incoherent Structure Factors (EISF). To be able to detect the dynamics of the polymer chain confined in a nanometric cylindrical pore of an AAO membrane, we have combined SANS, contrast matching and ZAC with neutron spin-echo. This strategy provides a direct measurement of the time and Q dependence of a *single* polymer chain (statistically averaged over all the chains of the sample). By interpreting the data within the framework of the reptation theory, we show that, in the conditions observed here (AAO pore radius $R_P = 90 \pm 10$ Å and the radius of gyration of the 35 kg/mol PEO $R_g = 69 \pm 1$ Å), confinement does not affect the dynamical property of the polymer chains nor the diameter of the reptation tube.

5 Conclusion

Inelastic and quasi-elastic neutron scattering scattering are central techniques to experimentally probe the dynamics of macromolecules (polymers or proteins). The dynamical information is measured in terms of the time or energy dependence of correlation functions of the atomic positions. The accessible time scale spans over decades from few fractions of picoseconds (time-of-flight techniques) to a few microseconds (Neutron spin-echo experiments). A specificity of the inelastic neutron techniques, is to directly and simultaneously probe the time and space correlation functions, so that the Q dependence of the system characteristic time(s) can be measured. The dynamics of the system can then be understood by challenging the experimental observables with physical models accounting for spatial and temporal quantities. Altogether, the results presented here have illustrated the capabilities of neutron scattering in achieving a real multiscale approach (1 ps to 0.6 μs and 0.1 to 40 nm) of a canonical soft-matter physical system: a polymer melt.

We have highlighted how it is possible to take advantage of strong isotopic effects (for example hydrogen/deuterium) to disentangle specific dynamical contributions of complex architectures. We have in particular shown how, in conjunction with H/D isotopic effects, high resolution QENS/NSE can be bridged to the Zero Average Contrast (ZAC) method to probe, in a non destructive way, the dynamics (hence potentially the viscosity) of a single polymer chain in bulk but also under severe nanometric confinement. This is a situation where more classical experimental approaches, for example rheology, would have been impossible.

Last, it should be noted that the formalisms of inelastic neutron scattering and NMR are very similar. We have shown in several instances that firm connections between QENS and NMR techniques can be drawn and help to apprehend more finely the physics of a polymer melt. We hope that these few tentative connections will encourage new comers to the field to extend the numerous NMR/QENS complementarities to different soft-matter related systems.

References

[1] T. Perring, *Lecture at the Oxford Neutron School* (2007)
[2] G.L. Squires, *Introduction to the Theory of Thermal Neutron Scattering* (Courier Corporation, 1978)
[3] L. Van Hove, Physical Review **95**, 249 (1954)
[4] F. Leclercq-Hugeux, M.V. Coulet, J.P. Gaspard, S. Pouget, J.M. Zanotti, Comptes Rendus Physique **8**, 884 (2007)
[5] V.F. Sears, Neutron News **3**, 26 (1992)
[6] J.M. Zanotti, Journal de Physique IV (Proceedings) **130**, 87 (2005)
[7] M. Bée, *Quasielastic Neutron Scattering* (Adam Hilger, 1988)
[8] F. Volino, A.J. Dianoux, Molecular Physics **41**, 271 (1980)
[9] V.F. Sears, Can. J. Phys. **44**, 1279 (1966)
[10] J.M. Zanotti, K. Lagrené, N. Malikova, P. Judeinstein, K. Panesar, J. Ollivier, S. Rols, M. Mayne-L'Hermite, M. Pinault, P. Boulanger, Eur. Phys. J. Spec. Top. **213**, 129 (2012)
[11] D. Klug, E. Whalley, E. Svensson, J. Root, V. Sears, Phys. Rev. B **44**, 841 (1991)
[12] G.P. Johari, O. Andersson, Phys. Rev. B **73**, 094202 (2006)
[13] M. Bastos, N. Alves, S. Maia, P. Gomes, A. Inaba, Y. Miyazaki, J.M. Zanotti, Physical Chemistry Chemical Physics **15**, 16693 (2013)
[14] M.C. Rheinstädter, C. Ollinger, G. Fragneto, F. Demmel, T. Salditt, Phys. Rev. Lett. **93**, 108107 (2004)
[15] J. Ollivier, J.M. Zanotti, *Diffusion inélastique de neutrons par temps de vol* (EDP Sciences, 2010), pp. 379–423
[16] M. Doi, S.F. Edwards, *The Theory of Polymer Dynamics* (Clarendon Press, 1988)
[17] M. Rubinstein, R.H. Colby, *Polymer Physics* (OUP Oxford, 2003)
[18] P.G. de Gennes, The Journal of Chemical Physics **55**, 572 (1971)

[19] R.L. Jones, S.K. Kumar, D.L. Ho, R.M. Briber, T.P. Russell, Nature **400**, 22080 (1999)
[20] H. Bodiguel, C. Fretigny, Phys. Rev. Lett. **97**, 266105 (2006)
[21] S. Sen, Y. Xie, S.K. Kumar, H. Yang, A. Bansal, D.L. Ho, L. Hall, J.B. Hooper, K.S. Schweizer, Physical review letters **98** (2007)
[22] F. Brochard, P.G. de Gennes, Journal de Physique Lettres pp. 399–401 (1979)
[23] A.N. Semenov, Phys. Rev. Lett. **80**, 1908 (1998)
[24] F. Li, L. Zhang, R.M. Metzger, Chem. Mater. **10**, 2470 (1998)
[25] K. Lagrené, J.M. Zanotti, M. Daoud, B. Farago, P. Judeinstein, Physical Review E **81** (2010)
[26] D. Richter, M. Monkenbusch, A. Arbe, J. Colmenero, *Neutron Spin Echo in Polymer Systems* (Springer, 2005)
[27] J.M. Zanotti, L.J. Smith, D.L. Price, M.L. Saboungi, Journal of Physics: Condensed Matter **18**, S2391 (2006)
[28] K. Schmidt-Rohr, H.W. Spiess, *Multidimensional Solid-state NMR and Polymers* (Academic Press, 1994)
[29] O. Borodin, G.D. Smith, Macromolecules **33**, 2273 (2000)
[30] M. Appel, G. Fleischer, Macromolecules **26**, 5520 (1993)
[31] C.K. Hall, E. Helfand, The Journal of Chemical Physics **77**, 3275 (1982)
[32] R. Dejean de La Batie, F. Lauprêtre, L. Monnerie, Macromolecules **21**, 2045 (1988)
[33] Q. Berrod, F. Ferdeghini, P. Judeinstein, N. Genevaz, R. Ramos, A. Fournier, J. Dijon, J. Ollivier, S. Rols, D. Yu et al., Nanoscale **8**, 7845 (2016)
[34] K. Lagrené, Ph.D. thesis, Paris Sud (2008), `https://tel.archives-ouvertes.fr/tel-00622540`
[35] J. Lal, S. Sinha, L. Auvray, Journal de Physique II pp. 1597 – 1615 (1997)
[36] M. Krutyeva, J. Martin, A. Arbe, J. Colmenero, C. Mijangos, G.J. Schneider, T. Unruh, Y. Su, D. Richter, The Journal of Chemical Physics **131**, 174901 (2009)
[37] J. Martín, M. Krutyeva, M. Monkenbusch, A. Arbe, J. Allgaier, A. Radulescu, P. Falus, J. Maiz, C. Mijangos, J. Colmenero et al., Phys. Rev. Lett. **104**, 197801 (2010)

AUTHOR INDEX